W0079146

Information Retrieval for Music and Motion

Meinard Müller

Information Retrieval for Music and Motion

With 136 Figures, 41 in color and 26 Tables

Meinard Müller
Institut für Informatik III
Universität Bonn
Römerstr. 164
53117 Bonn, Germany
meinard@cs.uni-bonn.de

Library of Congress Control Number: 2007932401

ACM Classification: H.3, H.5, I.3, J.5

ISBN 978-3-540-74047-6 Springer Berlin Heidelberg New York

Springer is a part of Springer Science+Business Media
springer.com

Typesetting by the author
Production: SPi, India
Cover design: KünkelLopka, Heidelberg

Printed on acid-free paper 45/3180/SPi 5 4 3 2 1 0

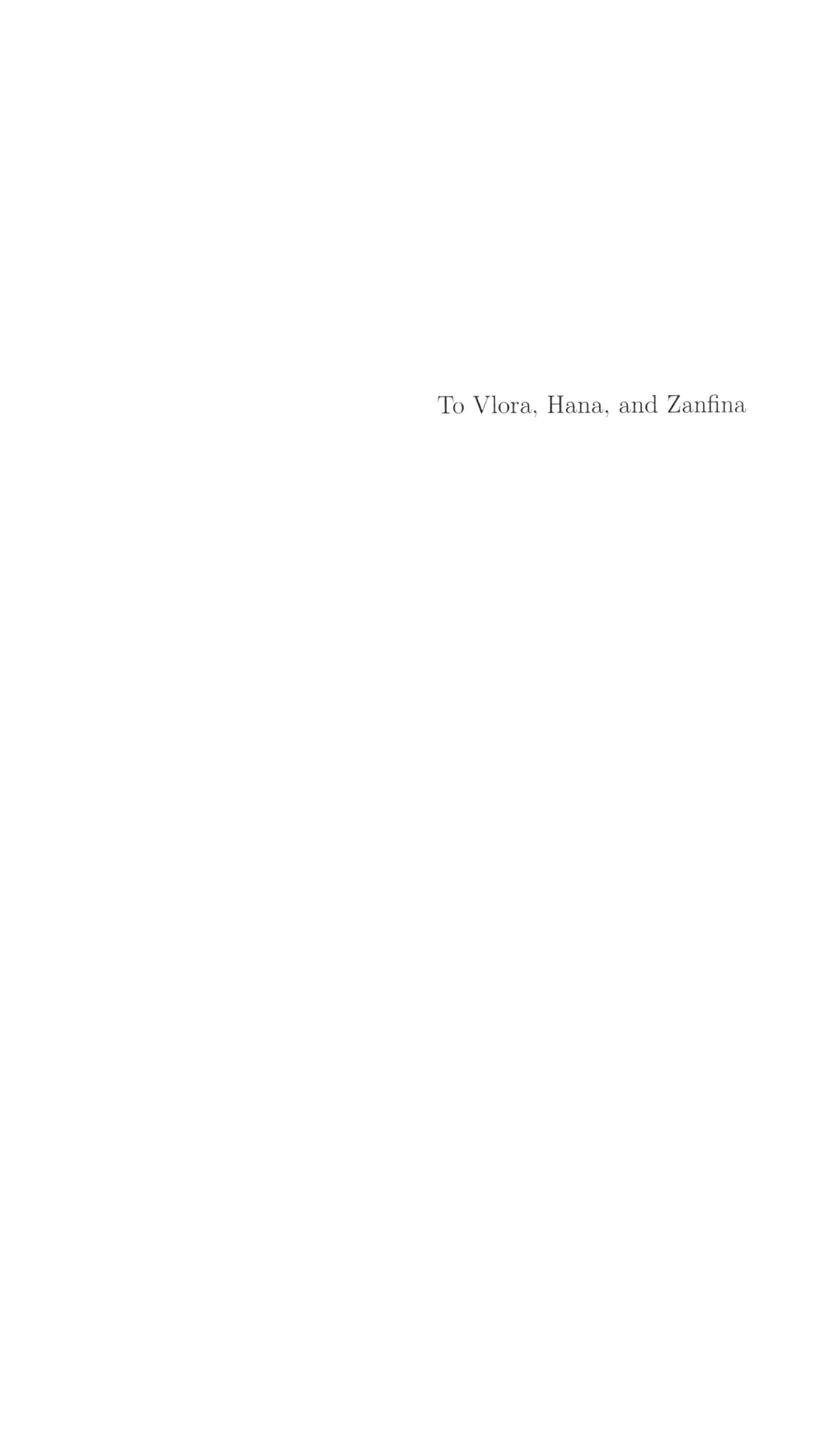

To Vlora, Hana, and Zanfina

Preface

Recent years have seen enormous advances in computerization and digitization as well as a corresponding growth in the use of information technology allowing users to access and experience multimedia content on an unprecedented scale. In this context, great efforts have been directed toward the development of techniques for searching and extracting useful information from huge amounts of stored data. In particular for textual information, powerful search engines have been implemented that provide efficient browsing and retrieval within billions of textual documents. For other types of multimedia data such as music, image, video, 3D shape, or 3D motion data, traditional retrieval strategies rely on textual annotations or metadata attached to the documents. Since the manual generation of descriptive labels is infeasible for large datasets, one needs fully automated procedures for data annotation as well as efficient *content-based* retrieval methods that only access the raw data itself without relying on the availability of annotations. A general retrieval scenario, which has attracted a large amount of attention in the field of multimedia information retrieval, is based on the *query-by-example paradigm*: given a query in form of a data fragment, the task is to automatically retrieve all documents from the database containing parts or aspects similar to the query. Here, the notion of similarity, which strongly depends on the respective application or on a person's perception, is of crucial importance in comparing the data. Frequently, multimedia objects, even though similar from a structural or semantic point of view, may reveal significant spatial or temporal differences. This makes content-based multimedia retrieval a challenging research field with many yet unsolved problems.

The present monograph introduces concepts and algorithms for robust and efficient information retrieval by means of two different types of multimedia data: waveform-based music data and human motion data. For both domains, music and motion, semantically related objects typically exhibit a large range of variations concerning temporal, spatial, spectral, or dynamic properties. In this book, we will study fundamental strategies for handling object deformations and variability in the given data with a view to real-world retrieval and browsing applications. Here, one important principle, which is applicable to general multimedia data, is to already absorb variations that are to be left unconsidered in the searching process at the feature level. This strategy makes it possible to use relatively strict and efficient matching techniques.

According to the two types of multimedia data to be considered, this monograph is organized in two parts. In Part I, we will discuss in depth several current problems in music information retrieval. In particular, we describe general strategies as well as efficient algorithms for music synchronization, audio matching, and audio structure analysis. We also show how the analysis results can be used in an advanced audio player to facilitate additional retrieval and browsing functionality. Then, in Part II, we will systematically introduce a general and unified framework for motion analysis, retrieval, and classification. Here, important aspects concern the design of suitable features, the notion of similarity used to compare data streams, as well as data organization.

Even though conceptually interrelated, the two parts of this monograph are kept independent, each giving a self-contained account of recent advances in information retrieval for the respective multimedia domain. Both parts have been organized in didactically prepared units: they start with introductory chapters covering the fundamentals required for the subsequent chapters and then present scientific contributions of the author. The detailed chapters at the beginning of each part give consideration to the interdisciplinary character of this work. Here, we also fix the notation, introduce a precise terminology, and supply rigorous mathematical foundations. In this monograph, we will encounter aspects from a multitude of research fields including information science, digital signal processing, audio engineering, musicology, and computer graphics.

This monograph is accessible to a wide audience, from students at the graduate level and lecturers to practitioners and scientists working in the above-mentioned research fields. Each part is suitable for use as stand-alone lecture notes for a graduate course in Computer Science. Here, the focus is on the study of fundamental algorithms and concepts for the analysis, classification, indexing, and retrieval of time-dependent data in the context of a specific multimedia domain. Important aspects concern the design of suitable features, the development of local and global similarity measures, as well as data organization. The general goal of the monograph is to highlight the interaction between modeling, experimentation, and mathematical theory while introducing the students to current research fields. Dividing the results into essentially independent chapters and including suitable recapitulations should allow a researcher to read chapters or individual sections of this monograph as self-contained units. Further notes including references to the literature are provided at the end of each chapter. Motivating and domain-specific introductions of this monograph can be found in Chap. 1.

Acknowledgments

Many of the results presented in this monograph have been obtained in collaboration with different people. I would like to take the opportunity to express my gratitude to my collaborators and the many people who have influenced and helped me in writing this monograph. First of all, I owe special thanks to Michael Clausen for his mental, intellectual, and financial support. He allowed me great latitude within his research group in independently conducting research projects and pursuing my own research goals. Particularly, I enjoyed the open and inspiring atmosphere in our many discussions. For his generous financial support, I want to express my gratitude to Armin B. Cremers, the head of the Department of Computer Science III and director of the Bonn-Aachen International Center for Information Technology (B-IT), University of Bonn.

Starting in 1998, the *Multimedia Signal Processing Group* headed by Michael Clausen conducted a six-year research project (MiDiLiB) on content-based indexing, retrieval, and compression of data in digital music libraries funded by the Deutsche Forschungsgemeinschaft (DFG). The research presented in Part I of this monograph can be seen as a kind of continuation of this project. In this context, I want to thank Frank Kurth for a close and fruitful collaboration in the field of music information retrieval over the last couple of years. In particular, he has been the driving force in launching and developing the SyncPlayer system. I am also grateful to Christian Fremerey and David Damm, who not only did most of the programming but also made important conceptual contributions to the SyncPlayer framework [73, 114, 115]. For an early collaboration, I want to thank Vlora Arifi, who pioneered the work on music synchronization [4, 5]. I also thank Henning Mattes for programming a very efficient and robust multiscale version of DTW used in our audio synchronization procedure [142].

The work on content-based motion retrieval has been initiated and encouraged in the year 2003 by Andreas Weber and Bernd Eberhardt. I want to express my gratitude to both of them. Bernd Eberhardt from HDM Stuttgart also supplied us with extensive motion capture material from which we assembled our motion database [149]. Most of the results in content-based motion retrieval [143, 145–147] are joint work with Tido Röder, whom I esteem in particular for his excellent technical skills (doing all the animations and videos). Starting from scratch, we jointly studied the foundations of computer animation and motion representations, while building up a comprehensive software framework for our motion retrieval research. I will never forget our "extreme" programming sessions, which were not only highly productive but also a lot of fun. Thank you very much! I also thank Bastian Demuth for his support in building up a prototype of a motion retrieval system [56].

In the process of writing this monograph, I have had the opportunity to teach much of the material as graduate courses in the Department of

Computer Science of the University of Bonn. I want to thank the students for their comments and valuable feedback.

There are many more people to whom I am obliged to express my gratitude for their help, support, stimulation, and encouragement. I will confine myself to only mentioning their names in alphabetical order: Rolf Bardeli, Jochen Bomm, Harun Celebi, Daniel Cremers, Martina Doelp, Hendrik Ewe, Thomas Fuchs, Michael Gleicher, Masataka Goto, Masakazu Jimbo, Reinhard Klein, Björn Krüger, Peter Lachart, Hans-Georg Müller, and Bodo Rosenhahn. Finally, I thank my three ladies Vlora, Hana, and Zanfina for being extremely supportive and for reminding me of the really important things in life.

Bonn,
June 2007

Meinard Müller

Contents

1 Introduction

In this chapter, we provide motivating and domain-specific introductions of the information retrieval problems raised in this book. Sect. 1.1 covers the music and Sect. 1.2, the motion domain. These two sections also include an outline of the two parts, provide a summary of all chapters, and discuss general literature relevant to music information retrieval and motion retrieval, respectively. Finally, in Sect. 1.3, we reveal the conceptual relations between the two parts. In particular, we point out the general concepts for content-based information retrieval, which apply to both music and motion domains, and even beyond.

1.1 Music Information Retrieval

For music, there is a vast amount of digitized data as well as a variety of associated data representations, which describe music at various semantic levels. Typically, digital music collections contain a large number of relevant digital documents for one musical work, which are given in various digital formats and in multiple realizations. For example, in the case of Beethoven's Fifth Symphony, a digital music library may contain the scanned pages of some particular score edition. Or the score may be given in a digital music notation file format, which encodes the page layout of sheet music in a machine-readable form. Furthermore, the library may contain various CD recordings such as the interpretations by Karajan and Bernstein, some historical recordings by Furthwängler and Toscanini, Liszt's piano transcription of Beethoven's Fifth played by Glenn Gould, as well as a synthesized version of a corresponding MIDI file. Different interpretations of Beethoven's Fifth often reveal large variations regarding tempo, dynamics, articulation, tuning, or instrumentation.

As illustrated by the Beethoven example, there are various digital manifestations of a musical work differing in format and content. In the field of *music information retrieval* (MIR), great efforts have been directed toward the development of technologies that allow users to access and explore music in all its different facets. For example, during playback of some CD recording, a digital music player of the future presents the corresponding musical score while highlighting the current playback position within the score. On demand, additional information about melodic and harmonic progression or rhythm and tempo is automatically presented to the listener. A suitable user interface displays the musical structure of the current piece of music and allows the user to directly jump to any key part within the recording without tedious fast-forwarding and rewinding. Furthermore, the listener is equipped with a Google-like search engine that enables him/her to explore the entire music collection in various ways: the user creates a query by specifying a certain note constellation or some harmonic or rhythmic pattern by whistling a melody or simply by selecting a short passage from a CD recording; the system then provides the user with a ranked list of available music excerpts from the collection that are musically related to the query. For example, querying a twenty-second excerpt of a Bernstein interpretation of the theme of Beethoven's Fifth, the system will return all other corresponding music clips in the database. This includes the repetition of the theme in the exposition or in the recapitulation within the same interpretation as well as the corresponding excerpts in all recordings of the same piece interpreted by other conductors. An advanced search engine is also capable of automatically identifying the theme even in the presence of significant variations, thus handling arrangements such as Liszt's piano transcription, synthesized versions, or rhythmically accompanied pop versions of Beethoven's Fifth.

Even though significant progress has been made in the development of advanced music players, there are still many yet unsolved problems in content-based music browsing and retrieval, which are due to the heterogeneity and complexity of music data. Here, *content based* means that in the comparison of music data, the system makes use of only the raw data, rather than relying on manually generated metadata such as keywords or other symbolic descriptions. While text-based retrieval of music documents using the composer's name, the opus number, or lyrics can be handled by means of classical database techniques, purely content based music retrieval constitutes a difficult research problem. How should a retrieval system be designed, if the user's query consists of a whistled melody fragment or a short excerpt of some CD recording? How can (symbolic) score data be compared with the content of (waveform based) CD recordings? What are suitable notions of similarity that capture certain (user specified) musical aspects while disregarding admissible variations concerning, e.g., the instrumentation or articulation? How can the musical structure, reflected by repetitive and musically related patterns, be automatically derived from a CD recording? These questions reflect only a small fraction of current MIR research topics that are closely related to automatic music analysis.

1.1.1 Outline of Part I

The first part of this monograph is kept self-contained and can be used as a basis for a graduate course in music information retrieval. The first few chapters cover the required fundamentals on music representations, digital signal processing, feature design, and dynamic time warping. The subsequent chapters then give an account on state-of-the-art techniques for recent MIR problems including music synchronization, audio matching, and audio structure analysis. We now give an overview of the main topics discussed in Part I.

In Chap. 2, we discuss basic properties of symbolic and physical music representations, which differ fundamentally in their respective structure and content. One general question is how different realizations of the same underlying piece of music can be transformed into a simplified and unified form – a procedure that is commonly referred to as feature extraction. Here, the goal is to reduce the music data to relevant key aspects (thus suppressing irrelevant details and variations) and to find a common platform making the comparison of music data algorithmically feasible. In this monograph, we mainly deal with waveform-based audio data as can be found in CD recordings. Due to the complexity of audio data, the extraction of musically interpretable parameters from a waveform constitutes a difficult problem. The single most important tool in the analysis of such data is the Fourier transform, which breaks up an audio signal into its constituting frequencies. We review the Fourier transform for continuous-time, periodic, as well as discrete-time signals. Furthermore, we discuss how the respective Fourier coefficients can be efficiently approximated by means of the discrete Fourier transform (DFT). As a second important tool from digital signal processing we review the concept of digital convolution filters. Such filters can be applied to modify the frequency content of an audio signal in some desired way – for example, using a suitable bandpass filter, one may remove all frequency components outside a specified frequency band.

Using a suitable filter bank consisting of a set of bandpass filters, where each filter corresponds to a piano note, an audio signal can be decomposed into a pitch representation. From such a representation, as explained in Chap. 3, one can derive pitch and onset features that indicate the presence of certain musical notes realized at certain physical onset times within the audio signal. As a further important feature representation, we discuss chroma-based audio features that strongly correlate to the harmonic progression of the audio signal. Here, the chroma correspond to the 12 traditional pitch classes of the equal-tempered scale. Chroma features, which show a high degree of robustness to variations in dynamics, timbre, and articulation as well as to local tempo deviations, turn out to be a powerful midlevel representation for comparing and relating music data in various realizations and formats. For example, to compare two different CD recordings of the same piece of music, one common strategy is to transform the two audio streams into sequences of chroma feature vectors. To deal with temporal variations, one needs a concept that brings the feature vectors of the two sequences into temporal correspondence. An important technique for computing such a correspondence or alignment between general sequences is known as dynamic time warping (DTW). In Chap. 4, we give a detailed, self-contained account on DTW and discuss several of its variants.

In the subsequent chapters, we address three important MIR problems: music synchronization, audio matching, and audio structure analysis. The common goal of these tasks is to automatically link several types of music representations, thus coordinating the multiple information sources related to a single piece of music. In Chap. 5, we introduce DTW-based algorithms, which automatically link two data streams, possibly given in different formats, representing the same piece of music. For example, in the case that one data stream represents the score and the other a CD recording of the same underlying piece, the *score-audio synchronization* task amounts to associating the note events given by the score with their physical occurrences in the audio file. This can also be seen as an automated annotation of the audio data stream by the score parameters. The basic goal of *audio matching*, as introduced in Chap. 6, can be described as follows: Consider an audio database containing several CD recordings for one and the same piece of music interpreted by various musicians. Then, given a short query audio clip of one interpretation, the goal is to automatically retrieve the corresponding excerpts from the other interpretations. In our approach, we use coarse audio features derived from a chroma representation to cope with possible variations in sound and articulation as well as a robust matching procedure to handle variations in tempo. Finally, in Chap. 7, we deal with a further prominent MIR problem referred to as *audio structure analysis*. Here, the goal is to automatically extract the repetitive structure or, more generally, the musical form of the underlying piece of music. Opposed to previous approaches, which are based on the constant tempo assumption, our approach to structure analysis allows for identifying musical repetitions even in the presence of significant temporal variations.

In conclusion, Chap. 8 describes our SyncPlayer system, which is an advanced audio player for multimodal presentation, browsing, and retrieval of music data. Integrating the proposed content-based analysis techniques, the SyncPlayer makes use of the automatically generated annotations and temporal linking structures to provide advanced functionalities such as automatic tracking of the score position in a CD recording, navigation within a single recording (intradocument navigation) or between different performances (interdocument navigation), or to extend score-based music retrieval for directly accessing audio recordings.

1.1.2 Further Notes

In the last decade, music information retrieval (MIR) has become active and multidisciplinary research field. Central MIR problems concern music information handling and retrieval, automated music recognition and classification, the design and extraction of musically relevant audio features, or the development of novel user interfaces. Due to the diversity and richness of music, MIR research brings together experts from a multitude of research fields ranging from information science, audio engineering, computer science, musicology, music theory, library science to law and business. For a general account on the multidisciplinary and multifaceted challenges of MIR research, we refer to the overview article by Downie [58]. Recently, Pardo [159] has edited a series of short overview articles, which outline current MIR research problems from a more technical point of view. The tutorial by Orio [157] summarizes fundamental issues on music representations, user interaction, music processing, and specifications of MIR systems. A detailed account on various music analysis problems from a multidisciplinary viewpoint, including aspects from psychoacoustics and music perception, can be found in the Ph.D. thesis by Scheirer [186]. The book edited by Klapuri and Davy [104] deals with the central MIR problem of automatic *music transcription* comprising prominent subproblems such as rhythm analysis, fundamental frequency analysis, source separation, and musical instrument classification. The book also covers important signal processing and pattern recognition methods from a statistical point of view. The comprehensive book by Mazzola [132] deals with mathematical music theory and introduces a conceptual basis for music composition, analysis, and performance. Even though this book presumes a thorough understanding of deep mathematical concepts, it gives a unified approach to a large number of musical aspects and constitutes a rich source of ideas worth considering in MIR research. Finally, we refer to the annual International Conference on Music Information Retrieval (ISMIR), which constitutes a multidisciplinary platform for researchers involved in work on accessing digital musical materials. The proceedings of this conference, which are available online and can be accessed via the conference homepage [97], contain a broad spectrum of research papers and reflect the state of the art in MIR research.

1.2 Motion Retrieval

Generally speaking, motion capturing is referred to as the process of recording real moving objects and creating an abstract three-dimensional digital representation of these motions. The resulting *motion capture data* or simply *mocap data* constitutes the basis for applications in various fields such as gait analysis, rehabilitation, physical therapy, biomechanical research, animal science, and sports performance analysis. With increasing importance, mocap data have also been used in computer animation to create realistic motions for both movies and video games. Here, one typically proceeds in several steps: after planning the motion capture shoots, the motions are performed by live actors and captured by the mocap system. The resulting data are cleaned and suitably processed using editing and blending techniques, and finally the motions are mapped to the animated characters.

Present motion capture systems can track and record human motions at high spatial and temporal resolutions, which makes it possible to capture even subtle movements and nuances of motion patterns. This is the reason why mocap-based animations usually look very realistic and natural compared with traditional hand-animated 3D models, where it is too time consuming and difficult to accurately represent motion details. On the downside, the life-cycle of a motion clip in an animation production is very short. Typically, a motion clip is captured, incorporated in a single 3D scene, and then never used again. Furthermore, the adjustment of once captured motion data to fit certain constraints and user-specified needs is a difficult problem. Therefore, it often seems easier to record new data rather than trying to manipulate the data. However, motion capturing is an expensive process – high-quality mocap systems easily cost more than one hundred thousand dollars including digital video cameras and software. For efficiency and cost reasons, the reuse of mocap data, as well as methods for modifying and adapting existing motion clips, is gaining in importance. Here, an active field of research is the application of editing, morphing, and blending techniques for the creation of new, realistic motions from prerecorded motion clips. Such techniques depend on motion capture databases covering a broad spectrum of motions in various characteristics. Only in the last few years, larger collections of motion materials have become publicly available.

Prior to reusing and processing motion capture material, one has to solve the fundamental problem of *identifying* and *extracting* suitable motion clips scattered in a given database. Traditional approaches to motion retrieval rely on manually generated annotations, where the motions to be identified are roughly described in words such as "a kick of the right foot followed by a punch." Retrieval is then performed at the metadata level. Since the manual generation of descriptive labels is infeasible for large data sets, one needs efficient *content-based* retrieval methods that access only the raw data. Here, a typical query mode is based on the *query-by-example paradigm*, which has attracted a large amount of attention in the field of information retrieval: Given a query in form of a data fragment, the task is to automatically retrieve all documents from the database containing parts or aspects similar to the query.

The crucial point in content-based motion retrieval is the notion of *similarity* used to compare different motions. Typically, the variations may concern the spatial as well as the temporal domain. For example, the kicks shown in Fig. 1.1 describe the same kind of motion even though they differ considerably with respect to motion speed as well as the direction and height of the kick. Due to these spatio temporal motion variations, content-based retrieval of motion capture data constitutes a difficult and time-consuming problem. Recent approaches to motion retrieval apply techniques such as dynamic time warping, which, however, are not applicable to large data sets due

Fig. 1.1. (*Top*) Seven poses from a side kick sequence. (*Bottom*) Corresponding poses for a front kick. The two kicking motions, even though being similar in some semantic sense, exhibit significant spatial and temporal differences

to their quadratic space and time complexity. In view of more efficient and robust motion retrieval strategies, the following observation is of fundamental importance: opposed to other data formats such as image or video, motion capture data format is based on a kinematic chain, which represents the human skeleton. This underlying model can be exploited by looking for geometric relations between specified body points of a pose, where the relations possess an explicit semantic meaning. For example, suppose the user is looking for all right foot kicking motions contained in the database. Even though there may be large variations between different kicking motions, all such motions share some common characteristics: first the right knee is stretched, then bent, and finally stretched again, while the right foot is raised during this process. Afterward, the right knee is again bent and then stretched, while the right foot drops to the floor again. In other words, by only considering the two simple boolean relations "right knee bent or not" and "right foot raised or not" in the temporal context, one can exclude all motions in the database that do not reveal the characteristic progression of relations as described above. With this strategy, one can typically cut down the search space very efficiently from several hours to a couple of minutes of motion capture data, which can then by analyzed and processed by more refined techniques.

1.2.1 Outline of Part II

In the second part of this monograph, we systematically introduce a general and unified framework for motion analysis, retrieval, and classification based on relational features. One main idea is to transform the motions into boolean feature sequences that show a high degree of invariance toward spatio temporal variations. Doing so, one can then adopt standard indexing methods based on inverted lists to facilitate efficient content-based motion retrieval. To keep Part II self-contained, we start with some fundamentals on motion capture data and their representations. The subsequent chapters then give an account of state-of-the-art techniques that represent the author's research output in the field of motion retrieval. The only material needed from Part I is covered in Chap. 4, which introduces the concept of dynamic time warping. We now give an overview of the main topics discussed in Part II.

In Chap. 9, for motion capture data, we describe a data model that is used throughout this monograph and then discuss some general similarity aspects that are crucial in view of motion comparison. Furthermore, we formally introduce the concept of kinematic chains, which are generally used to model flexibly linked rigid bodies such as robot arms or human skeletons. Kinematic chains are parameterized by joint angles, which in turn can be represented in various ways. We describe and compare three important angle representations based on rotation matrices, Euler angles, and quaternions. Each of these representations has its strengths and weaknesses depending on the respective analysis or synthesis application.

To compare two different motion sequences, one needs a concept of measuring similarity that can deal with spatial as well as temporal variations. One common strategy is to bring the frames of the two motion sequences into temporal correspondence and then to measure the frame-wise spatial differences under this correspondence. An important technique to time-align general sequences is known as dynamic time warping (DTW). For a detailed account on DTW, we refer to Chap. 4. In Chap. 10, we describe some DTW-based methods for motion comparison and retrieval based on suitable local cost measures.

As it turns out, the technique of dynamic time warping is computationally expensive and therefore prohibitive for large data sets. In Chap. 11, we study a different approach to motion analysis that is based on relational features assuming only the values zero and one. Such features express the presence or absence of certain relations among body parts. The main idea is to label individual frames of motion data according to such binary classifiers and to segment the data by merging adjacent frames with identical labels. Motion comparison can then be performed at the classifier and segment level. This qualitative approach leads to notions of motion similarity that, in many cases, are more intuitive and robust than purely numerical similarity measures. A further crucial advantage is that the boolean feature vectors are well suited for indexing the motion data using inverted lists. This affords very efficient content-based motion retrieval even for huge mocap databases, see Chap. 12.

In Chap. 13, we deal with the question of how to capture and learn the essence of an entire motion class. We introduce the concept of motion templates (MTs) as one possible solution by which the spatio temporal characteristics of a class of semantically related motions can be captured in an explicit and semantically interpretable matrix representation. The key property of MTs is that the variable aspects of a motion class can be automatically masked out in the comparison with unknown motion data. This facilitates automatic motion annotation as well as robust and efficient retrieval even in the presence of large spatio temporal variations, as discussed in Chap. 14.

1.2.2 Further Notes

We now give some references that may serve the reader as pointers to the literature on motion synthesis and analysis. Further links to the literature are given at the end of each chapter.

In view of massively growing multimedia databases of various types and formats, efficient methods for indexing and content-based retrieval have become an important issue. Vast literature exists on content-based database retrieval including text [218], image, and video data [11], as well as 3D models [74]. For the music scenario, Clausen and Kurth [42] give a unified approach to content-based retrieval; their group theoretical concepts generalize to other domains as well. The problem of indexing large time series databases has also attracted great interest in the database community, see, e.g., Keogh [100] and Last et al. [117] and references therein.

The reuse of motion capture data via editing and morphing techniques has been a central topic in data-driven computer animation for a decade, starting with [22,217]. Since then, many different methods have been suggested to create new, realistic motions from prerecorded motions; see, for example, [6, 35, 75, 106, 107, 169, 224] and the references therein. Only recently, motion capture data have become publicly available on a larger scale (e.g., CMU [44]), reinforcing the demand for efficient indexing and retrieval methods. Due to possible spatio temporal variations, the difficult task of identifying similar motion segments still bears open problems. Most of the previous approaches to motion comparison are based on features that are semantically close to the raw data, using 3D positions, 3D point clouds, joint angle representations, or PCA-reduced versions thereof, see [70, 93, 103, 107, 183, 220]. One problem of such features is their sensitivity toward pose deformations, as may occur in semantically related motions. To achieve more robustness, Müller et al. [146] introduce relational features and adaptive temporal segmentation, absorbing spatio temporal variations already at the feature level. A similar strategy has been used by Liu et al. [123], who transform motions into sequences of cluster centroids, which absorb spatio temporal variations. Motion comparison is then performed on these sequences.

Automatic motion annotation and classification are closely related to the retrieval problem and are important tasks in view of motion reuse. Arikan and Forsyth [6] propose a semi automatic annotation procedure for motion data using support vector machine classifiers. Rose et al. [180] group similar example motions into "verb" classes to synthesize new, user-controlled motions by suitable interpolation techniques. Several approaches for automatic classification and recognition of motion patterns are based on Hidden Markov models (HMMs), which are also a flexible tool to capture spatio temporal variations, see, e.g., [19, 87, 216]. Temporal segmentation of motion data can be viewed as another form of annotation, where consecutive, semantically related frames are organized into groups, see, e.g., [12, 67].

1.3 General Concepts

Even though the two types of multimedia data discussed in this monograph bear no immediate semantic relationship, music and motion share – from an abstract point of view – some common properties. First of all, both kinds of data are subject to relatively strong model assumptions. For example, most Western music is based on the traditional equal-tempered chromatic scale implying the occurrences of distinguished frequency components in an audio recording. This assumption can be exploited by extracting, e.g., pitch-based audio features that allow for some direct, musically relevant interpretation. Similarly, motion capture data are based on an explicit model in the form of a kinematic chain, which represents the human skeleton. This underlying model can be exploited by looking for geometric relations between specific body points of a pose, where the relations possess an explicit semantic meaning. For example, using the two boolean relations that check whether the "right knee is bent or not" and whether the "right foot is raised or not," one can capture important characteristics of a right foot kicking motion. As a first general strategy, we extract features from the raw data that closely correlate to semantic aspects of the underlying data while showing a high degree of invariance to irrelevant deformations and variations.

A further common property of music and motion data is their temporal dimension. In both cases the data can be transformed into time-dependent feature sequences that reflect the changing characteristics in the raw data over time. For example, an audio recording can be transformed into a sequence of pitch-based feature vectors that closely relate to the harmonic progression of the underlying piece of music. Similarly, a combination of simple boolean relations in the temporal context is often sufficient to characterize specific classes of similar motions. As a second general strategy, we exploit the temporal order of the extracted features to identify similar music and motion fragments.

Finally, note that for both domains, music and motion, semantically related objects exhibit a large range of variations concerning temporal, spatial, spectral, or dynamic properties. To handle deformations and variability in the objects, our approach is to simultaneously employ various invariance and fault-tolerance mechanisms at different conceptual levels. Firstly, by employing deformation-tolerant features, we already absorb a high degree of the undesired variations at the feature level. Secondly, to compare features or short-feature sequences, we introduce enhanced local cost measures that are suited to handle local temporal and spatial variations. Thirdly, by using global similarity measures that are based on mismatch, fuzzy, and time-warping concepts, we add another degree of robustness and fault tolerance at a more global level.

In addition to deformation tolerance and robustness, the question of efficiency is of fundamental importance in content-based multimedia retrieval, in particular in view of large amounts of data that have to be processed and searched. To speed up computations, we employ various methods ranging from data reduction and clustering techniques, over multiscale (coarse-to-fine) and preselection strategies, to index-based search and retrieval procedures. As it turns out, an overall procedure for content-based multimedia retrieval typically suffers from the trade-off between the capability of handling object deformations on the one hand and retrieval efficiency on the other hand. For example, time-warping strategies are powerful in coping with temporal deformations but are generally expensive with respect to computation time and memory. In contrast, index-based strategies may afford efficient data retrieval that scales to large data sets, however, at the expense of being rather inflexible to variations. Here, the usage of fault-tolerance mechanisms such as mismatch or fuzzy search restores flexibility to some extent, but at the price of increased computational cost. As a final general strategy, our decision for a particular choice or combination of methods will be led by the applicability and practicability of the overall retrieval procedures in real-world application scenarios.

In the following, we point out some of the conceptual relations between the two parts. As mentioned before, one general principle to cope with data variations is the usage of relatively coarse but semantically meaningful features that show a high degree of invariance to certain deformations. By exploiting available model assumptions, we introduce pitch/chroma-based features for the music domain (Chap. 3) and relational features for the motion domain (Chap. 11). To achieve robustness to local temporal variations, the respective feature sequences are further processed by considering short-time statistics of certain energy distributions (a kind of temporal smearing, see Sect. 3.3) and by applying adaptive segmentation techniques (a kind of temporal clustering, see Sect. 11.2).

To compare features or short sequences of features, we introduce various local cost (or equivalently, distance or similarity) measures. In many applications, one has to evaluate local cost measures for a large number of feature pairs. As a result, the evaluation becomes a time-critical task. Here, the idea is to cope with variations at the feature level, which then enables the use of relatively simple and efficiently computable local cost measures. For real-valued feature vectors, we often resort to local cost measures that are based on ordinary inner vector products (Sects. 5.2.2, 6.1.2, 7.2, 10.1.1, and 14.1), whereas for discrete-valued features, we also apply exact matching techniques (Sects. 6.4.1 and 12.1.2). To introduce further degrees of robustness and fault tolerance, we enhance local cost measures in various ways. Firstly, by incorporating contextual information into the local cost measure one can improve structural properties of the resulting cost matrix, which can largely simplify the subsequent extraction of global information. This principle has been applied in Sect. 5.2.5 to stabilize data alignment and in Sect. 7.2 to afford robust structure analysis. Secondly, we discuss a data-adaptive local cost measure, which disregards the class inconsistencies previously learned from training data (Sect. 14.1). Thirdly, exact feature matching is softened by introducing the concept of fuzzy matching. This technique is used for index-based audio matching (Sect. 6.4.2) as well as for index-based motion retrieval (Sect. 12.1.3). Semantically motivated local cost (similarity) measures are discussed in Sect. 5.3.2 for the music domain and in Sect. 10.1 for the motion domain. A further modification of local cost measures is discussed in Sect. 5.2.2 in the context of audio synchronization.

Based on a suitable local cost measure, dynamic time warping (DTW) is a well-known technique for aligning feature sequences and for handling local and global tempo variations. The DTW concept plays an important role throughout this monograph. In Chap. 4, we give a general introduction to DTW, summarize some of its variants, and discuss efficiency issues. General DTW-based methods for motion retrieval and comparison are presented in Chap. 10. An iterative multiple alignment procedure, which uses DTW as the main ingredient, is used in deriving a motion template (MT) representation expressing the essence of an entire class of semantically related motions (Chap. 13). These motion templates, in turn, facilitate robust motion retrieval based on a local variant of DTW (Chap. 14). In the music domain, we employ dynamic time warping for various synchronization tasks where different manifestations of the same piece of music are automatically linked (Chap. 5). In particular, we present an efficient multiscale DTW algorithm (Sect. 4.3) for audio synchronization (Sect. 5.2). In view of robustness and efficiency, several strategies have been suggested to replace or combine DTW-based techniques with methods based on local uniform scaling. We use linear scaling techniques in the context of audio matching to retrieve musically similar audio excerpts from a music database (Sect. 6.1.3). Similarly, in the context of audio structure analysis linear scaling techniques are employed to enhance the structural properties of self-similarity matrices (Sect. 7.2). This allows us to use a simple greedy strategy – instead of reverting to more cost-intensive DTW-based techniques – to extract local warping paths (Sect. 7.3).

In view of large amounts of data, efficient pruning methods are required, in particular in the retrieval stage, to cut down the search space to a small subset that still contains the relevant data. To this end, we adopt efficient indexing methods based on inverted files – a well-known index structure used in classical text retrieval, which is particularly suited for exact matching problems. Here, our contribution is to find suitable combinations of deformation-invariant features and tolerance mechanisms that enable us to use this relatively rigid indexing method in the music (Sect. 6.4) and in the motion domain (Chap. 12) despite large variations in the raw data. Introducing various types of fault tolerance, we show how the same index structures can be used for fuzzy search (Sects. 12.1.3, and 6.4.2), adaptive fuzzy search (Sect. 12.1.4), and keyframe search (Sects. 12.3.2 and 14.4).

To prove the practicability and efficiency of our analysis and retrieval methods, all algorithms presented in this monograph have been implemented and tested on comprehensive data collections. In particular, we conducted our experiments on real-world data (opposed to synthetic data generated under laboratory conditions). Our music retrieval algorithms have been tested on a corpus of more than one thousand audio recordings, reflecting a wide range of classical music (Sect. 6.2). For our motion experiments, we systematically recorded several hours of motion capture data containing a variety of motions performed by different actors (Sect. 13.4.1). These data were supplemented with motion data obtained from the CMU mocap database [44]. Several retrieval scenarios are described in detail including a motion retrieval system based on the query-by-example paradigm (Fig. 12.2, Sect. 12.2) and a system for efficient audio matching (Fig. 6.9, Sect. 6.4.3). In an ongoing software project, we are implementing a comprehensive system (SyncPlayer system) for multimodal inter- and intradocument browsing and retrieval in complex and inhomogeneous music collections by integrating various music analysis and retrieval techniques presented in this monograph. In Chap. 8, we outline the functionalities of the current version of the SyncPlayer.

Part I

Analysis and Retrieval Techniques for Music Data

2

Fundamentals on Music and Audio Data

In the first part of this monograph, we discuss content-based analysis and retrieval techniques for music and audio data. To account for the interdisciplinary character of this research field, we start in this chapter with some fundamentals on music representations and digital signal processing. In particular, we summarize basic facts on the score, MIDI, and audio format (Sect. 2.1). We then review various forms of the Fourier transform (Sect. 2.2) and give a short account of digital convolution filters (Sect. 2.3). Doing so, we hope to refine and sharpen the understanding of the required basic signal transforms. This will be essential for the design as well as for the proper interpretation of musically relevant audio features, see Chap. 3.

2.1 Music Representations

Modern digital music libraries contain textual, visual, and audio data. Among these multimedia-based types of information, music data pose many problems, since musical information is represented in diverse data formats. These formats, depending upon particular applications, differ fundamentally in their respective structures and content. In this section, we concentrate on three widely used formats for representing music data: The symbolic *score format* contains information on the notes such as musical onset time, pitch, duration, and further hints concerning dynamics and agogics (Sect. 2.1.1). The purely physical *audio format* encodes the waveform of an audio signal as used for CD recordings (Sect. 2.1.2). Finally, the *MIDI format* may be thought of as a hybrid of the last two data formats that explicitly represents content-based information such as note onsets and pitches but may also encode agogic and dynamic subtleties of some specific interpretation (Sect. 2.1.3).

2.1.1 Score Representation

A *musical score*, also referred to as *sheet music*, gives a symbolic description of what we commonly refer to – in particular for Western classical music – as the "piece of music." The score encodes a musical work in a formal language and depicts it in a graphical–textual form, which allows a musician to create a performance by following the given instructions. Figure 2.1 shows the first five measure of Beethoven's Fifth in a piano reduction. In a score, the music is represented by note objects, which, in turn, are given in terms of attributes such as pitch, musical onset time, duration, dynamics, or articulation. The tempo is specified by textual notations such as *Allegro con brio* or *Andante con moto* for global tempo instructions and *accelerando* or *ritardando* for local tempo variations. Similarly, loudness and dynamics are described by terms such as *piano*, *forte*, *crescendo*, or *diminuendo*. The score can be seen as a rough guide to music performance that requires previous knowledge and profound experience of the musician to create the intended sound. Typically, there is a lot of space for interpretational freedom, which often leads to variations in tempo, dynamics, or articulation. For example, the duration of the fermata in the theme of Beethoven's Fifth may vary significantly between two different interpretations. Even at the note level there may be variations in the note execution implied by signifiers such as trills, arpeggios, or grace notes.

Many codes have been suggested in the literature to represent sheet music in a digital, machine-readable form, see [178, 188] for an overview. At this point, we present some aspects of the music file format *MusicXML*, which has been recently developed to serve as a universal translator for common Western music notation, see [151]. MusicXML textually describes how note objects, measures, staves, systems, and etc. appear in a printed score. As an example, Fig. 2.2 shows how a middle E flat as appearing in the theme of Beethoven's Fifth is encoded. To simplify the notation, we denote the middle E flat by E$^\flat$4 following the convention that the suffix 4 refers to the octave

Fig. 2.1. Score representation of the first five measures of Beethoven's Fifth in a piano reduction (from [152])

```
<note>
  <pitch>
    <step>E</step>
    <alter>-1</alter>
    <octave>4</octave>
  </pitch>
  <duration>2</duration>
  <type>half</type>
</note>
```

Fig. 2.2. Textual description of a middle E flat (E$^\flat$4) with a duration of 2 quarters and notated as a half note in the MusicXML format (The clef, stave, signs, and measure have to be defined at the beginning of the MusicXML file.)

containing notes from the middle C (C4) and the middle B (B4). Similarly, a suffix is attached for notes of the lower and higher octaves such that the lowest note on a piano is denoted by A0 and the highest by C8. Now, in the MusicXML encoding of the half note E$^\flat$4, the tags `<note>` and `</note>` mark the beginning and the end of a MusicXML note element. The pitch element, delimited by the tags `<pitch>` and `</pitch>`, consists of a step element E (denoting the letter name of the pitch), the alter element -1 (changing E to E flat) and the octave element 4 (fixing the octave). Thus this note is a middle E flat. The element `<duration>2</duration>` encodes the duration of the note measured in quarter notes. (Here, it has been specified at some earlier point in the MusicXML file that the value 1 corresponds to a quarter note.) Finally, the element `<type>half</type>` tells us that this note is notated as a half note.

There are various ways to generate digital score representations. First, one could manually input the sheet music in a format such as MusicXML, which, however, is tedious and error prone. *Music notation software*, also referred to as *scorewriters*, supports the user in the task of writing and editing digitized sheet music, where the note objects can conveniently be input and modified by a computer's keyboard, a mouse, or an electronic piano. Many of the current music notation programs, including the two most popular scorewriters *Finale* and *Sibelius*, support the interchange of files in the MusicXML format. Another common way of generating digital scores is to scan printed sheet music with a scanner, which converts the score into a set of digitized images. At this stage, the computer considers these images as a mere collection of pixels or color samples and does not grasp the semantics of the sheet music. Therefore, in the second step, the digital images have to be further translated into a standard encoding scheme–for example into MusicXML–that reflects the semantics of the score symbols such as the notes, rests, or clefs. The process of scanning and transforming a printed score into a computer-editable format is known as *optical music recognition*, also abbreviated as OMR. Even though current commercial OMR software such as [162,190,195] report on an accuracy of over 99% for most sheet music (at least when using clean scans), in practice most of the OMR products still offer unsatisfactory recognition results [25]. One problem is that even small artifacts in the scan may lead to interpretation problems (e.g., a dot may be mixed up with a staccato mark) and to incorrect recognition results, which then requires manual intervention for correction. So far, optical music recognition of handwritten scores is a major unresolved problem, see, e.g. [77,136] for possible approaches.

2.1.2 Sound, Waveform, and Audio Representation

From a physical point of view, a *sound* or *audio signal* is generated by some vibrating object such as the vocal chords of a singer, the vibrating string and sound board of a violin, or the diaphragm of a kettle drum. These vibrations cause displacements and oscillations of the particles in the air, which in turn causes local regions of compression and rarefaction in the air. The alternating pressure travels as a wave from its source through the air to the listener or a microphone, which can then be perceived as sound by the human ear or be converted into an electrical signal by the microphone. Graphically, the change in air pressure at a certain location can be represented by a *pressure–time plot*, also referred to as the *waveform* of the sound, which shows the deviation of the air pressure from the average air pressure (usually this deviation is measured in pascal). If the points of high and low air pressure repeat in some alternating and regular fashion, the resulting waveform is also called *periodic*, see Fig. 2.3. In this case, the *period* of the waveform is defined to be the time between two successive high pressure points (or, equivalently, between two low pressure points). The *frequency*, measured in hertz (Hz), is then the reciprocal of the period. For example, the sinusoidal waveform shown in Fig. 2.3 has a period of a quarter second and hence a frequency of 4 Hz.

The *harmonic sound* or *pure tone*, which corresponds to a sinusoidal waveform as indicated in Fig. 2.3, can be considered as the prototype of an acoustic realization of a musical note. The property of a sound that correlates to the perceived frequency is commonly referred to as the *pitch*. For example, the

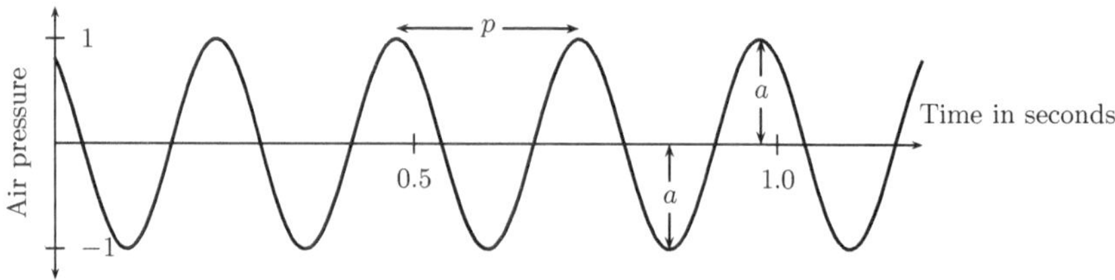

Fig. 2.3. Waveform of a periodic sound with a frequency of 4 Hz. The *horizontal arrow* indicates the period p, whereas the *vertical arrows* indicate the amplitude a

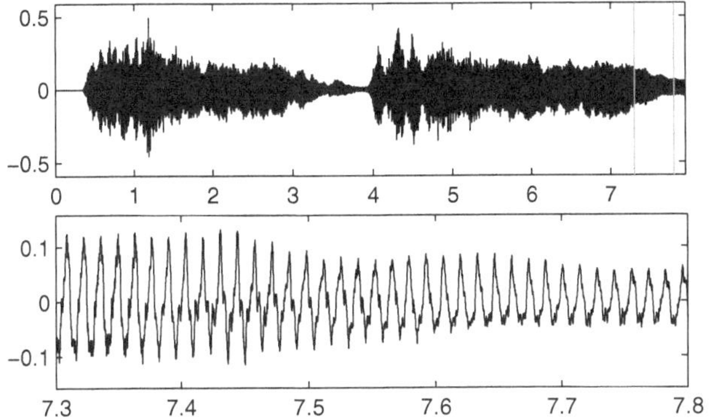

Fig. 2.4. (Top) Waveform of the first eight seconds of a Bernstein interpretation corresponding to the first five measures of Beethoven's Fifth as indicated in Fig. 2.1. (Bottom) Enlargement of the section between the seconds 7.3 and 7.8

middle A, also known as the concert pitch, has a frequency of 440 Hz. Since a slight change in frequency need not lead to a perceived change in pitch, one usually associates an entire range of frequencies with a single pitch. It is a well-known fact that the human sensation of pitch is logarithmic in nature, see [225]. For example, the perceived distance between the pitches of the A3 (220 Hz) and A4 (440 Hz) is the same as the perceived distance between the pitches A4 (440 Hz) and A5 (880 Hz). The interval between two sounds with half or double the frequency is referred to as an *octave*. The close relation between sounds separated by one octave, also referred to as *octave equivalency*, is closely related to our sensation of harmony and lays the foundation of the music notation based on the chromatic scale as used in traditional Western music. We will exploit this fact in music comparison by using sound representations where the frequencies are considered up to octave equivalence, see Sect. 3.3.

When playing a single note on an instrument, the resulting sound or *musical tone* is far from being a simple pure tone with a well-defined frequency. Intuitively, a musical tone can be regarded as a superposition of pure tones – so-called *harmonics* or *overtones* – whose frequencies differ by an integer multiple from a certain *fundamental frequency*.[1] Furthermore, a musical tone usually contains nonperiodic components such as noise-like components and transients, which typically appear during the attack phase of most instruments (e.g., when striking a piano key). Nevertheless, a musical tone conveys the sensation of a pitch, which corresponds to the fundamental frequency. As an example, consider Fig. 2.4 showing in its upper part the waveform of the first eight seconds of a Bernstein interpretation of Beethoven's Fifth. The lower part shows an enlargement of the section between the seconds 7.3 and 7.8,

[1] By definition, the first harmonic corresponds to the fundamental frequency, the second harmonic to the first overtone, and so on.

which reveals the almost periodic nature of the sound signal. The waveform within these 500 ms corresponds to the sound of a decaying D, which is played by the orchestra in unison in the fourth and fifth measure, see Fig. 2.1. Indeed one counts 37 periods within this section corresponding to a fundamental frequency of 74 Hz – the frequency of D2.

A further important aspect of music concerns the *dynamics*, which, in the physical context, refers to the intensity or loudness of a sound. At this point, we only give some intuitive description of these concepts and refer to the literature for a precise definition, see, e.g. [225]. Sound *intensity* refers to the energy of the sound per time and area unit, which closely correlates to the amplitude of the waveform. For a periodic waveform as shown in Fig. 2.3, the *amplitude* is defined as the magnitude of the maximal deviation of the air pressure from the average air pressure during one period. The *loudness* is – similar to the relation between pitch and frequency – the subjective psychological correlate of amplitude or sound intensity. Actually, research in psychoacoustics, i.e., the study of subjective human perception of sounds, has shown that loudness depends on many factors including frequency, timbre, amplitude, and duration. For the sake of simplicity, we may assume that the amplitude of the waveform relates to our perception of loudness. For arbitrary waveforms, one can consider the *envelope* as illustrated by Fig. 2.5b: the area enclosed by the envelope of the waveform within a certain time interval reflects the loudness (local energy) of the sound. For example, from Fig. 2.5b one can directly read off the decay of loudness of the sustained E flat (seconds 1.2 to 4) and the sustained D (seconds 5 to 7.9), which appear as fermata in the two

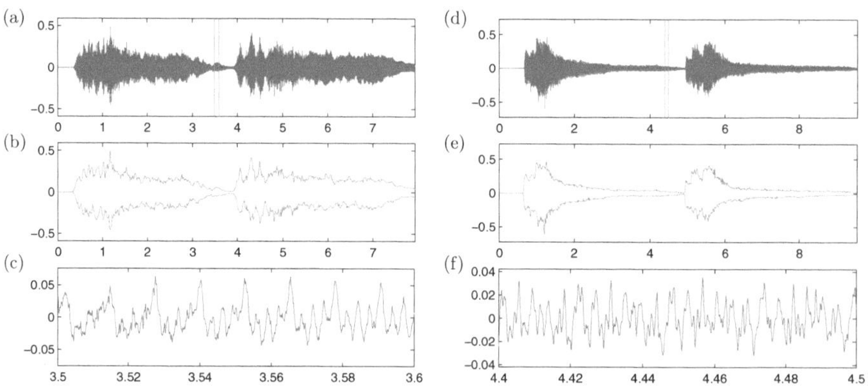

Fig. 2.5. (**a**) Waveform of a Bernstein interpretation of the first five measures of Beethoven's Fifth. (**b**) Envelope of the waveform shown in (a). (**c**) Enlargement of the section between the seconds 3.5 and 3.6. (**d**) Waveform of a piano interpretation by Glen Gould of the same five measures. (**e**) Envelope of the waveform shown in (d). (**f**) Enlargement of the section between the seconds 4.4 and 4.5

fate motives. The decrease of loudness during the two fermatas becomes even more apparent in the piano version played by Glen Gould, see Fig. 2.5e.

Besides pitch, loudness, and duration, there is another fundamental aspect of sound referred to as *timbre* or *tone color.* The timbre allows the listener to distinguish the musical tone of a violin, an oboe, or a trumpet even if the tone is played at the same pitch and with the same loudness. In particular, the timbre depends on the proportion of the harmonics' intensities as well as on the noise-like sound components. For details we refer to [189].

The waveform representation, as opposed to the score representation, encodes all information needed to reproduce the acoustic realization of a specific musical interpretation. This includes the temporal, dynamic, and tonal microdeviations that make the music seem alive. However, in waveform note parameters such as onset times, neither pitches nor note durations are given explicitly. As mentioned above, even a single note of the score becomes a complex sound when played on an instrument producing several harmonics as well as noise components and vibrations. Therefore, sound analysis and comparison is difficult to handle on the basis of waveform representations. This is also illustrated in Fig. 2.5c, which shows two rather distinct waveforms even though they belong to the same pitch E flat–once played by an orchestra and once played by a piano. The complexity of a waveform representation dramatically increases when considering polyphonic orchestral music, where the components of various musical tones interfere with each other and intermingle irrecoverably. This renders the task of determining note parameters from the waveform of a polyphonic orchestral piece a difficult one. Except for very simple music, the automatic conversion of a music performance into score notation by a computer–a task Mozart was capable of after listening to a polyphonic cantata only once–is still a largely open problem despite decades of research, see [104]. In Chap. 3 we discuss how one can, at least, extract note-related features from waveform representations, which then facilitates a comparison of various music and waveform representations at the feature level.

The waveform as introduced so far is *analog* in the sense that it is continuous in both time and amplitude, which leads to an infinite number of values associated with the waveform. Since a computer can only store and process a finite number of values, one has to convert the waveform into some *discrete* or *digital* representation–a process commonly referred to as *digitization*. The digitization of waveforms consists of two steps called *sampling* and *quantization*, see Fig. 2.6 for an illustration. In the first step, the waveform is read or *sampled* at uniform time intervals. Then, in the second step, the value of the waveform at each sampled point is constrained or *quantized* to a discrete set of values. For example, a compact disc (CD) stores digitized sound data at a sampling rate of 44 100 Hz, i.e., the waveform is sampled at 44 100 points per second. The resulting values are then quantized to a set of 65 536 possible values, which are then encoded by a 16-bit coding scheme. Note that the digitization of the waveform is a *lossy* transformation in the sense that one loses information in this process. Therefore, it is generally not

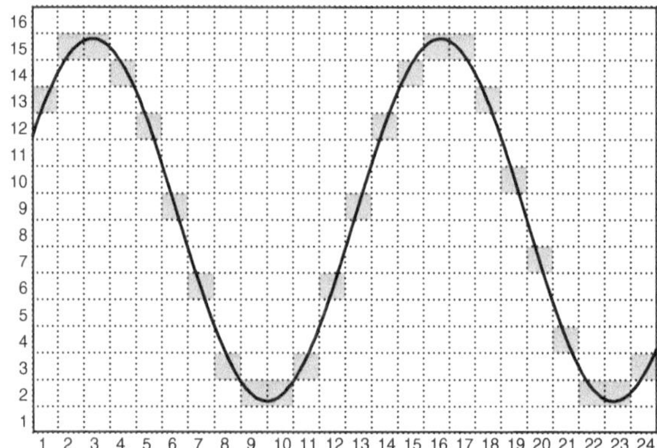

Fig. 2.6. Analog waveform (*black curve*) and digitized representation (the digitized values are indicated by the *gray boxes*). In this example, the waveform is sampled at 24 points and the quantization of the values employs a 4-bit (16 possible values) coding scheme

possible to reconstruct the original waveform from the digital representation. The introduced errors are known as *aliasing* and *quantization errors*, which may introduce audible sound artifacts such as harsh buzzing sounds or noise. For digital representations as used for CDs, however, the chosen sampling rate as well as the quantization resolution is chosen in such a way that the degradation of the waveform is not noticeable by the human ear. For further details, we refer to the literature, see, e.g., [164].

In conclusion of this section, we have seen that the physical realization of music in terms of sounds is a complex issue. We have briefly discussed the waveform representation and some fundamental properties of musical sounds such as pitch, loudness, and timbre–properties that also depend on perceptional aspects as studied in psychoacoustics, see [225]. In the subsequent sections, we only require an intuitive understanding of musical sound and its waveform representation. For the rest of this chapter, we simply use the term *audio representation* to refer to the physical domain of music, be it the musical sound of a performance in its analog form or the digitized waveform of a CD recording.

2.1.3 MIDI Representation

The *MIDI representation* of music may be thought of as a hybrid of the score and the audio representation: it can encode important information in the notes of the score as well as agogic and dynamic niceties of a specific interpretation. However, MIDI is quite limited especially in representing the timbre of a sound. In this section, we give a short overview of MIDI, which is used as the most common *symbolic* digital music interchange format today.

MIDI stands for *Musical Instrument Digital Interface* and has originally been developed as an industry standard to get digital electronic musical instruments from different manufacturers to work and play together, see [134]. Actually, it was the advent of MIDI in 1981–1983 that caused a rapid growth of the electronic musical instrument market. MIDI allows a musician to remotely and automatically control an electronic instrument or a digital synthesizer in real time. As an example, let us consider a digital piano, where the musician pushes down a key of the piano keyboard to start a sound and controls the intensity of the sound by the velocity of the keystroke. Releasing the key stops the sound. Instead of physically pushing and releasing the piano key, the musician may also trigger the instrument to produce the same sound by transmitting suitable MIDI messages, which encode the note-on, the velocity, and the note-off information. These MIDI messages may be automatically generated by some other electronic instrument or may be fed in via a computer. It is an important fact that MIDI does not represent musical sound directly, but only represents performance information encoding the instructions about how an instrument has been played or how music is to be produced.

The original MIDI standard was later augmented to include the file format specification *Standard MIDI File* (SMF), which describes how MIDI data should be stored on a computer. This file format allows users to exchange MIDI data regardless of the computer operating system and has provided a basis for an efficient Internet-wide distribution of music data in the SMF format, including numerous Web sites devoted to the sale and exchange of such data. A MIDI file contains a list of MIDI messages together with timestamps, which are required to determine the timing of the messages. Further information (called metamessages) is relevant to software that processes MIDI files.

For our purpose, the most important MIDI messages are the note-on and the note-off commands, which correspond to the start and the end of a note, respectively. Intuitively, one may think of a note-on and a note-off message to consist, among others, of a MIDI note number, a number for the key velocity, a channel specification, as well as a timestamp. The *MIDI note number* is an integer between 0 and 127 and encodes the pitch of a note. Here, MIDI pitches are based on the equal-tempered scale as used in Western classical music. Similar to an acoustic piano, where the 88 keys of the keyboard correspond to the musical pitches A0 to C8, the MIDI note numbers encode, in increasing order, the musical pitches C0 to G$^\sharp$9, see also Fig. 2.7. For example, the middle C denoted by C4 has the MIDI note number 60, whereas the concert pitch A4 has the MIDI note number 69. The *key velocity* is also an integer between 0 and 127, which basically controls the intensity of the sound–in the case of a note-on event it determines the volume whereas in the case of a note-off event it controls the decay during the release phase of the tone. The exact interpretation of the key velocity, however, depends on the respective instrument or synthesizer. The *MIDI channel* is an integer between 0 and 15. Intuitively speaking, this number prompts the synthesizer to use the

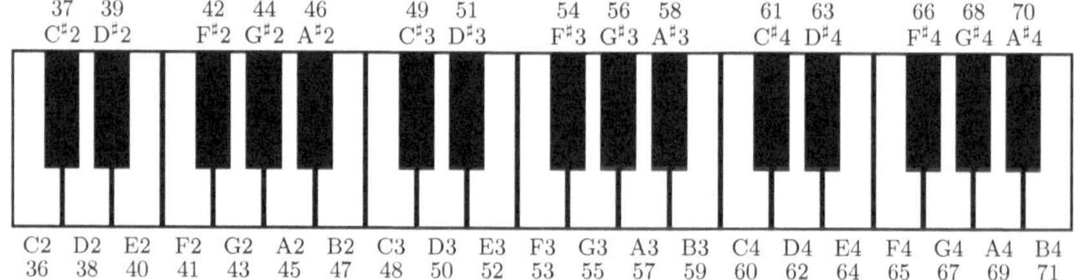

Fig. 2.7. Portion of a piano keyboard with the keys labeled by the pitch names and the MIDI note numbers. For example, the pitch A4 corresponds to the middle A (concert pitch) and has the MIDI note number 69

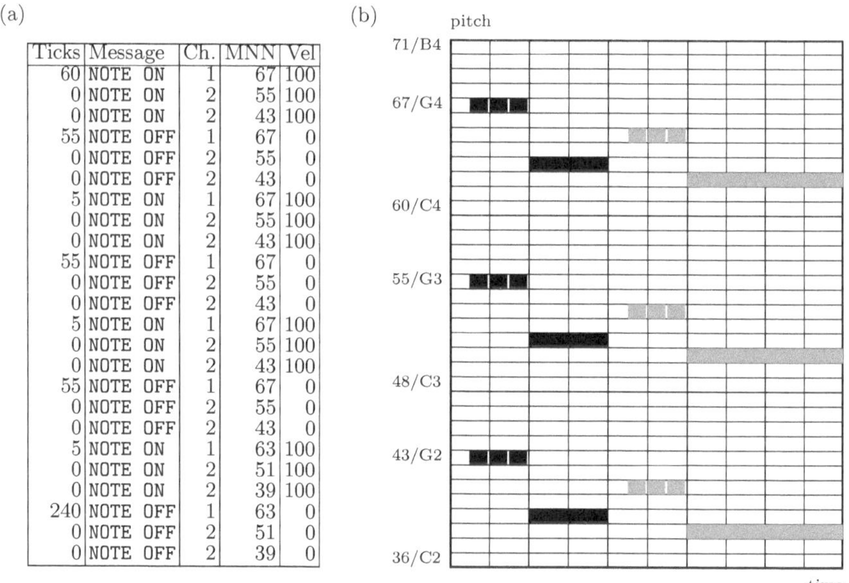
(a)

Ticks	Message	Ch.	MNN	Vel
60	NOTE ON	1	67	100
0	NOTE ON	2	55	100
0	NOTE ON	2	43	100
55	NOTE OFF	1	67	0
0	NOTE OFF	2	55	0
0	NOTE OFF	2	43	0
5	NOTE ON	1	67	100
0	NOTE ON	2	55	100
0	NOTE ON	2	43	100
55	NOTE OFF	1	67	0
0	NOTE OFF	2	55	0
0	NOTE OFF	2	43	0
5	NOTE ON	1	67	100
0	NOTE ON	2	55	100
0	NOTE ON	2	43	100
55	NOTE OFF	1	67	0
0	NOTE OFF	2	55	0
0	NOTE OFF	2	43	0
5	NOTE ON	1	63	100
0	NOTE ON	2	51	100
0	NOTE ON	2	39	100
240	NOTE OFF	1	63	0
0	NOTE OFF	2	51	0
0	NOTE OFF	2	39	0

Fig. 2.8. (**a**) MIDI encoding (simplified) of the first 12 notes shown in Fig. 2.1. (**b**) Piano roll representation of the score shown in Fig. 2.1. The first 12 notes corresponding to the notes in (a) are indicated by the *black bars*

instrument that has previously assigned to the respective channel number. Note that each channel, in turn, supports polyphony, i.e., multiple simultaneous notes. Finally, the *timestamp* is an integer value that represents how many clock pulses or *ticks* to wait before the respective note-on command is executed. Before we comment in more detail on the timing concept employed by MIDI, we illustrate the MIDI representation by means of the Beethoven example. Figure 2.8a shows a (simplified and tabular) MIDI encoding of the first fate motive corresponding to the first 12 notes of the score indicated in

Fig. 2.1. In this example, the notes of the right hand are assigned to channel 1 and the notes of the left hand to channel 2. The note-on and note-off events of MIDI files are often visualized by the *piano roll representation* as shown in Fig. 2.8b. Here, each horizontal bar corresponds to a note, where the vertical location of the bar indicates the pitch, and the start and end points in horizontal direction reflect the time interval during which the note is on. The piano roll representation as discussed here does not encode the velocity and channel information.

An important feature of the MIDI format is that it can handle musical as well as physical onset times and note durations. Similar to the score format, MIDI expresses timing information in terms of musical entities rather than using absolute time units such as microseconds. To this end, MIDI subdivides a quarter note into basic time units referred to as *clock pulses* or *ticks*. The number of pulses per quarter note (PPQN) is to be specified at the beginning in the so-called *header* of a MIDI file and refers to all subsequent MIDI messages. A common value, as used in the example of Fig. 2.8a, is 120 PPQN, which determines the resolution of the timestamps associated to note events. As mentioned above, a timestamp indicates how many ticks to wait before a certain MIDI message is executed relative to the previous MIDI message. For example, the first note-on message with the MIDI note number 67 is executed after 60 ticks, corresponding to the eighth rest at the beginning of Beethoven's Fifth. The second and third note-on messages are executed at the same time as the first one, encoded by the tick values zero. Then after 55 ticks, the MIDI note 67 is switched off by the note-off and so on.

Similar to the score representation, MIDI also allows for encoding and storing absolute timing information, however, at a much finer resolution level and in a more flexible way. To this end, one can include additional tempo messages that specify the number of microseconds per quarter note. From the tempo message, one can compute the absolute duration of a tick. For example, having 600 000 µs per quarter note and 120 PPQN, each tick corresponds to 50 000 µs. Furthermore, one can derive from the tempo message the number of quarter notes played in a minute–a measurement musicians refer to as *beats per minute* (BPM). For example, the 600 000 µs per quarter note correspond to 100 BPM. While the number of pulses per quarter note is fixed throughout a MIDI file, the absolute tempo information may be changed by inserting a tempo message between any two note-on or other MIDI messages. This makes it possible to account for not only global tempo information but also for local tempo changes such as accelerandi, ritardandi, or fermatas.

In this section, we have only scratched the surface of MIDI and its functionality. For a detailed description of the MIDI standard and related representations we refer to the literature, see [96, 134, 188] and the references therein. Even though hundreds of codes for representing symbolic music have been developed, the MIDI format is still the only symbolic music interchange format in wide use today–despite the fact that MIDI was originally designed to solve problems in electronic music performance and is limited in terms of the musical aspects it represents. As is noted in [80], MIDI is not capable of distinguishing between a middle D sharp and a middle E flat, both of which have the MIDI note number 63. Also, information on the representation of beams, the stem directions, or clefs cannot be encoded by MIDI. Furthermore, MIDI does not define a note element explicitly; rather, notes are bounded by note-on and note-off events (or note-on events with velocity 0). Rests are not represented at all and must be inferred from the absence of notes. Here, it is the objective of score representations such as MusicXML to explicitly describe all aspects to reproduce sheet music without ambiguities and missing data.

2.2 Fourier Transform

The Fourier transform is the most important mathematical tool in audio signal processing. It maps a time-dependent function f into a frequency-dependent function $\hat{f}$, which reveals the spectrum of frequency components that compose the original function. Loosely speaking, a function and its Fourier transform are two sides of the same information:

- The function f displays the time information and hides the information about frequencies. Intuitively, a signal corresponding to a musical recording shows *when* notes are played (change of the air pressure) but not *which* notes are played.
- The Fourier transform $\hat{f}$ displays information about frequencies and hides the time information. Intuitively, the Fourier transform of a music recording indicates which notes are played, but it is extremely difficult to figure out when they are played.

This fact is also illustrated in Fig. 2.9. In this section, we give a short overview of several variants of the Fourier transform and their interrelations. In Sect. 2.2.1, we start with a mathematical description of continuous-time as well as discrete-time signals and introduce the corresponding spaces of finite-energy signals. Such signals possess a Fourier representation, which can be thought of as a weighted superposition of sinusoids of various frequencies (Sect. 2.2.2). Computing the Fourier transform involves the evaluation of integrals or infinite sums. In practice, one has to approximate the Fourier transform by finite sums. This can be done efficiently by means of the well-known fast Fourier transform (FFT), see Sect. 2.2.3. References to the literature will be given in Sect. 2.2.4.

2.2.1 Signals and Signal Spaces

A *signal* can be defined as a function that conveys information about the state of behavior of a physical system. For example, a signal may describe the

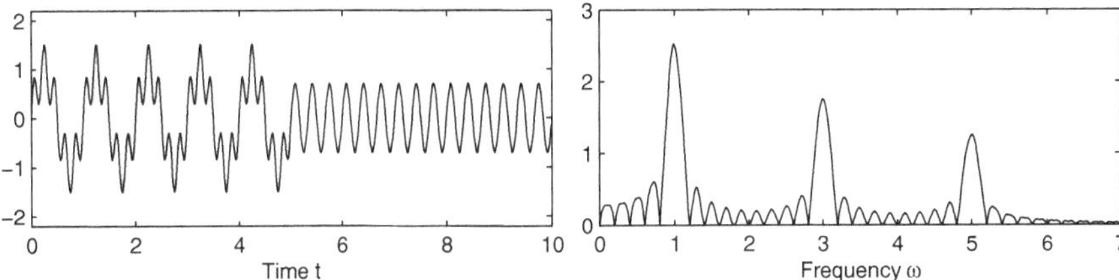

Fig. 2.9. A signal f (*left*) and the absolute value $|\hat{f}|$ of its Fourier transform (*right*). The two peaks at $\omega = 1$ and $\omega = 5$ of $|\hat{f}|$ correspond to the superposition of two sinusoidals constituting the first part of f, whereas the peak at $\omega = 3$ to the sinusoidal constituting the second part of f. The signal f is zero outside the interval $[0, 10]$. The ripples in the spectrum basically come from the phenomena known as *destructive interference*, where many different frequency components are needed to produce the compact support of f

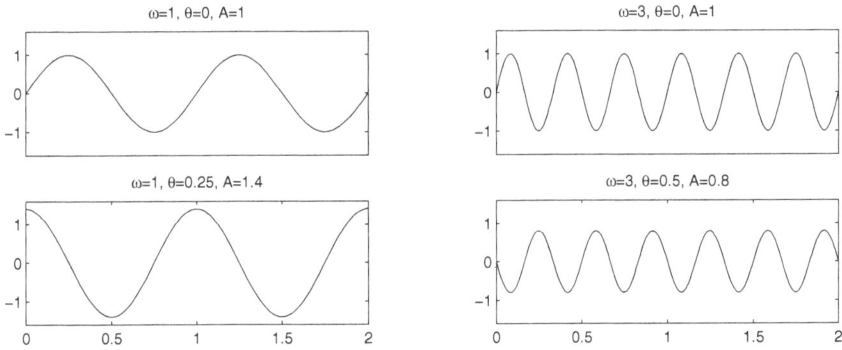

Fig. 2.10. The sinusoid $f(t) = A \sin(2\pi(\omega t - \varphi))$ displayed for $t \in [0, 2]$ and for various values A, ω, and φ

time-varying sound pressure at some place, the motion of a particle through some space, the distribution of light on a screen representing an image, or the sequence of images as in the case of a video signal. In the following, we only consider the case of audio signals as discussed in Sect. 2.1.2. Graphically, such a signal may be represented by its *waveform*, which depicts the amplitude of the air pressure over the time, see Fig. 2.4. In the following, we discuss two different kinds of signals: continuous-time and discrete-time signals.

Mathematically, a *continuous-time* (CT) or *analog* signal is a function $f : \mathbb{R} \to \mathbb{R}$, where the domain $\mathbb{R}$ represents the time axis and the range $\mathbb{R}$ the amplitude of the sound wave. As an example, consider the CT signal f defined by $f(t) := A \sin(2\pi(\omega t - \varphi))$, $t \in \mathbb{R}$, for fixed real parameters A, ω, and φ, see Fig. 2.10. This function is commonly referred to as *sinusoid* of *amplitude* A, *frequency* ω, and *phase* φ. From a musical point of view, A represents the intensity and ω the pitch of the signal. A CT signal is called *periodic* of *period* $\lambda \in \mathbb{R}_{>0}$ if $f(t) = f(t+\lambda)$ holds for all $t \in \mathbb{R}$. For example, the above sinusoid is a periodic signal of period $\lambda = 1/\omega$.

In contrast to a CT signal, a *discrete-time* (DT) signal is defined only on a discrete subset of the time axis. By means of a suitable encoding, one often assumes that this discrete set is a suitable subset I of the set $\mathbb{Z}$ of integers. Then a DT signal is defined to be a function $x : I \to \mathbb{R}$, where the domain I corresponds to points in time. Since one can extend any DT signal from the domain I to the domain $\mathbb{Z}$ simply by setting all values to zeros for points in $\mathbb{Z} \setminus I$, we may assume $I = \mathbb{Z}$. In the following discussion, we often use the symbols f and g to denote CT signals and the symbols x and y to denote DT signals. For the time parameter we often use the symbol t in the CT case and n in the DT case. In view of the Fourier transform, one typically extends the notion of a signal to *complex-valued* CT and DT functions replacing the range $\mathbb{R}$ by $\mathbb{C}$. Note that any real-valued function can be regarded as complex-valued function simply by setting the imaginary part to zero.

A typical procedure to transform a CT signal into a DT signal is known as *equidistant sampling*. Let $f : \mathbb{R} \to \mathbb{C}$ be a CT signal, then the *T-sampling* of f with respect to a real number $T > 0$ is defined to be the DT signal $x : \mathbb{Z} \to \mathbb{C}$ with $x(n) := f(T \cdot n)$. The number T is referred to as *sampling period* and the inverse $1/T$ as *sampling rate*, which is the number of samples per second measured in hertz (Hz). For example, common sampling rates for speech and music signals are 8 kHz for telephony, 32 kHz for digital radio, and 44.1 kHz for CD recordings. Note that the sampling rate is crucial for the quality of the signal. The side effects that are introduced by sampling are known as *aliasing* and are discussed later. Generally speaking, the CT world gives the "right" interpretation of physical phenomena, whereas the DT world is used to do the actual computations.

Phenomena such as superposition or amplification of signals can be modeled by means of suitable operations on the *space* of signals. Let D denote either the domain $\mathbb{R}$ of CT signal or the domain $\mathbb{Z}$ of DT signals. Then the *superposition* of two signals $f : D \to \mathbb{C}$ and $g : D \to \mathbb{C}$ is the sum $f + g$ defined pointwise by $(f + g)(t) := f(t) + g(t)$ for $t \in D$. Similarly, the *amplification* of a signal f by a real factor λ is the scalar multiple λf, which is also defined pointwise. Figure 2.11 gives an example of a superposition of amplified signals. We see in Sect. 2.2.2, how the Fourier transform can be regarded as a kind of reverse operation, which decomposes a given signal into elementary signals with an explicit physical interpretation. The space $\mathbb{C}^D := \{f | f \colon D \to \mathbb{C}\}$ with the above addition and scalar multiplication defines a complex vector space of infinite dimension.

In view of certain signal manipulations such as the Fourier transform or filtering operations, the spaces $\mathbb{C}^{\mathbb{R}}$ or $\mathbb{C}^{\mathbb{Z}}$ are far too large. One therefore defines suitable subspaces that guarantee certain signal properties and imply the feasibility of the desired signal manipulations. Very important subspaces of $\mathbb{C}^{\mathbb{Z}}$ or $\mathbb{C}^{\mathbb{R}}$ are the so-called *Lebesgue spaces* $\ell^p(\mathbb{Z})$ and $L^p(\mathbb{R})$, respectively,

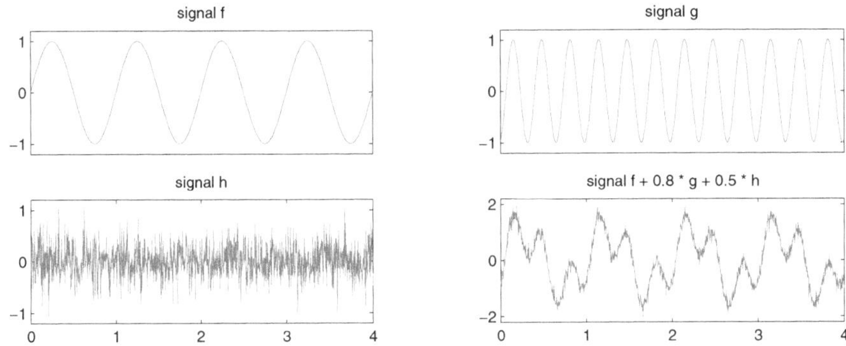

Fig. 2.11. The *lower right corner* shows the superposition of the sinusoids f, the amplified sinusoid g, and the amplified noise signal h

where $p \geq 1$ is either a real parameter or $p = \infty$. Furthermore, identifying the 1-periodic functions of $\mathbb{C}^{\mathbb{R}}$ with the space $\mathbb{C}^{[0,1]}$, one obtains a Lebesgue space $L^p([0,1])$. For a definition and a proof of the main properties of these spaces we refer to [68]. We also refer to the literature for notions such as *norm*, *inner product*, *Banach space*, and *Hilbert space*, see also Sect. 2.2.4 for further comments and references. Table 2.1 summarizes the definition of the Lebesgue spaces $\ell^2(\mathbb{Z})$, $L^2(\mathbb{R})$, and $L^2([0,1])$. These spaces are of particular interest due to the existence of an inner product and the induced Hilbert space structure. By the inner product one can generalize geometric concepts such as *angles* and *orthogonality* as known from the case of finite-dimensional Euclidean spaces to the case of infinite-dimensional function spaces. For a given signal f, the quantity $\|f\|^2$ is also referred to as the *energy* of f. Note that the Lebesgue spaces for $p = 2$ consist exactly of the signals of finite energy.

2.2.2 Fourier Representations

The basic idea of the Fourier representation is to represent a signal as a weighted superposition of independent elementary frequency functions that possess an explicit physical interpretation. Each of the weights expresses to which extend the corresponding elementary function contributes to the original signal, thus revealing a certain aspect of the signal.

At this point, we do not deal with the mathematical intricacies concerning convergence or integrability–we make some remarks on the technical details in Sect. 2.2.4. We start our discussion with the case of 1-periodic CT signals and the space $L^2([0,1])$. Note that any constant as well as any $\frac{1}{k}$-periodic function for an integer $k \in \mathbb{N}$ is 1-periodic too. The sinusoidal $t \mapsto \sqrt{2}\sin(2\pi kt)$ may be regarded as the archetype of a $\frac{1}{k}$-periodic function, which represents a pure tone of k Hz having unit energy, i.e., having an $L^2([0,1])$-norm of one.

Table 2.1. Overview of the Lebesgue spaces $L^2(\mathbb{R})$, $L^2([0,1])$, and $\ell^2(\mathbb{Z})$ and their respective Fourier representation and Fourier transform

Signal space	$L^2(\mathbb{R})$	$L^2([0,1])$	$\ell^2(\mathbb{Z})$
Inner product	$\langle f\|g\rangle = \int_{t\in\mathbb{R}} f(t)\overline{g(t)}\mathrm{d}t$	$\langle f\|g\rangle = \int_{t\in[0,1]} f(t)\overline{g(t)}\mathrm{d}t$	$\langle f\|g\rangle = \sum_{n\in\mathbb{Z}} x(n)\overline{y(n)}\mathrm{d}t$
Norm	$\\|f\\|_2 = \langle f\|f\rangle^{\frac{1}{2}}$	$\\|f\\|_2 = \langle f\|f\rangle^{1/2}$	$\\|x\\|_2 = \langle x\|x\rangle^{1/2}$
Definition	$L^2(\mathbb{R}) :=$ $\{f:\mathbb{R}\to\mathbb{C} \mid \\|f\\|_2<\infty\}$	$L^2([0,1]) :=$ $\{f:[0,1]\to\mathbb{C} \mid \\|f\\|_2<\infty\}$	$L^2(\mathbb{Z}) :=$ $\{f:\mathbb{Z}\to\mathbb{C} \mid \\|x\\|_2<\infty\}$
Elementary frequency function	$\mathbb{R}\to\mathbb{C}$ $t\mapsto \mathrm{e}^{2\pi\mathrm{i}\omega t}$	$[0,1]\to\mathbb{C}$ $t\mapsto \mathrm{e}^{2\pi ikt}$	$\mathbb{Z}\to\mathbb{C}$ $n\mapsto \mathrm{e}^{2\pi\mathrm{i}\omega n}$
Frequency parameter	$\omega\in\mathbb{R}$	$k\in\mathbb{Z}$	$\omega\in[0,1]$
Fourier representation	$f(t) = \int_{\omega\in\mathbb{R}} c_\omega \mathrm{e}^{2\pi\mathrm{i}\omega t}\mathrm{d}\omega$	$f(t) = \sum_{k\in\mathbb{Z}} c_k \mathrm{e}^{2\pi ikt}$	$x(n) = \int_{\omega\in[0,1]} c_\omega \mathrm{e}^{2\pi\mathrm{i}\omega n}\mathrm{d}\omega$
Fourier transform	$\hat{f}:\mathbb{R}\to\mathbb{C}$ $\hat{f}(\omega) = \int_{t\in\mathbb{R}} f(t)\mathrm{e}^{-2\pi\mathrm{i}\omega t}\mathrm{d}t$ $c_\omega = \hat{f}(\omega)$	$\hat{f}:\mathbb{Z}\to\mathbb{C}$ $\hat{f}(k) = \int_{t\in[0,1]} f(t)\mathrm{e}^{-2\pi ikt}\mathrm{d}t$ $c_k = \hat{f}(k)$	$\hat{x}:[0,1]\to\mathbb{C}$ $\hat{x}(\omega) = \sum_{n\in\mathbb{Z}} x(n)\mathrm{e}^{-2\pi\mathrm{i}\omega n}$ $c_\omega = \hat{x}(\omega)$

Strictly speaking, one needs additional technical assumptions and modifications in the above definitions as discussed in Sect. 2.2.4

More generally, all phase-shifted versions $t \mapsto \sqrt{2}\cos(2\pi(kt-\varphi))$ have this interpretation. The idea of the Fourier transform is to represent any 1-periodic function in terms of such sinusoidals. In fact, it can be shown that any real-valued signal $f \in L^2([0,1])$ can be written as

$$f(t) = d_0 + \sum_{k\in\mathbb{N}} d_k\sqrt{2}\cos(2\pi(kt-\varphi_k)) \tag{2.1}$$

for suitable amplitudes $d_k \in \mathbb{R}_{\geq 0}$ and phases $\varphi_k \in [0,1)$. This superposition exhibits the frequency content of f as follows: the coefficient d_k reflects the contribution of the corresponding sinusoidal of k Hz, whereas the coefficient φ_k shows how the sinusoidal has to be shifted to best "explain" or "match" the original signal.

For each frequency parameter $k \in \mathbb{N}$, instead of using one sinusoidal of arbitrary phase, one may use two sinusoidals of different, but fixed phase. In particular, using the two sinusoidals $t \mapsto \sqrt{2}\cos(2\pi kt)$ and $t \mapsto \sqrt{2}\sin(2\pi kt)$, one obtains the well-known *Fourier series*

$$f(t) = a_0 + \sum_{k\in\mathbb{N}} a_k\sqrt{2}\cos(2\pi kt) + \sum_{k\in\mathbb{N}} b_k\sqrt{2}\sin(2\pi kt) \tag{2.2}$$

for suitable coefficients $a_0, a_k, b_k \in \mathbb{R}$, $k \in \mathbb{N}$. These coefficients are also referred to as *Fourier coefficients* of f. From the equality $\cos(2\pi(kt - \varphi_k)) = \cos(2\pi kt)\cos(2\pi\varphi_k) + \sin(2\pi kt)\sin(2\pi\varphi_k)$ it follows that the superposition in (2.1) can be derived from the Fourier coefficients by setting $d_0 = a_0$, $d_k = \left(a_k^2 + b_k^2\right)^{1/2}$, and $\varphi_k = (1/2\pi)\arccos\left(a_k/d_k\right)$, $k \in \mathbb{N}$. Furthermore, the set $\left\{1, \sqrt{2}\cos(2\pi k\,\cdot), \sqrt{2}\sin(2\pi k\,\cdot) | k \in \mathbb{N}\right\}$ can be shown to form an orthonormal (ON) basis of the Hilbert space $L^2([0,1])$. Thus, the Fourier coefficients are given by the inner products of the signal f with the basis functions of the ON basis:

$$a_0 = \langle f|1\rangle = \int_{t\in[0,1]} f(t)\mathrm{d}t, \tag{2.3}$$

$$a_k = \langle f|\sqrt{2}\cos(2\pi k\,\cdot)\rangle = \sqrt{2}\int_{t\in[0,1]} f(t)\cos(2\pi kt)\mathrm{d}t, \tag{2.4}$$

$$b_k = \langle f|\sqrt{2}\sin(2\pi k\,\cdot)\rangle = \sqrt{2}\int_{t\in[0,1]} f(t)\sin(2\pi kt)\mathrm{d}t. \tag{2.5}$$

The two real-valued sinusoids $t \mapsto \cos(2\pi kt)$ and $t \mapsto \sin(2\pi kt)$ may be combined to form a single complex-valued *exponential* function $\mathbf{e}_k : [0,1] \to \mathbb{C}$ defined by

$$\mathbf{e}_k(t) := \mathrm{e}^{2\pi\mathrm{i}kt} := \cos(2\pi kt) + \mathrm{i}\ \sin(2\pi kt). \tag{2.6}$$

As often in mathematics, the transfer of a problem from the real into the complex world can have several advantages. Firstly, the concept of Fourier series can be naturally generalized from real-valued to complex-valued signals. Secondly, one obtains a concise and elegant formula of the Fourier representation. Thirdly, as we will see in a moment, the amplitude d_k and phase φ_k can be naturally expressed by a single complex Fourier coefficient. Again, one can show that the set $\{\mathbf{e}_k | k \in \mathbb{Z}\}$ forms an ON basis of $L^2([0,1])$. The resulting expansion of a signal $f \in L^2([0,1])$ with respect to this basis leads to the equality

$$f(t) = \sum_{k\in\mathbb{Z}} c_k \mathbf{e}_k(t) = \sum_{k\in\mathbb{Z}} c_k \mathrm{e}^{2\pi\mathrm{i}kt}, \tag{2.7}$$

which is also referred to as (complex) *Fourier series*. The corresponding (complex) *Fourier coefficients* $c_k \in \mathbb{C}$ are given by

$$c_k = \langle f|\mathbf{e}_k\rangle = \int_{t\in[0,1]} f(t)\overline{\mathrm{e}^{2\pi\mathrm{i}kt}}\mathrm{d}t = \int_{t\in[0,1]} f(t)\mathrm{e}^{-2\pi\mathrm{i}kt}\mathrm{d}t. \tag{2.8}$$

Now, for a real-valued signal f, one obtains $c_{-k} = \overline{c_k}$ for $k \in \mathbb{Z}$ (this, however, does not hold for a complex-valued signal). In other words, for real-valued signals the coefficients with negative indices are redundant. Furthermore, one easily shows the identities $a_0 = c_0$, $a_k = \sqrt{2}\,\mathrm{Re}(c_k)$, and $b_k = -\sqrt{2}\,\mathrm{Im}(c_k)$, $k \in \mathbb{N}$. Recall that the complex number c_k can be represented in polar coordinates $c_k = |c_k|\mathrm{e}^{2\pi\mathrm{i}\gamma_k}$, where $|c_k| \in \mathbb{R}_{\geq 0}$ is the absolute value (distance from the origin) of c_k and $\gamma_k \in [0,1)$ the phase (angle) of c_k. From this, one obtains

$$d_k = \left(a_k^2 + b_k^2\right)^{1/2} = \left(2\,\mathrm{Re}(c_k)^2 + 2\,\mathrm{Im}(c_k)^2\right)^{1/2} = \sqrt{2}\,|c_k|, \tag{2.9}$$

$$\begin{aligned}\varphi_k &= \frac{1}{2\pi}\arccos\left(\frac{a_k}{d_k}\right) = \frac{1}{2\pi}\arccos\left(\frac{\mathrm{Re}(c_k)}{|c_k|}\right)\\ &= \frac{\arccos(\cos(2\pi\gamma_k))}{2\pi} = \gamma_k.\end{aligned} \tag{2.10}$$

In other words, the amplitude d_k (up to a constant normalization factor) and the phase φ_k in the expansion (2.1) coincide with the absolute value and phase of the complex Fourier coefficient c_k, respectively. The map $\hat{f}:\mathbb{Z}\to\mathbb{C}$ defined by $\hat{f}(k) := c_k$ is also referred to as the *Fourier transform* of f. Furthermore, it can be shown that the Fourier transform is an energy preserving map or *isometry* between the Hilbert spaces $L^2([0,1])$ and $\ell^2(\mathbb{Z})$. In other words, if $f \in L^2([0,1])$, then $\hat{f} \in \ell^2(\mathbb{Z})$ and $\|f\|_{L^2([0,1])} = \|\hat{f}\|_{\ell^2(\mathbb{Z})}$.

The general idea of the Fourier representation carries over from the case of periodic to the case of nonperiodic signals in $L^2(\mathbb{R})$. In the nonperiodic case, however, the integer frequency parameters $k \in \mathbb{Z}$ do not suffice to "describe" a signal. Instead, one defines an exponential function $\mathbf{e}_\omega : [0,1] \to \mathbb{C}$, $textrme_\omega(t) := e^{2\pi i\omega t}$, for any parameter $\omega \in \mathbb{R}$. Then, replacing summation by integration one obtains the following nonperiodic analog of the Fourier series:

$$f(t) = \int_{\omega\in\mathbb{R}} c_\omega \mathbf{e}_\omega(t)\mathrm{d}\omega = \int_{\omega\in\mathbb{R}} c_\omega \mathrm{e}^{2\pi i\omega t}\mathrm{d}\omega \tag{2.11}$$

for $t \in \mathbb{R}$, where c_ω is defined by

$$c_\omega = \int_{t\in\mathbb{R}} f(t)\overline{\mathbf{e}_\omega(t)}\mathrm{d}t = \int_{t\in\mathbb{R}} f(t)\mathrm{e}^{-2\pi i\omega t}\mathrm{d}t. \tag{2.12}$$

Strictly speaking, one needs some technical adjustments in these constructions, which, however, will not play any further role for the applications to be discussed, see Sect. 2.2.4 for further remarks. Basically, the (2.11) shows that a signal $f \in L^2(\mathbb{R})$ can be written as an infinitesimal superposition of the elementary frequency functions $\mathbf{e}_\omega$. The numbers c_ω have the same interpretation as the Fourier coefficients c_k. The frequency-dependent function $\hat{f}:\mathbb{R}\to\mathbb{C}$ defined by $\hat{f}(\omega) := c_\omega$ is also called *Fourier transform* of f. Again, it can be shown that the Fourier transform is energy preserving. In other words, if $f \in L^2(\mathbb{R})$, then $\hat{f} \in L^2(\mathbb{R})$ and $\|f\|_{L^2(\mathbb{R})} = \|\hat{f}\|_{L^2(\mathbb{R})}$.

Finally, the case of DT signals can be regarded to be dual to the case of periodic CT signals. As we will see, the Fourier transform of a DT signal is a periodic CT signal, whereas the Fourier transform of a periodic CT signal is a DT signal. More precisely, let $x \in \ell^2(\mathbb{Z})$ be a DT signal of finite energy. Then the *Fourier representation* of x is given by

$$x(n) = \int_{\omega\in[0,1]} c_\omega \mathbf{e}_\omega(n)\mathrm{d}\omega = \int_{\omega\in[0,1]} c_\omega \mathrm{e}^{2\pi i\omega n}\mathrm{d}\omega \tag{2.13}$$

for $n \in \mathbb{Z}$, where c_ω is defined by

$$c_\omega = \sum_{n\in\mathbb{Z}} x(n)\overline{\mathbf{e}_\omega(n)} = \sum_{n\in\mathbb{Z}} x(n)\mathrm{e}^{-2\pi\mathrm{i}\omega n}. \tag{2.14}$$

In other words, the signal x can be represented as (infinitesimal) superposition of the 1-sampled elementary frequency functions $\mathbf{e}_\omega$, where only the frequencies $\omega \in [0,1]$ are needed. Intuitively, the restriction of the frequency parameters to the set $[0,1]$ can be explained as follows: for integer frequency parameter $k \in \mathbb{Z}$, one has $\mathbf{e}_\omega(n) = \mathbf{e}_{\omega+k}(n)$ for all sample points $n \in \mathbb{Z}$. In other words, two frequency functions with an integer difference in their frequency parameter coincide on the set of sampling points and cannot be distinguished as 1-sampled DT signals. This phenomenon is also known as *aliasing*. Analogous to the CT case, the map $\hat{f} : [0,1] \to \mathbb{C}$ defined by $\hat{f}(\omega) := c_\omega$ is also referred to as the *Fourier transform* of f. Furthermore, $\|x\|_{\ell^2(\mathbb{Z})} = \|\hat{x}\|_{L^2([0,1])}$. An overview of all three variants of the Fourier representation can be found in Table 2.1.

2.2.3 Discrete Fourier Transform

Computing the Fourier transform of signals involves the evaluation of integrals or infinite sums, which is, in general, computational infeasible. Also computing approximations of such integrals via Riemann sums can be computationally expensive. Therefore, to compute reasonable approximations of the Fourier transform at suitable frequency parameters, one has to employ fast algorithms–possibly at the expense of precision.

In most applications, one uses the *discrete Fourier transform* (DFT) for approximating the Fourier transform. Mathematically, the DFT of size N is a linear mapping $\mathbb{C}^N \to \mathbb{C}^N$ given by the $N \times N$ matrix

$$\mathrm{DFT}_N := \left(\Omega_N^{kj}\right)_{0\le k,j<N} = \begin{pmatrix} 1 & 1 & \cdots & 1 \\ 1 & \Omega_N & \cdots & \Omega_N^{(N-1)} \\ \vdots & \vdots & \ddots & \vdots \\ 1 & \Omega_N^{(N-1)} & \cdots & \Omega_N^{(N-1)(N-1)} \end{pmatrix}, \tag{2.15}$$

where we set $\Omega_N := \mathrm{e}^{-2\pi i/N}$. For an input vector $v := (v(0), v(1), \ldots, v(N-1))^{\mathrm{T}} \in \mathbb{C}^N$, the evaluation of the DFT is given by the matrix–vector product $\hat{v} := \mathrm{DFT}_N \cdot v$, where $\hat{v} = (\hat{v}(0), \hat{v}(1), \ldots, \hat{v}(N-1))^{\mathrm{T}} \in \mathbb{C}^N$ denotes the output vector. Note that the straightforward computation of the matrix–vector product requires $O(N^2)$ multiplications and additions. This is for most applications too slow–in many cases one has to deal with large $N \gg 10^5$. The important point is that there is an efficient algorithm, the so-called *fast Fourier transform* (FFT), which only requires $O(N \log N)$ multiplications and additions. The main idea of the FFT algorithm, which was originally found by Gauss in about 1805 and rediscovered by Cooley and Tukey in 1965, is based on a factorization of the DFT matrix into a sequence consisting of $O(\log N)$ sparse matrices each of which can be evaluated with $O(N)$ operations, see [41].

We now investigate, how one can approximate certain values of the Fourier transform by the entries $\hat{v}(k)$, $k \in [0:N-1]$. We start with the case of a DT signal $x \in \ell^2(\mathbb{Z})$ and define an input vector v by setting $v := (x(0), x(1), \ldots, x(N-1))$. From $\hat{v} = \mathrm{DFT}_N \cdot v$, we obtain

$$\hat{v}(k) = \sum_{j=0}^{N-1} x(j)\Omega_N^{kj} = \sum_{j=0}^{N-1} x(j)\mathrm{e}^{-2\pi\mathrm{i}(k/N)j} = \hat{x}\left(\tfrac{k}{N}\right) - \sum_{j\in\mathbb{Z}\setminus[0:N-1]} x(j)\,\mathrm{e}^{-2\pi\mathrm{i}(k/N)j}. \tag{2.16}$$

If the energy of x is concentrated in the interval $[0:N-1]$, i.e., if $x(j) \approx 0$ for $j \in \mathbb{Z} \setminus [0:N-1]$, then $\hat{v}(k)$ approximates $\hat{x}\left(\frac{k}{N}\right)$. In other words, the entry $\hat{v}(k)$ approximates the Fourier transform $\hat{x}$ at the frequency parameter $\omega = k/N$. The vector $\hat{v}$ can be regarded as an approximate $\frac{1}{N}$-sampling in the interval $[0,1]$ of the 1-periodic Fourier transform $\hat{x} \in L^2([0,1])$.

More generally, starting with a CT signal $f \in L^2(\mathbb{R})$, one obtains a DT signal x by T-sampling the signal f, i.e., $x(n) := f(T \cdot n)$ for $n \in \mathbb{Z}$. We first investigate the relation of the Fourier transforms $\hat{x}$ and $\hat{f}$:

$$\begin{aligned}\hat{x}(\omega) &= \sum_{j\in\mathbb{Z}} x(j)\,\mathrm{e}^{-2\pi\mathrm{i}\omega j} = \sum_{j\in\mathbb{Z}} f(Tj)\,\mathrm{e}^{-2\pi\mathrm{i}\omega j} \\ &\approx \int_{t\in\mathbb{R}} f(Tt)\mathrm{e}^{-2\pi\mathrm{i}\omega t}\mathrm{d}t = \frac{1}{T}\int_{t\in\mathbb{R}} f(t)\mathrm{e}^{-2\pi\mathrm{i}\omega t/T}\mathrm{d}t \\ &= \frac{1}{T}\hat{f}\left(\frac{\omega}{T}\right).\end{aligned} \tag{2.17}$$

Here, the approximation sign expresses that the value $\hat{x}(\omega)$ is a Riemann sum of the integral $(1/T)\hat{f}\,(\omega/T)$. The accuracy of the approximation very much depends on the properties of the integrand $f(Tt)\mathrm{e}^{-2\pi\mathrm{i}\omega t}$. Note that in the computation of the Riemann sum, where one sums over $j \in \mathbb{Z}$, the integrand is 1-sampled. Due to aliasing effects, at least arising from the factor $\mathrm{e}^{-2\pi\mathrm{i}\omega t}$, it is hopeless to expect any meaningful approximation results for $\omega \in \mathbb{Z}\setminus\left[-\frac{1}{2}, \frac{1}{2}\right]$. Actually, $\hat{x}$ is 1-periodic, whereas $\omega \mapsto (1/T)\hat{f}\,(\omega/T)$ is an $L^2(\mathbb{R})$ function, which approaches zero for $\omega \to \pm\infty$. Within the interval $\left[-\frac{1}{2}, \frac{1}{2}\right]$, however, in particular when approaching the frequency $\omega = 0$, the Riemann sum $\hat{x}(\omega)$ approximates the value $1/T\hat{f}\,(\omega/T)$ with increasing accuracy.

We now combine the results induced by the DFT approximation (2.16) and the Riemann approximation (2.17). As before, let $f \in L^2(\mathbb{R})$ be a CT signal, x the T-sampling of f, and $v := (x(0), x(1), \ldots, x(N-1))^{\mathrm{T}}$ for some $N \in \mathbb{N}$. Assuming that f is real valued, one can easily check that $\hat{f}(\omega) = \overline{\hat{f}(-\omega)}$, $\hat{x}(\omega) = \overline{\hat{x}(-\omega)}$, and $\hat{v}(k) = \overline{\hat{v}(N-k)}$. Therefore, we only have to consider positive frequency parameters. Then, we obtain from (2.16) and (2.17)

$$\hat{v}(k) \approx \hat{x}\left(\frac{k}{N}\right) \approx \frac{1}{T}\hat{f}\left(\frac{k}{N}\cdot\frac{1}{T}\right), \tag{2.18}$$

which gives meaningful approximations for $k = 0, 1, \ldots, \lfloor \frac{N}{2} \rfloor$. Furthermore, the coefficients $\hat{v}(k)$ for $k = N-1, N-2, \ldots, \lfloor \frac{N}{2} \rfloor + 1$ are redundant and correspond to negative frequencies. We illustrate this result by an example. Assuming a sampling rate $1/T = 50$ and $N = 100$, we obtain $\hat{v}(k) \approx 50 \cdot \hat{f}(k/2)$ for $k = 0, 1, \ldots, N/2$. In other words, the Fourier transform of f is approximated at the frequencies $\omega = 0, 1/2, 1, 3/2, \ldots, 25$, where the highest frequency $\omega = 1/2T = 25$ is the so-called *Nyquist frequency* with respect to the sampling rate $1/T$. This corresponds to an approximate $(1/TN)$-sampling of the frequency domain in the range zero up to the Nyquist frequency. Note that one can increase the accuracy of the approximations as well as the resolution of the frequency domain by suitably modifying the numbers N and T. We refer to Fig. 2.12 for an illustration.

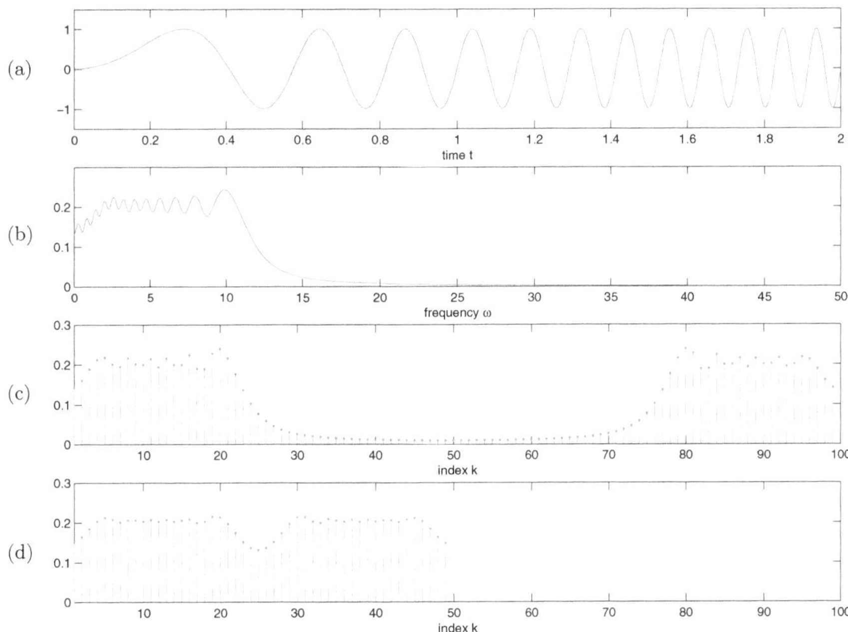

Fig. 2.12. (**a**) CT signal f shown on the time interval $[0, 2]$. The signal f is defined to be zero outside this interval. (**b**) Absolute value $|\hat{f}|$ of the Fourier transform shown on the frequency interval $[0, 50]$. (**c**) The absolute values $T \cdot |\hat{v}(k)|$ for $k \in [0 : N-1]$, where $\hat{v} = \mathrm{DFT}_N(v)$ with $N = 100$, $v(k) = f(Tk)$ for $k \in [0 : N-1]$, and $1/T = 50$. Note that $T \cdot \hat{v}(k) \approx \hat{f}(k/2)$ for $k = 0, \ldots, 50$. (**d**) Analog to (c) with $N = 50$ and $1/T = 25$. Note that $T \cdot \hat{v}(k) \approx \hat{f}(k/2)$ for $k = 0, \ldots, 25$. However, the accuracy of the approximations, in particular for higher frequencies, is much lower than in (c)

2.2.4 Further Notes

The Fourier transform is one of the most important transforms in particular in the field of digital signal processing (DSP). The basic definitions and main properties are discussed in nearly every introductory book on this subject. Exemplarily, we refer to the classical introductory book on *Signals and Systems* by Oppenheim et al. [156] on DSP. A summary of the main properties of the Fourier transform as needed in the subsequent chapters can also be found in [43]. In the book *Real Analysis* by Folland [95], one finds a mathematically rigorous introduction to Lebesgue theory as well as to the Fourier transform. A soft introduction to the main ideas of the time-frequency analysis is given in [68]. There also exists vast literature on the fast Fourier transform, which is also known as the *Cooley–Tukey algorithm*. Exemplarily, we refer to the original article by Cooley and Tukey [45], which works in case that the length N of the DFT is a power of 2. In [41], one finds various variants of the FFT, which, in combination, makes it possible to efficiently evaluate a DFT of arbitrary length $N \in \mathbb{N}$ with a time complexity of $O(N \log N)$.

So far, we have only scratched the topic of sampling and aliasing, which are of crucial importance for digital signal processing. In general, a CT signal has to be suitably approximated in order to describe it by a finite number of discrete parameters. Mathematically, the discrete set of parameters could be the Fourier coefficients (for periodic signals), the coefficients of polynomials (when representing a function by its Taylor series), or the values of a CT signal at a finite number of points in time, also referred to as *sampling*. In all cases there are certain requirements on the original CT signal, e.g., periodicity or differentiability, to guarantee certain bounds on the approximation error. In the case of sampling, theses requirements concern the frequency content of the original signal. The famous *sampling theorem* by Shannon says that an *Ω-bandlimited* signal $f \in L^2(\mathbb{R})$ (i.e., the Fourier transform $\hat{f}$ vanishes for $|\omega| > \Omega$ for a real number $\Omega > 0$) can be reconstructed perfectly from the T-sampling of f with $T := 1/2\Omega$, see [43, 156, 167]. Similarly, one may reduce the sampling rate of a DT signal x without loss of information if x does not contain any frequencies above half of the Nyquist rate. For example, the sampling rate may be reduced from 44.1 kHz to 4 000 Hz without any aliasing artifacts if one ensures that there are no frequency components above 2 000 Hz. For further details, we refer to the literature.

We close this section by discussing some technical details concerning the Lebesgue spaces and the Fourier transform. For further details and the proofs we refer to [68]. Firstly, in the definition of the Lebesgue spaces one has to postulate the Borel measurability of the functions for the integrals to be defined. Secondly, the spaces $L^p(\mathbb{R})$ and $L^p([0,1])$ are actually defined to be quotient spaces where two functions $f, g \in L^p(\mathbb{R})$ are identified if $\|f - g\|_p = 0$, i.e., if they differ only up to a null set. Thirdly, the equality in the Fourier representation and in the Fourier transform is just an equality in the L^2-sense, i.e., equality up to a null set. Under additional assumptions on f one also obtains pointwise equality. For example, if f is a continuously differentiable periodic CT signal, the Fourier series converges uniformly to f on the interval $[0, 1]$. Fourthly, the integral in the definition (2.12) of the Fourier transform of a signal $f \in L^2(\mathbb{R})$ does not exist in general. Therefore, one usually defines the Fourier transform for signals $f \in L^1(\mathbb{R}) \cap L^2(\mathbb{R})$ and then extends the definition to all signals $f \in L^2(\mathbb{R})$ using the so-called Hahn–Banach Theorem. Explicitly, the Fourier transform $\hat{f}$ can be defined as limit over increasing finite integration domains:

$$\hat{f}(\omega) := \lim_{N \to \infty} \int_{t \in [-N,N]} f(t) \mathrm{e}^{-2\pi \mathrm{i} \omega t} \mathrm{d}t. \tag{2.19}$$

Similarly, one has to define the Fourier representation, see (2.11). Finally, note that the exponential functions $\mathbf{e}_\omega : \mathbb{R} \to \mathbb{C}$ are not contained in $L^2(\mathbb{R})$ (as a $(1/\omega)$-periodic function they do not possess finite energy over $\mathbb{R}$). Therefore, the integral in (2.12) cannot be written as inner product as it has been possible for the Fourier series (2.8).

2.3 Digital Filters

The general term of a filter is used to denote any procedure that alters a signal's characteristics. In the audio context, various types of filters are applied to modify the waveform in some specified way and to change its spectral properties. Of particular interest are filters that boost or attenuate a signal's frequencies in certain frequency bands. The most important class of digital filters is based on the concept of convolution, which can be thought of as a general moving average (Sect. 2.3.1). As it turns out, all filters that satisfy certain linearity, invariance, and stability conditions can be expressed by means of a convolution with a suitable DT signal h. The Fourier transform of h is referred to as frequency response of the filter and exhibits important characteristics on the frequency selectivity of the underlying filter (Sect. 2.3.2). The frequency response can be used to specify certain filter characteristics as required for a specific application. For example, to reduce the sampling rate from 44.1 to 4 kHz, one needs an antialiasing filter that removes all frequencies above 2 000 Hz while retaining all frequencies below 2 000 Hz. Issues concerning filter specifications and filter design are discussed in Sects. 2.3.3 and 2.3.4. Further comments and references to the literature are found in Sect. 2.3.5.

2.3.1 Convolution Filters

Mathematically, a *filter* can be regarded as a mapping $I \to O$ that transforms an input signal $x \in I$ into an output signal $y \in O$, where I and O denote suitable signal spaces. In the following, we restrict ourselves to the case of discrete signals and consider, if not stated otherwise, the signal spaces $I = \ell^2(\mathbb{Z})$ and $O = \ell^2(\mathbb{Z})$. Some remarks on the CT case can be found in Sect. 2.3.5.

An important class of filters can be described by so-called *convolution*– a concept that constitutes an indispensable mathematical tool in functional analysis. Loosely speaking, the convolution of two DT signals x and y is a kind of multiplication that produces again a DT signal $x * y$ and that in a sense represents the amount of overlap between x and a reversed and translated version of y. The convolution of x and y at position $n \in \mathbb{Z}$ is defined by

$$(x * y)(n) := \sum_{k \in \mathbb{Z}} x(k)y(n-k). \tag{2.20}$$

Note that the sum in (2.20) could be infinite for general x and y. Therefore, the convolution $x * y$ exists only under suitable conditions on the DT signals. The easiest condition is that x and y have finitely many nonzero entries. To be more specific, we define the length $\ell(x)$ of such a signal x to be

$$\ell(x) := 1 + \max\{n | x(n) \neq 0\} - \min\{n | x(n) \neq 0\}. \tag{2.21}$$

Then, for two signals x and y of finite positive length the convolution $x * y$ exists and $\ell(x * y) = \ell(x) + \ell(y) - 1$. Another condition, also known as *Young inequality*, says that if $x \in \ell^1(\mathbb{Z})$ and $y \in \ell^p(\mathbb{Z})$, then $|(x * y)(n)| < \infty$ for all $n \in \mathbb{Z}$ and $\|x * y\|_p \leq \|x\|_1 \cdot \|y\|_p$. For a proof and further sufficient conditions, we refer to [68].

Fixing a DT signal $h \in \ell^1(\mathbb{Z})$, the Young inequality implies that one obtains an operator $C_h : \ell^2(\mathbb{Z}) \to \ell^2(\mathbb{Z})$ by setting $C_h(x) := h * x$ for $x \in \ell^2(\mathbb{Z})$. This operator, which is also referred to as *convolution filter*, has the following properties. Firstly, C_h is linear, i.e., $C_h(ax + by) = aC_h(x) + bC_h(y)$ for $x, y \in \ell^2(\mathbb{Z})$ and $a, b \in \mathbb{C}$. Secondly, C_h is invariant under time shifts, i.e., first shifting a signal in time and then applying C_h yields the same result as first applying C_h and then shifting the output signal. More precisely, let x^k, $x^k(n) := x(n-k)$ for $n \in \mathbb{Z}$, denote the *time shift* of a signal x by $k \in \mathbb{Z}$, then $C_h(x^k) = C_h(x)^k$. Thirdly, let $\delta \colon \mathbb{Z} \to \mathbb{C}$ denote the *impulse* signal with $\delta(0) = 1$ and $\delta(n) := 0$ for $n \in \mathbb{Z} \setminus \{0\}$, then $C_h(\delta) = h$, i.e., h can be recovered from C_h.

Now, the importance of convolution in the filter context stems from the following fact: all linear filters $T : \ell^2(\mathbb{Z}) \to \ell^2(\mathbb{Z})$ that are invariant under time shifts and satisfy a continuity condition[2] can be expressed as convolution filter. Indeed, defining $h := T(\delta)$, one can show that $T = C_h$. In this case, the DT signal h is also referred to as *impulse response* of the filter T and $h(n)$, $n \in \mathbb{Z}$, is called the nth *filter coefficient*. Furthermore, if $h \in \ell^1(\mathbb{Z})$, the filter T is called *stable*. In summary, there is a one-to-one correspondence between stable filters with the above properties and elements in $\ell^1(\mathbb{Z})$. Therefore, one often simply speaks of the *filter* $h \in \ell^1(\mathbb{Z})$ meaning the underlying convolution filter $T = C_h$. In the following, if not stated otherwise, all filters are be assumed to be of the form $T = C_h$.

---

[2] The continuity condition requires that all convergent sequences of input signals in $\ell^2(\mathbb{Z})$ are mapped to convergent sequences of output signals in $\ell^2(\mathbb{Z})$.

We close this section with some more definitions. A filter $T = C_h$ is called *FIR filter* (finite impulse response) if h has finite length and *IIR filter* (infinite impulse response) otherwise. If $h \neq 0$ is the impulse response of some FIR filter, then $\ell(h)$ is also called the *length* and $\ell(h) - 1$ the *order* of the FIR filter. Finally, a filter is called *causal* if $h(n) = 0$ for $n < 0$. The property of causality becomes important in the context of real-time signal processing applications, where one cannot observe the future values of the signal. For example, filtering an input signal x by a causal FIR filter $T = C_h$ of order N is described by

$$T(x)(n) = \sum_{\ell=0}^{N} h(\ell)x(n-\ell), \tag{2.22}$$

for filter coefficients $h(0), \ldots, h(N)$ with $h(0) \neq 0$ and $h(N) \neq 0$. The output signal $T(x)$ at point n only depends on the "past" samples $x(n-1), \ldots, x(n-N)$ and the "present" sample $x(n)$ of the input signal x.

2.3.2 Frequency Response

To characterize the properties of a convolution filter $T = C_h$, the Fourier transform $\hat{h}$ of the impulse response h, also called the *frequency response* of the filter T, plays a crucial role. In the filter context, one often uses capital letters G, H, X, Y to denote the Fourier transform of discrete filters $g, h \in \ell^1(\mathbb{Z})$ or DT signals $x, y \in \ell^2(\mathbb{Z})$.

One main property of the Fourier transform is that convolution in the time domain is transferred into pointwise multiplication in the spectral domain, see [68]. In other words, if $y = C_h(x) = h * x$ denotes the output signal for some input signal $x \in \ell^2(\mathbb{Z})$ and some filter $h \in \ell^1(\mathbb{Z})$, then

$$Y(\omega) = \hat{y}(\omega) = \widehat{h * x}(\omega) = \hat{h}(\omega) \cdot \hat{x}(\omega) = H(\omega) \cdot X(\omega). \tag{2.23}$$

The equality $Y(\omega) = H(\omega) \cdot X(\omega)$ can be interpreted in the following way. Recall from Sect. 2.2.2 that the value $X(\omega)$ encodes the contribution of the elementary frequency function $n \mapsto \mathbf{e}_\omega(n)$ to x. Hence, the filter h modifies this contribution according to the value $H(\omega)$. If $|H(\omega)| > 1$, the signal component of frequency ω is boosted, if $|H(\omega)| = 1$ it is left unchanged, and if $|H(\omega)| < 1$ it is attenuated. The phase of $H(\omega)$ encodes the relative shift of the corresponding elementary frequency function in the filtering process, see Sect. 2.3.3 for further details. Representing the complex value $H(\omega)$ in *polar coordinates*

$$H(\omega) = |H(\omega)| \cdot \mathrm{e}^{2\pi i \Phi_h(\omega)}, \tag{2.24}$$

the resulting function $\omega \mapsto |H(\omega)|$ is called the *magnitude response*, and the function $\omega \mapsto \Phi_h(\omega)$ is called the *phase response* of the filter h. We refer to Fig. 2.13 for an example. The magnitude response is often displayed in a scale measured in *decibel* (dB), which is suitable to study the behavior of values close to 0. In the decibel scale, a positive value $r \in \mathbb{R}_{>0}$ is transformed into the

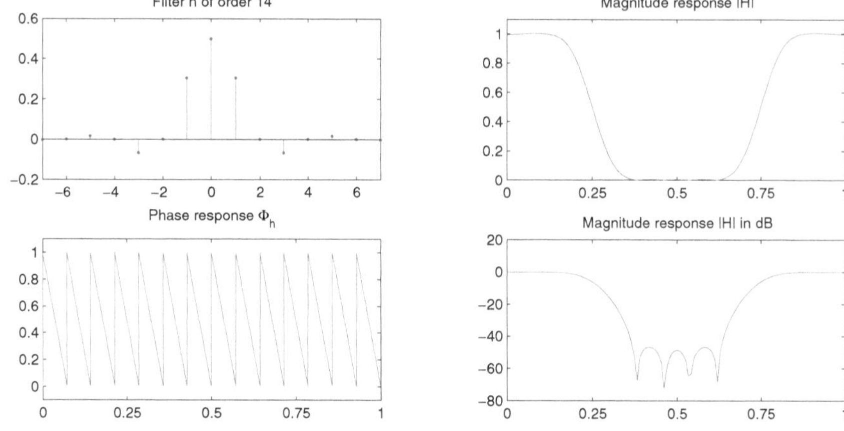

Fig. 2.13. A filter h of order 14, its phase response Φ_h, and its magnitude response $|H|$. Note that the values $\Phi_h(\omega)$ are defined only up to some integer value, hence normalized to the interval $[0,1)$. The characteristics of the magnitude response $|H|$ can be better seen in the decibel scale

value $\Delta(r) := 20\log_{10}(r)$. This transformation is monotonously decreasing, $\Delta(1) = 0$, and $\lim_{r\to 0}\Delta(r) = -\infty$. Note that $\Delta(2r) \approx 6 + \Delta(r)$, i.e., doubling the value corresponds to an increase of about 6 db.

We summarize some properties of the frequency response of a filter, cf. Sect. 2.2.2. Firstly, if h is a real-valued filter, then $\overline{H(\omega)} = H(-\omega)$, $|H(\omega)| = |H(-\omega)|$ (magnitude response is an even function), and $\Phi_h(-\omega) = -\Phi_h(\omega)$ (phase response is an odd function). Secondly, since the frequency response is a 1-periodic function, H is specified by its values on the interval $[0, \frac{1}{2}]$ for a real-valued filter h. Thirdly, spectral information concerning low frequencies is encoded by values $H(\omega)$ with $\omega \approx 0$, whereas spectral information concerning high frequencies is encoded by values $H(\omega)$ with $\omega \approx 1/2$. As an example, suppose x is the T-sampling with sampling rate $1/T = 44.1\,\mathrm{kHz}$ of a CT signal f. Recall from (2.17) that $X(\omega) \approx 1/T\hat{f}(\omega/T)$ for $\omega \in [0, \frac{1}{2}]$. Let h be a filter with a frequency response that has, for example, the values $H(0) = 0$ and $H(1/4) = 2$. Then by passing x through the filter h the low frequencies are cut off, whereas the frequencies around $\omega = 1/4$ are amplified by a factor of 2. Note that these frequencies correspond to frequencies around 11 025 Hz in the original CT signal f, which shows how the interpretation of the filter effects depends on the sampling rate.

According to the magnitude response $|H|$, one distinguishes various types of filters. In particular, let $0 \le \omega_0 < \omega_1 \le 1/2$ and let h be a real-valued filter with magnitude response

$$|H(\omega)| \approx \begin{cases} 1 \text{ if } \omega \in (\omega_0, \omega_1) \\ 0 \text{ if } \omega \in [0, \omega_0] \cup [\omega_1, \frac{1}{2}]. \end{cases} \tag{2.25}$$

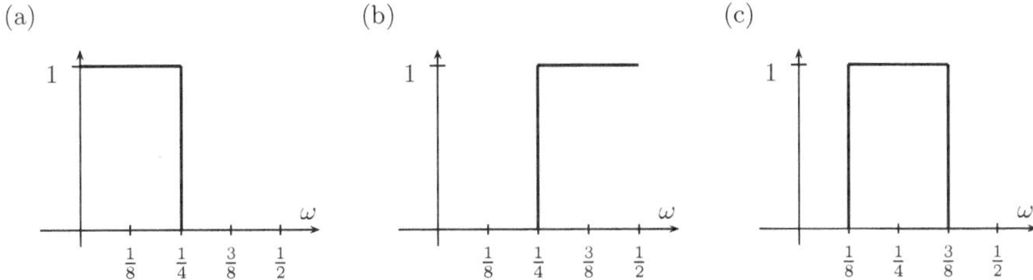

Fig. 2.14. Magnitude responses of (**a**) an ideal lowpass filter with cutoff frequency $\omega_1 = 1/4$, (**b**) an ideal highpass filter with cutoff frequency $\omega_0 = 1/4$, and (**c**) an ideal bandpass filter with cutoff frequencies $\omega_0 = 1/8$ and $\omega_1 = 3/8$

Then h is called a *lowpass filter* if $\omega_0 = 0$, a *highpass filter* if $\omega_1 = 1/2$, and a *bandpass filter* if $0 < \omega_0$ and $\omega_1 < 1/2$, see Fig. 2.14. These filters pass frequencies within the range (ω_0, ω_1), also referred to as *passband* while rejecting the frequencies outside that range, also referred to as *stopband*. The frequencies ω_0 and ω_1 are also called *cutoff frequencies*.

2.3.3 Filter Specifications

In the design of frequency-selective filters, the desired filter characteristics are specified in the frequency domain in terms of the magnitude and the phase response. One then determines the coefficients of an FIR or IIR filter that satisfies or closely approximates the desired frequency response specifications. There are many different theoretical methods for filter design which have been implemented and incorporated in numerous computer software programs such as [131]. We now summarize some of the common filter specifications in terms of a filter's magnitude and phase response.

Suppose, we want to design an ideal lowpass filter with cutoff frequency ω_0. Theoretically, one can design an ideal filter simply by inverting the frequency response as indicated in Fig. 2.14a. Let $\mathrm{sinc} : \mathbb{R} \to \mathbb{R}$ denote the *sinc function* as defined by

$$\mathrm{sinc}(t) := \begin{cases} \frac{\sin \pi t}{\pi t} & \text{for } t \neq 0, \\ 1 & \text{for } t = 0. \end{cases} \tag{2.26}$$

Then, a straightforward computation yields that the filter coefficients of such an ideal filter are given by $h(n) = 2\omega_0 \mathrm{sinc}(2\omega_0 n)$ for $n \in \mathbb{Z}$. This filter, however, has several drawbacks: It has infinitely many nonzero filter coefficients and it is neither causal nor stable. In view of an actual computer implementation, filtering with an ideal filter is not feasible. Therefore, one has to work with approximations that typically reveal the following phenomena: the frequency response H of an realizable filter shows *ripples* in the *passband* and *stopband*. In addition, H has no infinitely sharp cutoff from passband to stopband, i.e., H cannot drop from unity to zero abruptly and sharp discontinuities are smeared into a range of frequencies, see Fig. 2.15.

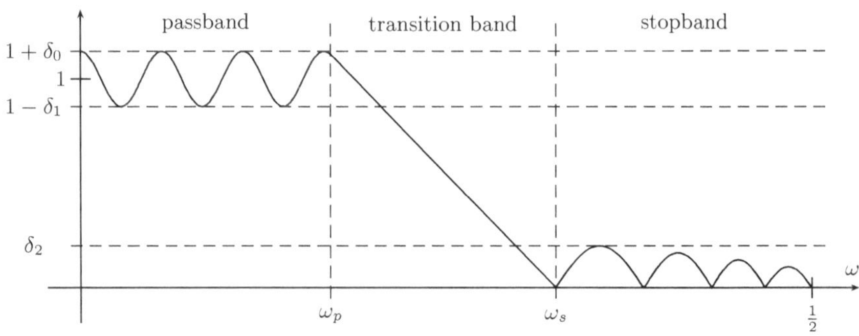

Fig. 2.15. Magnitude characteristics of physically realizable lowpass filters

In applications some degradations in the fequency response may be tolerable such as a small amount of ripples in the pass- or stopband. The following filter specifications are explained for the case of a lowpass filter and are illustrated by Fig. 2.15. The transition of the frequency response from passband to stopband is called *transition band* of the filter. The bandedge frequency ω_p defines the edge of the passband, while the frequency ω_s denotes the beginning of the stopband. The difference $\omega_\mathrm{s}-\omega_\mathrm{p}$ is referred to as *transition width*. Similarly, the width of the passband is called the *bandwidth* of the filter. For example, if the filter is lowpass with a passband edge frequency ω_p, its bandwidth is ω_p. If there are ripples in the passband of the filter, the maximal deviation of the ripples above and below 1 are denoted by δ_0 and δ_1, respectively. Then, the magnitude $|H|$ varies within the interval $[1-\delta_1, 1+\delta_0]$. The maximal value of ripples in the stopband of the filter is denoted as δ_2. In the same way, filter characteristics can be defined for a highpass filter. In the case of a bandpass filter, one has two stopbands as well as two corresponding transition bands to the left and right of the passband. Accordingly, one has to define parameters for each of these bands separately. In a typical filter design problem, given the above specifications and given a class of filters, a filter h is constructed that lies within this class and best fits the desired design requirements. Of course, the degree to which H approximates the specifications depends in particular on the order of the filter.

So far, we have only considered specifications that concern the magnitude response of a filter. Next, we discuss filter characteristics encoded by the phase response, see (2.24). To substantiate the interpretation of the phase response, we investigate the effect of a filter h on the 1-sampled elementary frequency function $n \mapsto \mathbf{e}_\omega(n) := \mathrm{e}^{2\pi\mathrm{i}\omega n}$:

$$(h * \mathbf{e}_\omega)(n) = \sum_{k\in\mathbb{Z}} h(k)\mathrm{e}^{2\pi\mathrm{i}\omega(n-k)} = H(\omega)\mathbf{e}_\omega(n) = |H(\omega)|\mathrm{e}^{2\pi\mathrm{i}(\omega n+\Phi_h(\omega))}. \tag{2.27}$$

Note that the phase induces a time shift in the elementary frequency function, which generally depends on the parameter ω. Thinking of a general signal as a superposition of amplified elementary frequency functions, such frequency-dependent delays generally destroy the delicate interaction of different frequency components, also referred to as constructive and destructive interference, and leads to strong and undesirable distortions in the output signal. Therefore, one often poses the requirement that $\Phi_h(\omega) = c\omega$ modulo 1 for some constant $c \in \mathbb{R}$. In this case one obtains $(h * e_\omega)(n) = |H(\omega)|e_\omega(n + c)$, where the delay c is independent of the frequency ω. For a general input signal, such a consistent delay in all frequency components simply leads to an overall delay in the filtering process, which is not considered a distortion. This observation leads to the following definitions. If $\Phi_h(\omega) = c\omega$ modulo 1 for some $c \in \mathbb{R}$, the filter h is said to be of *linear phase*, see Fig. 2.13 for an example. The function $\tau_h : [0, 1] \to \mathbb{R}$ defined by

$$\tau_h(\omega) := -\frac{\mathrm{d}\Phi_h}{\mathrm{d}\omega}(\omega) \tag{2.28}$$

is called the *group delay* of the filter h, where discontinuities in the phase that are due to integer ambiguities are left unconsidered. The value $\tau_h(\omega)$ can be interpreted as the time delay that a signal component of frequency ω undergoes in the filtering process. Obviously, the group delay is constant for a filter that has linear phase. In the next section, we give some examples and discuss an offline strategy that allows us to compensate phase distortions for filters with nonlinear phase.

2.3.4 Examples

We start our discussion with the case of FIR filters, which possess an impulse response of finite length. Note that any FIR filter can be made causal by suitably shifting the filter coefficients. It can be shown that the property of linear phase is equivalent to a certain symmetry or asymmetry condition on the filter coefficients. More precisely, let h be a causal FIR filter of order N, then h has linear phase if and only if $h(n) = \pm h(N - n)$ for $n = 0, 1, \ldots, N$, see [167, 197].

As a first example, suppose we want to design a causal FIR filter that approximates the ideal lowpass filter having a cutoff frequency ω_0 while having a linear phase. Recall from Sect. 2.3.3 that the ideal lowpass filter is obtained by 1-sampling the sinc function $t \mapsto 2\omega_0\mathrm{sinc}(2\omega_0 t)$, which is a symmetric function. To obtain a causal FIR filter h_N of order N with linear phase, one can simply take a shifted version of the middle part of these coefficients. Thus the nonzero coefficients are given by

$$h_N(n) = 2\omega_0\mathrm{sinc}\big(2\omega_0\left(n - \tfrac{N}{2}\right)\big), \quad 0 \le n \le N. \tag{2.29}$$

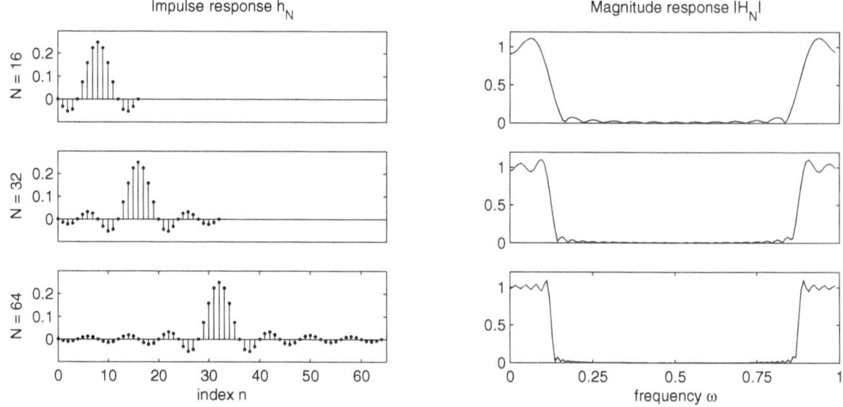

Fig. 2.16. Approximations of the ideal lowpass filter with cutoff frequency $\omega_0 = 1/8$

Fig. 2.16 shows the filters h_N for various orders in the case $\omega_0 = 1/8$. For even N, the frequency response of h_N is given by

$$H_N(\omega) = \mathrm{e}^{2\pi\mathrm{i}\omega N/2} \sum_{n\in[-\frac{N}{2}:\frac{N}{2}]} 2\omega_0 \cdot \mathrm{sinc}(2\omega_0 n)\mathrm{e}^{-2\pi\mathrm{i}n\omega}. \tag{2.30}$$

In other words, the magnitude response $|H_N|$ is the $(N/2)$-truncated Fourier series of the frequency response of the ideal lowpass filter. The phase response is given by $\Phi_{h_N}(\omega) = 2\pi\mathrm{i}\omega N/2$, thus resulting in a group delay τ_{h_N} of constant value $N/2$. This kind of filter design has the major disadvantage that it lacks a precise control of the filter characteristics as indicated in Fig. 2.15. In particular, to obtain a small transition width one generally needs a large number of nonzero filter coefficients. Also, one generally cannot guarantee the size of the ripples. As indicated in Fig. 2.16, there is a significant oscillatory overshoot of H_N at $\omega = \omega_0$ independent of the value N. This phenomenon, which is also known as *Gibbs phenomenon*, results from the nonuniform mean-square convergence of the Fourier series (2.30) and manifests itself in the design of FIR filters.

Other filter design methods have been suggested that provide total control of the filter specifications in terms of ω_p, ω_s, δ_1, and δ_2 as indicated in Fig. 2.15. In particular certain classes of IIR filters including Chebyshev and elliptic filters are suitable to cope with very strict filter specifications such as sharp transitions between pass- and stopband while using a relatively small number of filter parameters. As a disadvantage, however, these filters have nonlinear phase. Before continuing this discussion, we recall the concept of *recursive filtering*, which allows for realizing certain IIR filters despite infinitely many nonzero filter coefficients. Here, the main idea is to reuse certain output samples, which are previously produced in the filtering process, as input samples. A general recursive filter h is described by the *difference equation*

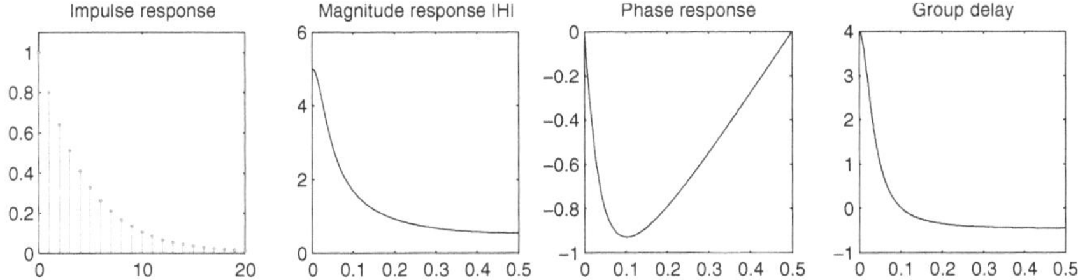

Fig. 2.17. The causal IIR filter specified by the forward filter coefficient $a_0 = 1$ and the backward filter coefficients $b_0 = 1$ and $b_1 = -0.9$

$$y(n) = -\sum_{k=1}^{M} a(k)y(n-k) + \sum_{k=0}^{N} b(k)x(n-k), \tag{2.31}$$

where the $a(k)$ are referred to as *forward filter coefficients*, the $b(k)$ as *feedback filter coefficients*, and $\max(N, M)$ the *order* of the filter. Setting $a(0) = 1$ and taking the Fourier transform of the equation $\sum_{k=0}^{M} a(k)y(n-k) = \sum_{k=0}^{N} b(k)x(n-k)$, one obtains $A(\omega)Y(\omega) = B(\omega)X(\omega)$. In other words, the frequency response H, if existent, is given by $H(\omega) = B(\omega)/A(\omega)$. For example, the filter specified by $a_0 = 1$, $b_0 = 1$, and $b_1 = -0.9$ is a causal IIR filter of order $N = 1$, where the impulse reponse h is given by $h(n) = 0.9^n$ for $n \geq 0$, see Fig. 2.17.

As a second example, we consider *elliptic filters*, which will be used in the applications described in the subsequent chapters. Elliptic filters are the most effective filters in the sense that they meet given magnitude response specifications with the lowest order of any filter type. This allows for designing filters with a small amount of ripples in the pass- and stopband as well as with a small transition width, see Fig. 2.18 for examples. However, elliptic filters reveal significant phase distortions, particularly near the band edges.

To compensate for a nonlinear phase, one can use a strategy known as *forward–backward filtering*. First, the input signal x is passed through a filter h to yield $y = h * x$. Time reversing y, one obtains a signal y^r, $y^r(n) = y(-n)$, which is again passed through the filter h to yield $z = h * y^r$. The time reversal z^r has precisely zero phase, whereas the magnitude is modified by the square of the filter's magnitude response. This fact can be seen immediately in the spectral domain using $H^r = \overline{H}$ and $Z^r = \overline{Z}$:

$$Z^r = \overline{Z} = \overline{HY^r} = \overline{H\overline{Y}} = \overline{H\overline{HX}} = \overline{H}HX = |H|^2 X. \tag{2.32}$$

Note that this strategy only works in an offline scenario, where the entire signal is available prior to filtering. We refer to Fig. 3.3 for an example.

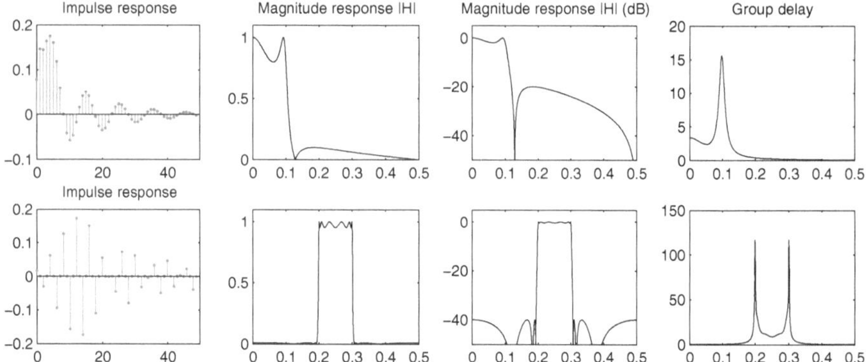

Fig. 2.18. (*Top*) Elliptic lowpass filter of order 4 satisfying the filter specifications $\omega_{\mathrm{p}} = 0.1$ and $\omega_{\mathrm{s}} = 0.15$ (bandedge frequencies), $\delta_0 = 0$ and $\delta_1 = 0.2$ (passband ripples), as well as $\delta_2 = 0.1$ corresponding to $-20\,\mathrm{dB}$ (stopband ripples), see Fig. 2.15. (*Bottom*) Elliptic bandpass filter of order 12 for $\omega_{\mathrm{p}} = 0.2$ and $\omega_{\mathrm{s}} = 0.19$ (left-hand side), $\omega_{\mathrm{p}} = 0.3$ and $\omega_{\mathrm{s}} = 0.31$ (right-hand side), $\delta_0 = 0$ and $\delta_1 = 0.05$, as well as $\delta_2 = 0.01$ corresponding to $-40\,\mathrm{dB}$.

2.3.5 Further Notes

Most of the concepts discussed in this section can be found in standard textbooks on digital signal processing such as [156, 167] and are implemented in signal processing software, see, e.g., [131].

When a filter is to be implemented in software, various important issues concerning efficiency and numerical stability have to be considered. For example, in realizing the recursion (2.31) there are many possible strategies each relying on different filter representations such as the direct form, the cascade form, or the lattice structure. Depending upon the strategy, there may be significant differences in the number of arithmetic operations required to perform the filtering. A second issue concerns the memory requirements for storing the filter parameters, the past inputs, past outputs, and any intermediately computed values. Thirdly, one has to be aware of numerical problems that arise from rounding errors when using finite representation of real numbers. For a detailed discussion of these issues, we refer to [167].

As discussed above, a linear time-invariant filter h can be regarded as a spectral shaping function that modifies the input signal spectrum $X(\omega)$ according to its frequency response $H(\omega)$, see [167]. Therefore, the desired filter characteristics are frequently specified in the spectral domain. Given a set of possibly contradicting or even physically unrealizable requirements, the filter design process can be regarded as an optimization problem where each requirement contributes a particular term to an error function which has to be minimized. Such optimization tasks constitute difficult mathematical problems and may involve computational expensive algorithms. Many of the techniques for designing digital IIR filter are based on the conversion of classical analog filters described by some linear constant-coefficient differential equation [167]. In particular, the design of elliptic filters is mathematically involved being based on elliptic rational functions, see [128]. However, once the filter design methods are available in signal processing software, they are easily manageable simply by specifying the desired filter characteristics.

In this section, we have mainly discussed the case of lowpass filters. Actually, there is a straightforward method to construct bandpass filters from lowpass filters based on so-called *modulation*, see [156]. For a given filter $h \in \ell^1(\mathbb{Z})$ and $\lambda \in \mathbb{R}$, the λ-modulated filter h_λ is defined by $h_\lambda(n) := \mathrm{e}^{-2\pi\mathrm{i}\lambda n}x(n)$, $n \in \mathbb{Z}$. Then, it is easy to see that $H_\lambda(\omega) = H(\omega+\lambda)$. A modulation in the time domain thus corresponds to a shift in the frequency domain. In case h has real coefficients, one can obtain a modulated filter with real coefficients by considering the real part $g_\lambda := \mathrm{Re}(h_\lambda)$. The filter g_λ is also referred to as *cosine modulation* of h and yields the frequency response $G_\lambda(\omega) = (1/2)(H(\omega+\lambda)+H(\omega-\lambda))$. We refer to Fig. 2.19 for an illustration of the modulation effects.

Finally, we remark that for certain applications it may be important to recover the original signal from certain filter outputs. For example, the process of inverting the effect of a convolution filter is known as *deconvolution*. If an

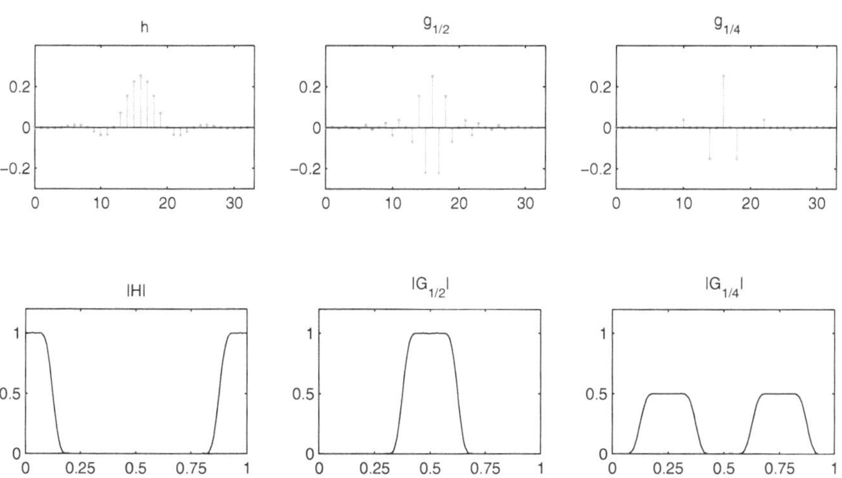

Fig. 2.19. (*Top*) Impulse responses of a lowpass filter h and cosine-modulated versions g_λ for $\lambda = 1/2$ and $\lambda = 1/4$. (*Bottom*) Corresponding magnitude responses

invertible convolution filter and its inverse are causal and stable, then they are also referred to as filters with *minimum phase*. For details we refer to the standard literature such as [156, 167, 197]. Dealing with audio analysis rather than synthesis applications, the invertibility of filters or filter banks will not play a crucial role in the following chapters.

3

Pitch- and Chroma-Based Audio Features

Automatic music processing poses a number of challenging questions because of the complexity and diversity of music data. As discussed in Sect. 2.1, one generally has to account for various aspects such as the data format (e.g., score, MIDI, audio), the instrumentation (e.g., orchestra, piano, drums, voice), and many other parameters such as articulation, dynamics, or tempo. To make music data comparable and algorithmically accessible, the first step in all music processing tasks is to extract suitable *features* that capture relevant key aspects while suppressing irrelevant details or variations. Here, the notion of *similarity* is of crucial importance in the design of audio features. In some applications and particularly in the case in music retrieval, one may be interested in characterizing an audio recording irrespective of certain details concerning the interpretation or instrumentation. Conversely, other applications may be concerned with measuring just the niceties that relate to a musician's individual articulation or emotional expressiveness.

As our main goal of this chapter, we introduce audio features and mid-level representations that are particularly useful in the music retrieval context. In Sect. 3.1, we describe how an audio signal can be decomposed into spectral bands, where each band corresponds to a pitch of the equal-tempered scale as used in Western music. The decomposition will be realized by a suitable multirate filter bank consisting of elliptic filters. The pitch representation of the audio signal can then be used as a basis for deriving audio features of various characteristics. As a first example, we introduce the STMSP (short-time mean-square power) features that measure the local energy content of each subband and indicate the presence of certain musical notes within the audio signal (Sect. 3.2). Taking the (discrete) derivative of the energy curves, one obtains onset features. Such features are particularly suitable for the analysis of piano music, where each keystroke inevitably leads to a sudden increase of energy in certain pitch bands [141]. In Sect. 3.3, we will derive a chroma representation from the pitch representation by suitably combining pitch bands that correspond to the same chroma. Here, the chroma refers to the 12 traditional pitch classes of the equal-tempered scale. The chroma features are much coarser than pitch features and show a high degree of robustness to variations in timbre and instrumentation. Adding a further degree of abstraction, we then introduce a new class of scalable and robust audio features by considering short-time statistics over energy distributions within the chroma bands. The resulting CENS (chroma energy normalized statistics) features strongly correlate to the short-time harmonic content of the underlying audio signal and absorb variations of properties such as dynamics, timbre, articulation, execution of note groups, and temporal microdeviations. Furthermore, due to their low temporal resolution, CENS features can be processed efficiently [139, 140]. Further notes and references to the literature can be found in Sect. 3.4.

3.1 Pitch Features

We now introduce some techniques for decomposing an audio signal into frequency bands that correspond to the musical notes. In the following, a musical note is identified with its corresponding MIDI pitch p, e.g., the note A4 with $p = 69$. The associated frequency, 440 Hz in case of A4, is also referred to as *center frequency* of the note. In the decomposition we consider the range A0 ($p = 21$) to C8 ($p = 108$) corresponding to the keys of a standard piano. For each pitch, we design a bandpass filter that passes all frequencies around the respective center frequency while rejecting all other frequencies.

Since a good signal analysis is the basis for our further procedure, the imposed filter requirements are stringent: to properly separate adjacent notes, the passbands of the filters should be narrow, the cutoffs should be sharp, and the rejection in the stopband should be high. In addition, the filter orders should be small to allow for efficient computations. Recall from Sect. 2.1.2 that the MIDI note number (the notes of the equal-tempered scale) depend on their center frequencies in some logarithmic fashion. Actually, let $f(p)$ denote the center frequency of the pitch $p \in [1:120]$, then one has the relation

$$f(p) = 2^{\frac{p-69}{12}} \cdot 440, \tag{3.1}$$

see Table 3.1 for examples. It follows that the bandwidth of a filter associated to a low pitch must be smaller than that of a high pitch. However, very small bandwidths and short transitions may lead to numerical problems in the filter design as well as in the filtering process. To alleviate the filter requirements, one has the work with different sampling rates. For example, one may use a sampling rate of 22 050 Hz for high pitches ($p = 93, \ldots, 108$), 4 410 Hz for medium pitches ($p = 57, \ldots, 92$), and 882 Hz for low pitches ($p = 21, \ldots, 56$). Adapting the sampling rates also takes into account that the time resolution naturally decreases for lower frequencies. Furthermore, a reduction of the sampling rate significantly speeds up the computations. The proposed partitioning of the pitches and the allocated sampling rates constitutes a compromise between resolution, numerical stability, and computational complexity.

Table 3.1. Filter specifications shown for the notes A3 ($p = 57$) to A4 ($p = 69$)

Note	p	$f(p)$ (CT)	ω_{p1} (CT)	ω_{p2} (CT)	Width (CT)	sr	ω_{p1} (DT)	ω_{p2} (DT)	Width (DT)	Q
A3	57	220.00	215.60	224.40	8.80	882	0.2444	0.2544	0.0100	25.0
A$^\sharp$3	58	233.08	228.42	237.74	9.32	882	0.2590	0.2696	0.0106	25.0
B3	59	246.94	242.00	251.88	9.88	882	0.2744	0.2856	0.0112	25.0
C4	60	261.63	256.39	266.86	10.47	4410	0.0581	0.0605	0.0024	25.0
C$^\sharp$4	61	277.18	271.64	282.73	11.09	4410	0.0616	0.0641	0.0025	25.0
D4	62	293.66	287.79	299.54	11.75	4410	0.0653	0.0679	0.0027	25.0
D$^\sharp$4	63	311.13	304.90	317.35	12.45	4410	0.0691	0.0720	0.0028	25.0
E4	64	329.63	323.04	336.22	13.19	4410	0.0733	0.0762	0.0030	25.0
F4	65	349.23	342.24	356.21	13.97	4410	0.0776	0.0808	0.0032	25.0
F$^\sharp$4	66	369.99	362.59	377.39	14.80	4410	0.0822	0.0856	0.0034	25.0
G4	67	392.00	384.16	399.84	15.68	4410	0.0871	0.0907	0.0036	25.0
G$^\sharp$4	68	415.30	407.00	423.61	16.61	4410	0.0923	0.0961	0.0038	25.0
A4	69	440.00	431.20	448.80	17.60	4410	0.0978	0.1018	0.0040	25.0
A$^\sharp$4	70	466.16	456.84	475.49	18.65	4410	0.1036	0.1078	0.0042	25.0

The third to sixth columns refer to specifications in the CT domain. The third column shows the center frequency $f(p)$ of pitch p, the forth and fifth columns indicate the cutoff frequencies ω_{p_1} and ω_{p_2} to the left and right, respectively, and the sixth column the bandwidth. The seventh column shows the sampling rate (sr) used in the filtering step. The next three columns indicate the cutoff frequencies and the bandwidth relative to this sampling rate. The last column shows the Q factor

A different partitioning of the pitches with other sampling rates may work equally well.

Because of their excellent cutoff properties, we use elliptic filters as described in Sect. 2.3.4. As a compromise between numerical stability in the filter design, filter order, and filter characteristics, each filter is implemented using an eighth-order elliptic filter with 1 dB passband ripple and 50 dB rejection in the stopband. The bandwidth of a filter is often specified by means of the so-called *quality factor* or *Q factor*, which denotes the ratio of center frequency to bandwidth. To separate the notes, we use a Q factor of $Q = 25$ and a transition band that has half the width of the passband. For example, for the concert pitch $p = 69$ (A4) with center frequency $f(p) = 440$ Hz, the bandwidth w is given by $w = \frac{f(p)}{Q} = \frac{440}{25} = 17.6$. From this we obtain the cutoff frequencies $\omega_{p_1} = 440 - 17.6/2 = 431.2$ and $\omega_{p_2} = 440 + 17.6/2 = 448.8$ for the left and right, respectively. Similarly, the stopband frequencies are given by $\omega_{s_1} = 440 - 17.6 = 422.4$ and $\omega_{s_2} = 440 + 17.6 = 457.6$ Hz. So far, the filter specifications are given in absolute terms as if we were dealing with CT signals. To obtain the filter specifications with respect to a specific sampling rate, one has to divide the entities by the sampling rate, see (2.17). For example, having a sampling rate of 4 410 Hz, the center frequency corresponds to $\omega = 440/4410 = 0.0998$ and the cutoff frequencies to $\omega_{p_1} = 0.0978$ and

$\omega_{p_2} = 0.1018$, and so on. The filter specifications for some of the pitches are shown in Table 3.1.

Generally, an array of bandpass filters that separates the input signal into several components is referred to as *filter bank*. The output obtained from applying a single filter to an input signal is called *subband* of the input signal. Furthermore, a *multirate* filter bank involves, by definition, several sampling rates. The filters specified above form a multirate filter bank with all filters having the same Q factor, see Fig. 3.1. In the following, this filter bank will be simply referred to as *pitch filter bank*.

Prior to applying the pitch filter bank, one has to provide the input signal with the required sampling rates. As an example, suppose that the input signal x_0 is given at a sampling rate of 22 050 Hz. After applying an antialiasing lowpass filter with cutoff frequency $\omega_p = 0.1$, as indicated by Fig. 3.2, the signal is downsampled by a factor of five to obtain a DT signal x_1 at 4 410 Hz. The signal x_1 is again passed through the same antialiasing filter and downsampled by a factor of five to obtain a DT signal x_2 at 882 Hz. To obtain a pitch representation of the original signal, each filter of the pitch filter bank is applied to one of the DT signals x_0, x_1, and x_2, according to the required

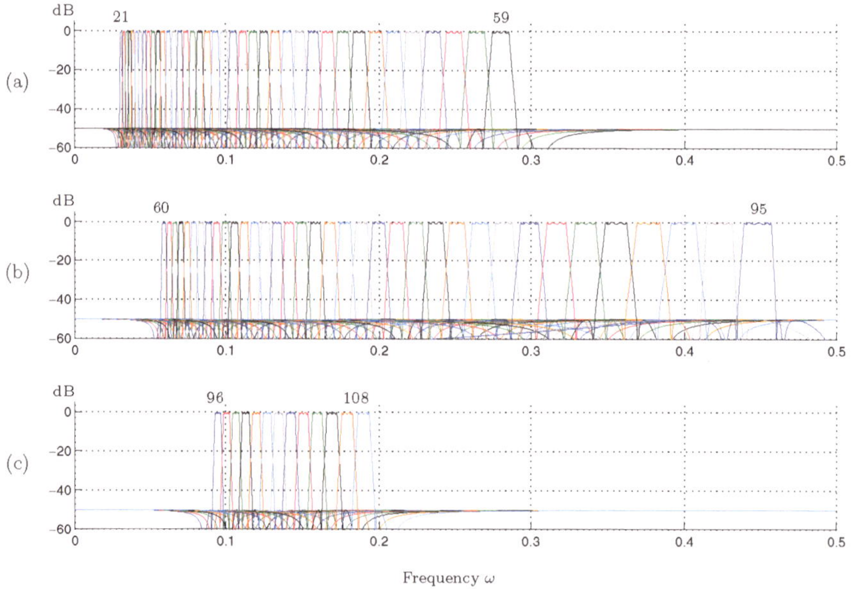

Fig. 3.1. Magnitude responses in dB for the elliptic filters of the multirate pitch filter bank. Each filter corresponds to a MIDI pitch $p \in [21 : 108]$. **(a)** Filters for $p \in [21 : 59]$ with respect to the sampling rate 882 Hz. **(b)** Filters for $p \in [60 : 95]$ with respect to the sampling rate 4 410 Hz. **(c)** Filters for $p \in [96 : 108]$ with respect to the sampling rate 22 050 Hz

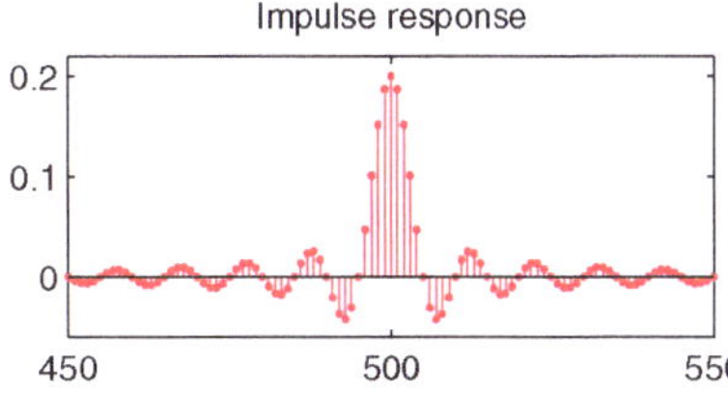

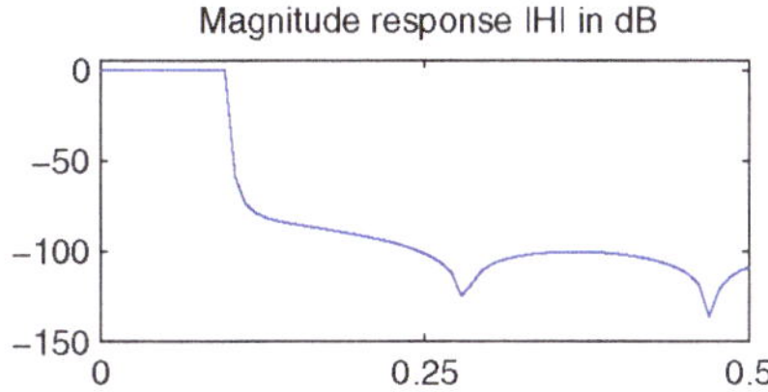

Fig. 3.2. Antialiasing filter of order $N = 1\,000$ with cutoff frequency $\omega_p = 0.1$, which is used prior to downsampling by a factor of five

sampling rate. The group delays are corrected by using forward–backward filtering as described in Sect. 2.3.4, see also Fig. 3.3.

The decomposition into pitch subbands provides musically meaningful and temporally accurate information about the distribution of a signal's frequency components, which can be used as a front-end signal processing step in music analysis. Because of the high-time resolution in the subbands, however, the memory requirements for such a pitch decomposition are generally much larger than for the original signal. Furthermore, the decomposition assumes a reasonable tuning of the instruments according to the equal-tempered scale. Because of the passband properties of the pitch filters, deviations of up to ± 25 cents[1] from the respective center frequency can be absorbed. Global deviations in tuning can be compensated by employing a suitably adjusted filter bank. However, phenomena such as strong string vibratos or pitch oscillation as is typical for, e.g., kettledrums lead to significant and problematic pitch smearing effects. Here, the detection and smoothing of such fluctuations, which is certainly not an easy task, may be necessary prior to the filtering step.

3.2 Local Energy (STMSP) and Onset Features

One of the most important tasks in the analysis of music recordings is the extraction of musically relevant audio features that indicate the presence of certain musical notes in the audio signal. To this end, we introduce audio features that measure the *local energy* or *short-time mean-square power* (STMSP) in each of the pitch subbands. More precisely, let x denote a subband signal and $w \in \mathbb{N}$ be a fixed window size, then the STMSP of x at $n \in \mathbb{Z}$ is defined as $\sum_{k \in [n - \lfloor \frac{w}{2} \rfloor : n + \lfloor \frac{w}{2} \rfloor]} |x(k)|^2$. To reduce the data rate by a factor of $d \in \mathbb{N}$, the STMSP is only evaluated every d samples. Note that the computation of the STMSP can be regarded as a convolution of the squared signal with a rectangular window of size w followed by downsampling by a factor d.

---
[1] The *cent* is a logarithmic unit to measure musical intervals. The interval between two adjacent notes of the equal-tempered scale is equal to 100 cents.

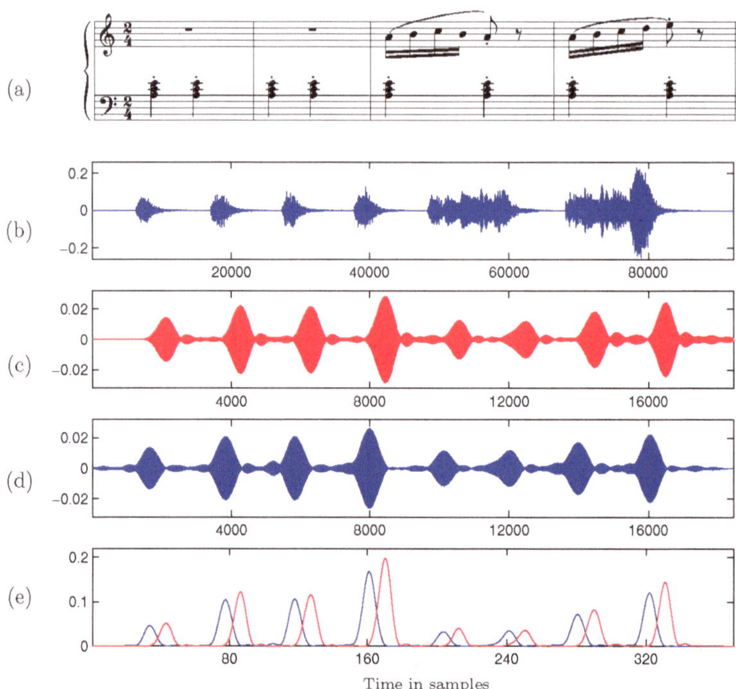

Fig. 3.3. **(a)** First four measures of Op. 100, No. 2 by Friedrich Burgmüller. **(b)** Audio representation of a corresponding piano recording (roughly 4.2 s) sampled at 22 050 Hz. **(c)** Output signal of the filter corresponding to the note C4 ($p = 60$). As input signal, the signal of (b) was downsampled to 4 410 Hz. Note the shift in the output signal due to the group delay of the filter. **(d)** Corrected output signal obtained by forward–backward filtering with the same filter and the same input signal as in (c). **(e)** STMSP (see Fig. 3.4 for a description) obtained from the signals (c) and (d)

As a variant, one often replaces the rectangular window by some other window function such as a triangular or the Hann window.

We compute the STMSP for each of the 88 pitch subbands. Note that the interpretation of the window size w and the downsampling factor d depends on the sampling rate of the respective subband. For example, the STMSP with $w = 501$ and $d = 250$ for a subband sampled at 22 050 Hz has a resolution of $250/22\,050 = 0.00113$ s per window, which corresponds to a feature sampling rate of 88.2 Hz. To obtain the same resolution for a subband sampled at 4 410 Hz one has to chose $w = 101$ and $d = 50$, and so on. To make the STMSPs of different subbands comparable with each other, one has to compensate for the different window sizes w. In our case, the different sampling rates of 22 050, 4 410, and 882 Hz are compensated in the energy computation by introducing an additional factor of 1, 5, and 25, respectively.

As illustration, we consider some of the energy curves obtained from a piano recording of a piece by Friedrich Burgmüller, see Fig. 3.3a. The note C4 ($p = 60$) appears every quarter beat in the left hand of the score, which is also reflected by STMSP of the corresponding subband as shown in Fig. 3.3e. This figure also illustrates that compensating the group delay is of great importance in view of temporal accuracy. Figure 3.4 shows the STMSPs for the pitches $p = 69$, $p = 71$, and $p = 72$. Even though the note A4 ($p = 69$) only appears three times in the right hand of the score, there is significant energy in this subband at the eight time positions corresponding to the quarter beat. This is due to the fact that the note A4 corresponds to the first overtone of the note A3, which is played every quarter beat by the left hand. Similarly, this holds for the note C5 ($p = 72$). The STMSP for the pitch $p = 71$ clearly reveals the three occurrences of the note $B4$ in the right hand of the score. Looking at the energy curves in the logarithmic decibel scale, it becomes evident that some of the signal's energy is spread over large parts of the pitch spectrum. As it turns out, the nonperfect behavior of our bandpass filters only causes marginal artifacts. The main reason for the energy spread is due to the fact that striking a single piano key already generates a complex sound. This sound not only consists of the fundamental pitch and its harmonics but also comprises inharmonicities that are spread over the entire spectrum and are caused by the keystroke (mechanical noise) as well as transient and resonance

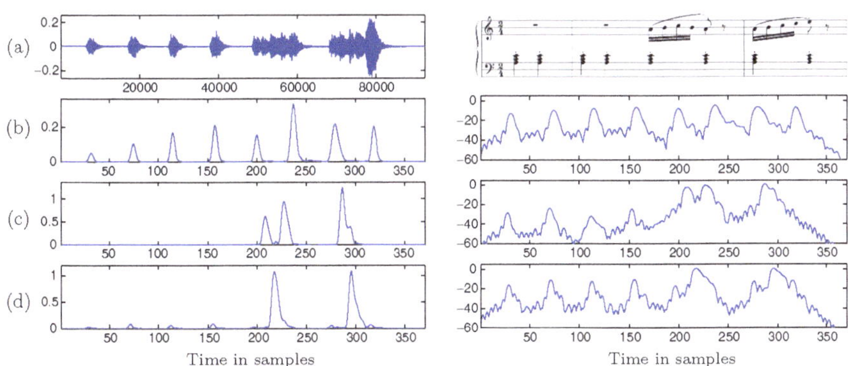

Fig. 3.4. (a) Audio recording (sampling rate 22 050 Hz) and score as in Fig. 3.3. **(b)** STMSP for the subband (sampling rate 4 410 Hz) corresponding to the note A4 ($p = 69$) using a window size $w = 101$ and a downsampling factor $d = 50$ resulting in a time resolution of 11.3 ms. The right side shows the STMSP in the logarithmic decibel scale. **(c), (d)** STMSPs for the notes B4 ($p = 71$) and C5 ($p = 72$), respectively, computed as in (b)

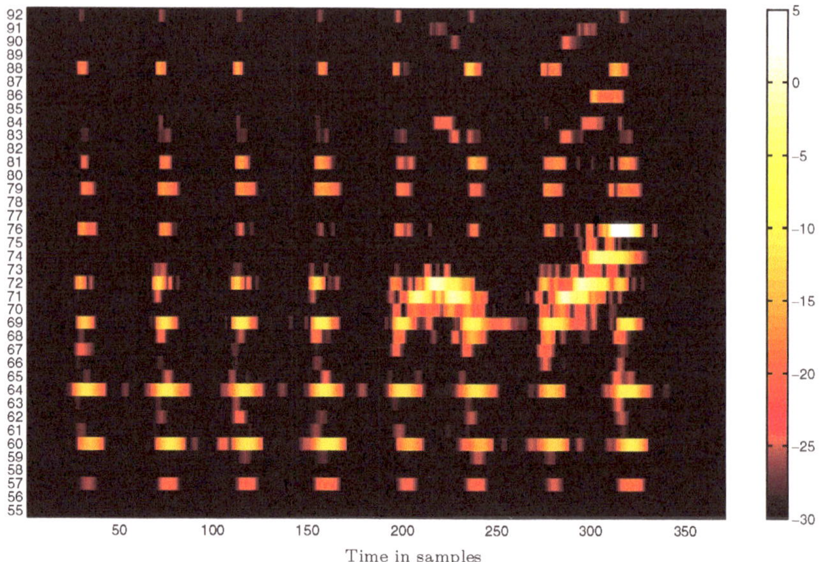

Fig. 3.5. Time-pitch plot for the audio recording as shown in Fig. 3.3b. The rows correspond to MIDI pitches and the columns to time measured in samples with a time resolution of 11.3 ms. Each row shows the STMSP of the specified pitch, where the values are color coded according to the decibel scale. For example, the curve shown in on the right hand side of Fig. 3.4b corresponds to the row labelled by $p = 69$

effects, see [16,66]. This effect can also been seen in Fig. 3.11. A compact representation of the local energy distribution in the subbands is obtained by a time-pitch plot as shown in Fig. 3.5, where the energy level is suitably color coded. In Sect. 3.4, we will discuss how to obtain a similar pitch representation by suitably pooling the spectral coefficients obtained from a short time Fourier transform (STFT).

For many instruments such as the piano or the guitar, the sound of a note is loudest immediately after it is played and then fades with time. Striking a piano key results in a sudden energy increase[2]. This increase may not be significant relative to the entire signal's energy – in particular if the keystroke is soft and the generated sound is masked by the remaining signal. However, the energy increase relative to the pitch bands corresponding to the fundamental pitch and harmonics of the respective note may still be substantial. This observation suggests the following general feature extraction procedure

[2] This phase is also referred to as *attack phase* of the sound.

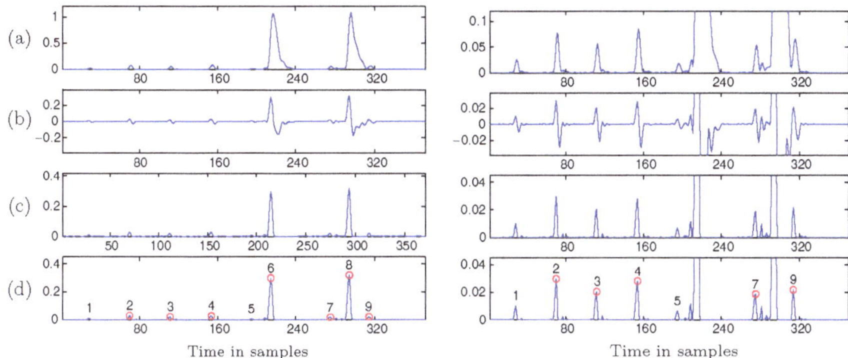

Fig. 3.6. The figures to the right are enlarged versions (in vertical direction) of the figures to the left. **(a)** STMSP for pitch $p = 72$ as shown in Fig. 3.4. **(b)** First-order difference signal. **(c)** Half-wave rectified difference signal. **(d)** The peaks chosen by the peak picking strategy are indicated by the *red circles*

to detect candidates for note onsets, see [141, 185]. For a given audio signal, we compute the STMSP for each pitch $p \in [21 : 108]$. Let x denote such a local energy curve for a fixed pitch, then the first-order difference x' (the discrete time derivative) of x is defined by $x'(n) := x(n) - x(n-1)$, $n \in \mathbb{Z}$. This difference is further processed by taking only the positive values of the signal and setting the negative values to zero – a process, which is also referred to as *half wave rectification*. Altogether, for each note p, we obtain a rectified difference signal, which will be referred to as *onset signal* OS_p. The procedure is illustrated by Fig. 3.6.

The local maxima or *peaks* of the onset signal OS_p indicate the positions of locally maximal energy increases in the respective band. Such peaks are good candidates for onsets of piano notes that have the fundamental pitch or some harmonics corresponding to p. For example, the pitches p, $p-12$, $p-19$, $p-24$, or $p-28$ all have some harmonics corresponding to the pitch p. In practice, one often has to cope with "bad" peaks that are not generated by onsets. In particular, resonance and beat effects (caused by the interaction of strings) may lead to additional peaks in the onset signals. Furthermore, a note played in forte may generate peaks in subbands that do not correspond to harmonics (e. g., caused by mechanical noise). The distinction of such "bad" peaks and onset peaks is often impossible and the peak picking strategy becomes a delicate problem. Since in general the "bad" peaks are less distinct than the "good" ones, we use local thresholds (local averages) to discard the peaks below these thresholds.

In Fig. 3.6d, the numbers indicate the musically significant peaks of the onset signal OS_{72}, whereas the circles indicate the peaks that were chosen by the peak picking strategy. The peaks 6 and 8 correspond to the note C5 ($p = 72$) played two times in the right hand. The other labelled peaks correspond to the first harmonics of the note C4 ($p = 60$) played in the left hand. The peaks 1 and 5 were rejected by our local threshold constraints. This also holds for the "bad" (unlabelled) peaks close to the peaks 6 and 8. After a suitable conversion into absolute timing positions, we obtain a list of peaks for each piano note p. Each peak is specified by a triple (p, t, s) where p denotes the pitch corresponding to the subband, t the time position in seconds within the audio file, and s the size expressing the significance of the peak (corresponding to the velocity of the note). The representation, where only these peaks are stored, will be referred to as *onset representation* of the audio signal. One important property of this representation is that onset features have a straightforward musical interpretation and are directly comparable with MIDI data (Sect. 2.1.3). We will show how this property can be used for efficient and accurate synchronization of MIDI and audio data (Sect. 5.3). Finally, note that the onset representation is sparse compared to the original audio signal. Experiments have shown that the amount of data for a mono audio signal sampled at 22 050 Hz requires roughly 30–50 times more memory than the corresponding onset representation.

3.3 Chroma and CENS Features

It is a well-known phenomenon that human perception of pitch is periodic in the sense that two pitches are perceived as similar in "color" if they differ by an octave. On the basis of this observation, a pitch can be separated into two components, which are referred to as *tone height* and *chroma*, see [14, 191]. In the following, we only consider the musical notes of the equal-tempered scale. Then, using the MIDI pitch notation as introduced in Sect. 2.1, the tone height refers to the octave number and the chroma to the respective pitch spelling attribute contained in the set $\{\mathrm{C}, \mathrm{C}^\sharp, \mathrm{D}, \ldots, \mathrm{B}\}$. Note that in the equal-tempered scale, different pitch spellings such $\mathrm{C}^\sharp$ and $\mathrm{D}^\flat$ refer to the same chroma. A *pitch class* is defined to be the set of all pitches that share the same chroma. For example, the pitch class corresponding to the chroma C is the set $\{\ldots, \mathrm{C0}, \mathrm{C1}, \mathrm{C2}, \mathrm{C3}, \ldots\}$ consisting of all pitches separated by an integral number of octaves. For simplicity, we will not distinguish between a chroma and its respective pitch class.

A chroma-based audio representation can be easily obtained from a pitch representation simply by adding up the subbands that correspond to the same pitch class. More precisely, we first decompose an audio signal into 88 pitch subbands, compute the STMSP for each subband, and then add up all STMSPs that belong to the same pitch class. This results in an STMSP for the respective chroma. For example, to compute the STMSP of the chroma C, we add up the STMSPs of the pitches C1, C2, …, C8

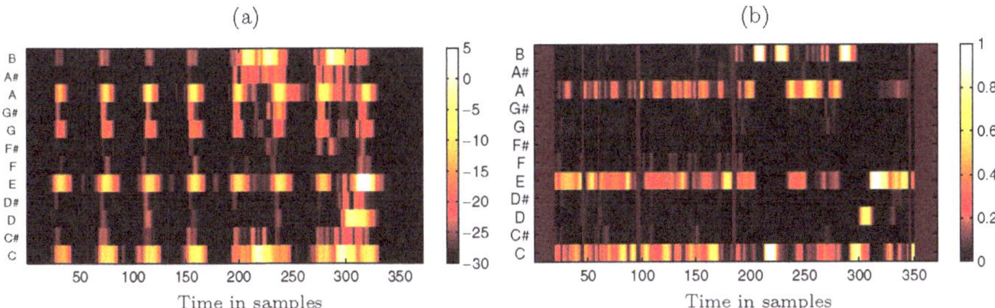

Fig. 3.7. (a) Time-chroma plot (with logarithmic decibel scale) of the chroma representation obtained from the pitch representation shown in Fig. 3.5. **(b)** Normalized time-chroma plot of (a)

(MIDI pitches $24, 36, \ldots, 108$). This yields a real 12-dimensional vector $v = (v(1), v(2) \ldots, v(12)) \in \mathbb{R}^{12}$ for every window, where $v(1)$ corresponds to chroma C, $v(2)$ to chroma C$^\sharp$, and so on. The resulting representation will be referred to as *chroma representation* of the audio signal, which is shown in Fig. 3.7a for our Burgmüller example. Recall that the timbre of a sound strongly relates to the energy distribution in the harmonics. Therefore, due to the octave equivalence, chroma features show a high degree of robustness to variations in timbre. Furthermore, chroma features account for the close octave relationship in both melody and harmony as prominent in Western music and are ideal for the analysis of music that is characterized by prominent harmonic progression, see [14].

In the remainder of this section, we introduce several modifications of the chroma representation to add further degrees of robustness. To absorb difference in the sound intensity or dynamics, we introduce the normalized chroma representation. To this end, one replaces a chroma vector v by $v/\|v\|_1$, which expresses the relative distribution of the signal's energy within the 12 chroma bands, see Fig. 3.7b. Here, $\|v\|_1 := \sum_{i=1}^{12} |v(i)|$ denotes the ℓ^1-norm of v. To avoid random energy distributions occurring during passages of very low energy (e.g., passages of silence before the actual start of the recording or during long pauses), we replace a chroma vector v by the uniform distribution in case $\|v\|_1$ falls below a certain threshold.

In view of possible variations in local tempo, articulation, and note execution, the local chroma energy distribution features are still too sensitive. Furthermore, the applications described in the Chaps. 5–7 will require a flexible and computationally inexpensive procedure to adjust the feature resolution. Therefore, we further process the chroma features by introducing a second, much larger statistics window and consider *short-time statistics* concerning the chroma energy distribution over this window. First, a given audio signal is transformed into a feature sequence $(v_1, v_2, \ldots, v_N)$ consisting of chroma vectors $v_n \in [0,1]^{12}$, $n \in [1:N]$. Then, defining the quantization function $\tau : [0,1] \to \{0,1,2,3,4\}$ by

$$\tau(a) := \begin{cases} 0 \text{ for } & 0 \leq a < 0.05, \\ 1 \text{ for } & 0.05 \leq a < 0.1, \\ 2 \text{ for } & 0.1 \leq a < 0.2, \\ 3 \text{ for } & 0.2 \leq a < 0.4, \\ 4 \text{ for } & 0.4 \leq a \leq 1, \end{cases} \tag{3.2}$$

we quantize each chroma energy distribution vector $v_n = (v_n(1), \ldots, v_n(12)) \in [0,1]^{12}$ by applying τ to each component of v_n, yielding

$$\tau(v_n) := (\tau(v_n(1)), \ldots, \tau(v_n(12))). \tag{3.3}$$

Intuitively, this quantization assigns a value of 4 to a chroma component $v_n(i)$ if the corresponding chroma class contains more than 40% of the signal's total energy and so on. The chroma components below a 5% threshold are excluded from further consideration, which introduces some robustness to noise. The thresholds are chosen in a logarithmic fashion to account for the logarithmic sensation of sound intensity, see [225]. For example, the vector $v_n = (0.02, 0.5, 0.3, 0.07, 0.11, 0, \ldots, 0)$ is transformed into the vector $\tau(v_n) := (0, 4, 3, 1, 2, 0, \ldots, 0)$.

In a subsequent step, we convolve the sequence $(\tau(v_1), \ldots, \tau(v_N))$ component-wise with a Hann window of length $\mathbf{w} \in \mathbb{N}$. This again results in a sequence of 12-dimensional vectors with nonnegative entries, representing a kind of weighted statistics of the energy distribution over a window of $\mathbf{w}$ consecutive vectors. In a last step, this sequence is downsampled by a factor of $\mathbf{d}$. The resulting vectors are normalized with respect to the ℓ^2-norm (Euclidean norm). The resulting features are referred to as $\text{CENS}^{\mathbf{w}}_{\mathbf{d}}$ (chroma energy normalized statistics). These features are elements of the following set of vectors:

$$\mathcal{F} := \{v = (v = (v(1), \ldots, v(12)) \in [0,1]^{12} \mid \|v\|_2 = 1\}, \tag{3.4}$$

where $\|v\|_2 := (\sum_{i=1}^{12} v(i)^2)^{\frac{1}{2}}$. The entire procedure for the CENS computation is summarized in Fig. 3.8.

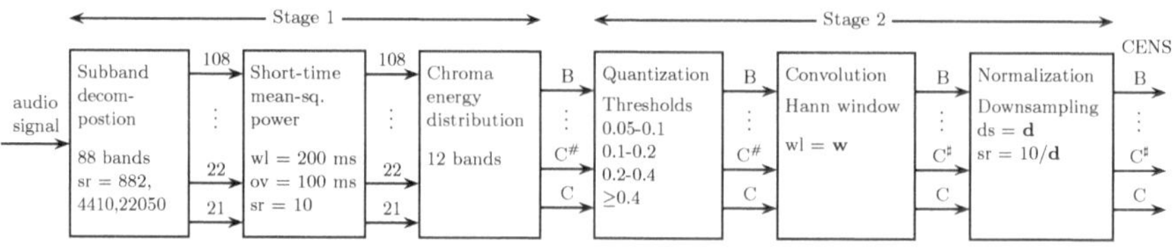

Fig. 3.8. Two-stage CENS feature design (wl = window length, ov = overlap, sr = sampling rate, ds = downsampling factor)

In the following, we will use a 200 ms rectangular window with an overlap of half the window size to compute the STMSP for each of the 88 subbands, see Sect. 3.2. This results in a resolution of 100 ms per feature or, equivalently, in a feature sampling rate of 10 Hz. Then, using $\mathbf{w} = 41$ and $\mathbf{d} = 10$ in the CENS computation, one obtains one CENS vector per second, each considering roughly 4 100 ms of the original audio signal. Figure 3.9 shows the resulting sequence of CENS feature vectors for our Burgmüller example. As this example demonstrates, CENS sequences correlate closely with the smoothed harmonic progression of the underlying audio signal, while absorbing variations in other parameters. First, the normalization makes the CENS features invariant to variations in dynamics. Second, using chroma instead of pitches introduces a high degree of robustness to variations in timbre. Third, taking statistics over relatively large windows not only smoothes out local time deviations as may occur for articulatory reasons but also compensates for different realizations of note groups such as trills or arpeggios. As an example, we consider Beethoven's Fifth in an orchestral version conducted by Bernstein and in some piano version played by Glen Gould (Fig. 2.5). Even though these two interpretations exhibit significant variations in articulation and instrumentation, the resulting CENS sequences are similar (Fig. 3.10). As we will see in Chap. 6, such CENS sequences, which establish a temporal relation between different feature vectors, become powerful in characterizing harmony-based music, Finally, note that once the signal has been decomposed into the STMSP subbands (with a fixed feature sampling rate that represents the finest required resolution level for a specific application), one may adjust

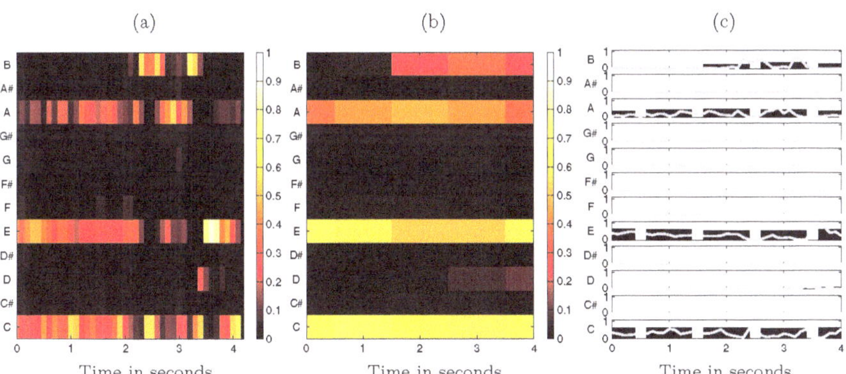

Fig. 3.9. **(a)** Normalized chroma representation for the Burgmüller example (Fig. 3.3) using a 200 ms rectangular window with an overlap of half the window size (10 features per second). **(b)** Resulting CENS sequence for $\mathbf{w} = 41$ and $\mathbf{d} = 10$. Each column represents a 12-dimensional feature vector (1 feature per second). **(c)** The light curves represent the chroma energy distributions as in (a), whereas the dark bars represent the CENS features as in (b)

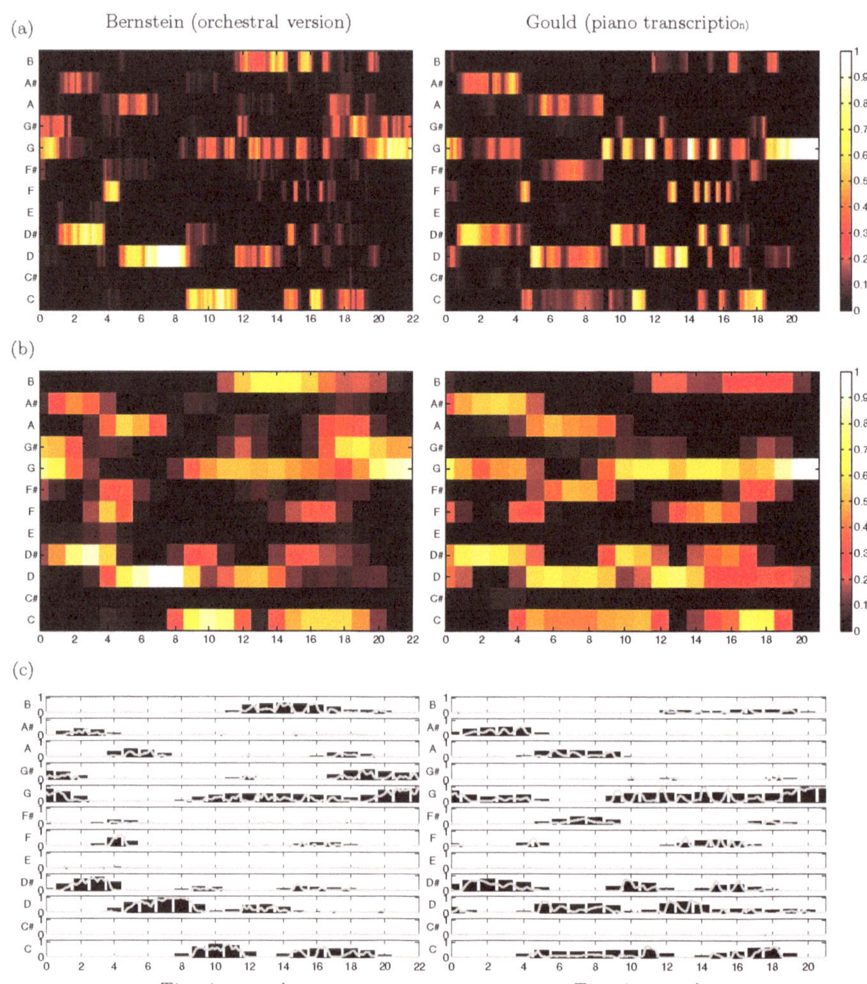

Fig. 3.10. Normalized chroma representations **(a,c)** and CENS sequences **(b,c)** as in Fig. 3.9 for Beethoven's Fifth (first 21 measures) in an orchestral version conducted by Bernstein (*left*) and in some piano version played by Glen Gould (*right*)

the granularity and sampling rate of the CENS features simply by modifying the parameters **w** and **d** without repeating the cost-intensive spectral audio decomposition. Furthermore, one can adjust the thresholds and values of the quantization function τ to mask out or enhance certain aspects of the audio signal, thus making the CENS features insensitive to noise components or boosting the strongest chroma subband, which often correlates to the melody.

3.4 Further Notes

The decomposition of an audio signal into suitable subbands is a major step in most applications of content-based audio analysis. In this context, a large number of audio representations have been suggested, the most prominent being the classical *short-time Fourier transform* (STFT), which is also referred to as *windowed Fourier transform* (WFT). The STFT, which was introduced by Dennis Gabor in the year 1946, is used to represent the frequency and phase content of windowed frames of the audio signal. For details and further references we refer to [43,98,104]. Considering the energy of the windowed frequency content one obtains the *spectrogram*, which results in a time-frequency representation of the original signal [131]. From this representation one can obtain a pitch representation by suitably pooling the Fourier coefficients. To this end, one associates each coefficient of the spectrogram to the pitch whose center frequency is closest to the frequency expressed by the Fourier coefficient, and then adds up the associated coefficients [14]. However, this approach is problematic since the STFT computes spectral coefficients that are *linearly* spread over the spectrum, see Sect. 2.2.3. This may lead to a frequency resolution that is not sufficient to separate musical notes of low frequency. For example, suppose the audio signal is sampled at 22 050 Hz, then using a window size of 2 048 samples in the STFT will yield (approximated) Fourier coefficients for the frequencies $k \cdot 10.77\,\mathrm{Hz}$, $k = 0, 1, \ldots, 1\,024$, see (2.18). This resolution would not be sufficient to resolve the notes below $F3$ ($p = 53$), where the difference of the center frequencies of adjacent pitches is below 10 Hz. One possible strategy to increase the frequency resolution for the lower notes is to use several STFTs that work at different sampling rates and with different window sizes (similarly to using a multirate filter bank), see, e. g., [83]. The filter bank techniques discussed in this chapter can be seen as a flexible alternative to the STFT for decomposing an audio signal into pitch subbands. Similar filter bank approaches can be found in the literature, see [17, 185]. However, the filter characteristics may differ significantly and depend upon the respective application. For example, in our approach the filters were chosen in view of excellent cutoff properties and efficient processing. In other applications it may be important that the original signal can be recovered from the subbands, which typically comes at the expense of long filter responses, see [17].

The design of musically relevant audio features that are robust to certain variations is of fundamental importance in music information retrieval and there is a large number of approaches suggested in the literature, see for example [104, 186]. One prominent example are the mel-frequency cepstral coefficients (MFCC), which are perceptually based spectral features originally designed for speech processing applications, see [170]. Similarly to the pitch decomposition, the frequency bands are positioned logarithmically on the so-called *mel scale*, which approximates the response of the human auditory system. In the music context, MFCCs have turned out to be useful in capturing timbral characteristics and are used in particular for music classification tasks, see [104, 125]. The idea of describing a pitch by means of its tone height and chroma is the foundation of how musical notes in Western music are notated, see [191]. In recent years, due to their explicit musical meaning as well as their robustness to timbral deformations, chroma features have been extensively used in music analysis tasks, see [14, 104]. As another general strategy used in this chapter, we considered certain (short-time) statistics to add further degrees of robustness. Global statistics such as pitch histograms have previously been applied to fields such as music genre classification, see, e.g., [204]. The combination of using chroma features and local statistics has first been proposed in [140].

As has been mentioned earlier, the analysis of music recordings constitutes a difficult problem – in view of musical variations between different realizations as well as the complexity of musical sounds. As a final note, we emphasize that even the sound resulting from playing a single note on a piano is a complex mixture of different frequencies. This fact is illustrated by Fig. 3.11, which shows the pitch decomposition of a chromatic scale starting with the note A0 ($p = 21$) and ending with A4 ($p = 69$). Naturally, the sounds for higher pitches exhibit a much cleaner pitch spectrum and sharper onsets than the ones for lower pitches, the latter being noisy and smeared over time. The energy distribution in the harmonics not only depends on the respective instrument,

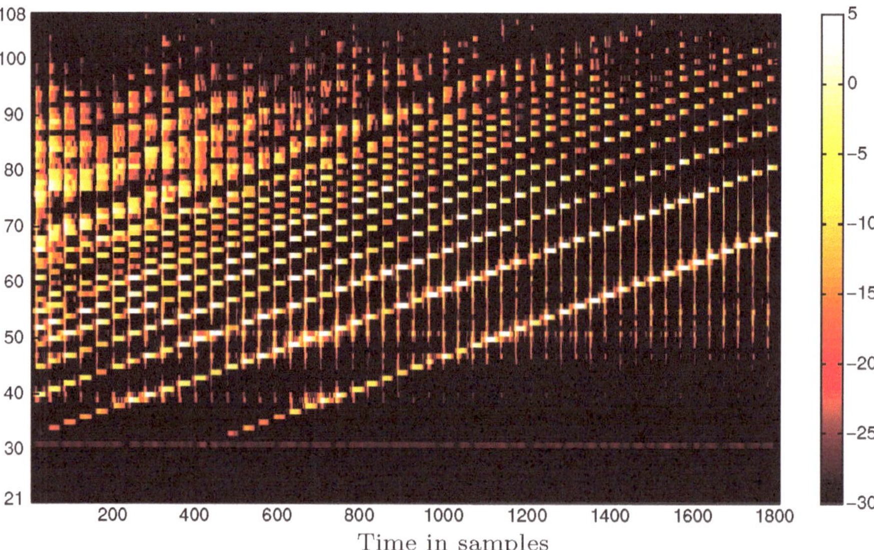

Fig. 3.11. Time–pitch plot (logarithmic scale) for a piano recording of the chromatic scale starting with note A0 ($p = 21$) and ending with A4 ($p = 69$). The rows correspond to MIDI pitches and the columns to time measured in samples with a time resolution of 46.4 ms

but also on the sound intensity and tone height. For lower notes, the signal's energy is often contained in the higher harmonics, while the listeners still have the perception of a low sound. Furthermore, the frequency content of a sound significantly depends on the microphone's frequency response, which is often poor for very low frequencies as was the case in our recording of the piano notes. Furthermore, Fig. 3.11 shows the noise in the subband $p = 31$ (with center frequency $f(p) = 49.0\,\mathrm{Hz}$), which stems from the alternating current at 50 Hz. Finally, observe the typical note attacks that are visible in Fig. 3.11 as vertical lines spreading over several octaves in the pitch representation.

4

Dynamic Time Warping

Dynamic time warping (DTW) is a well-known technique to find an optimal alignment between two given (time-dependent) sequences under certain restrictions (Fig. 4.1). Intuitively, the sequences are warped in a nonlinear fashion to match each other. Originally, DTW has been used to compare different speech patterns in automatic speech recognition, see [170]. In fields such as data mining and information retrieval, DTW has been successfully applied to automatically cope with time deformations and different speeds associated with time-dependent data.

In this chapter, we introduce and discuss the main ideas of classical DTW (Sect. 4.1) and summarize several modifications concerning local as well as global parameters (Sect. 4.2). To speed up classical DTW, we describe in Sect. 4.3 a general multiscale DTW approach. In Sect. 4.4, we show how DTW can be employed to identify all subsequence within a long data stream that are similar to a given query sequence (Sect. 4.4). A discussion of related alignment techniques and references to the literature can be found in Sect. 4.5.

4.1 Classical DTW

The objective of DTW is to compare two (time-dependent) sequences $X := (x_1, x_2, \ldots, x_N)$ of length $N \in \mathbb{N}$ and $Y := (y_1, y_2, \ldots, y_M)$ of length $M \in \mathbb{N}$. These sequences may be discrete signals (time-series) or, more generally, feature sequences sampled at equidistant points in time. In the following, we fix a *feature space* denoted by $\mathcal{F}$. Then $x_n, y_m \in \mathcal{F}$ for $n \in [1:N]$ and $m \in [1:M]$. To compare two different features $x, y \in \mathcal{F}$, one needs a *local cost measure*, sometimes also referred to as *local distance measure*, which is defined to be a function

$$c : \mathcal{F} \times \mathcal{F} \to \mathbb{R}_{\geq 0}. \tag{4.1}$$

Typically, $c(x, y)$ is small (low cost) if x and y are similar to each other, and otherwise $c(x, y)$ is large (high cost). Evaluating the local cost measure for

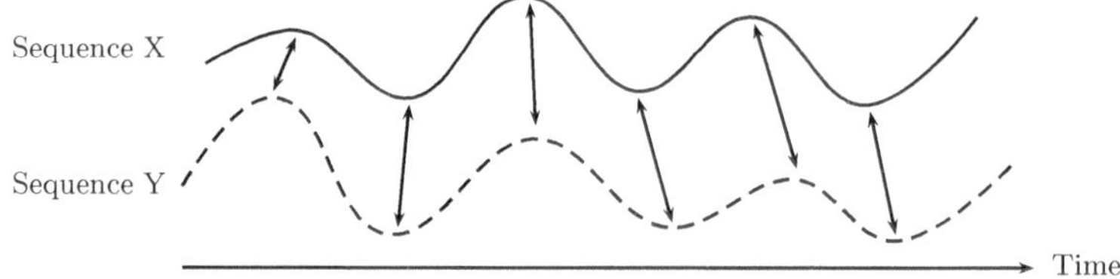

Fig. 4.1. Time alignment of two time-dependent sequences. Aligned points are indicated by the *arrows*

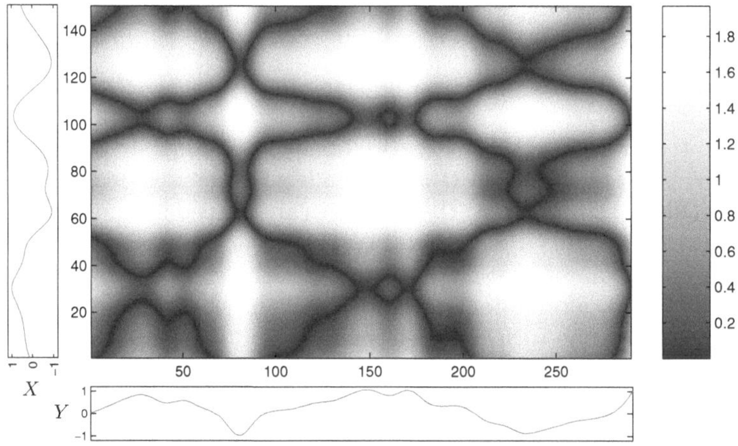

Fig. 4.2. Cost matrix of the two real-valued sequences X (*vertical axis*) and Y (*horizontal axis*) using the Manhattan distance (absolute value of the difference) as local cost measure c. Regions of low cost are indicated by *dark colors* and regions of high cost are indicated by *light colors*

each pair of elements of the sequences X and Y, one obtains the *cost matrix* $C \in \mathbb{R}^{N \times M}$ defined by $C(n, m) := c(x_n, y_m)$, see Fig. 4.2. Then the goal is to find an alignment between X and Y having minimal overall cost. Intuitively, such an optimal alignment runs along a "valley" of low cost within the cost matrix C, see Fig. 4.4 for an illustration. The next definition formalizes the notion of an alignment.

Definition 4.1. *An* (N, M)-warping path *(or simply referred to as* warping path *if N and M are clear from the context) is a sequence $p = (p_1, \ldots, p_L)$ with $p_\ell = (n_\ell, m_\ell) \in [1 : N] \times [1 : M]$ for $\ell \in [1 : L]$ satisfying the following three conditions.*

(i) Boundary condition: $p_1 = (1, 1)$ *and* $p_L = (N, M)$.
(ii) Monotonicity condition: $n_1 \leq n_2 \leq \ldots \leq n_L$ *and* $m_1 \leq m_2 \leq \ldots \leq m_L$.
(iii) Step size condition: $p_{\ell+1} - p_\ell \in \{(1, 0), (0, 1), (1, 1)\}$ *for* $\ell \in [1 : L - 1]$.

Note that the step size condition (iii) implies the monotonicity condition (ii), which nevertheless has been quoted explicitly for the sake of clarity. An (N, M)-warping path $p = (p_1, \ldots, p_L)$ defines an alignment between two sequences $X = (x_1, x_2, \ldots, x_N)$ and $Y = (y_1, y_2, \ldots, y_M)$ by assigning the element x_{n_ℓ} of X to the element y_{m_ℓ} of Y. The boundary condition enforces that the first elements of X and Y as well as the last elements of X and Y are aligned to each other. In other words, the alignment refers to the entire sequences X and Y. The monotonicity condition reflects the requirement of faithful timing: if an element in X precedes a second one this should also hold for the corresponding elements in Y, and vice versa. Finally, the step size condition expresses a kind of continuity condition: no element in X and Y can be omitted and there are no replications in the alignment (in the sense that all index pairs contained in a warping path p are pairwise distinct). Figure 4.3 illustrates the three conditions.

The *total cost* $c_p(X, Y)$ of a warping path p between X and Y with respect to the local cost measure c is defined as

$$c_p(X, Y) := \sum_{\ell=1}^{L} c(x_{n_\ell}, y_{m_\ell}). \tag{4.2}$$

Furthermore, an *optimal warping path* between X and Y is a warping path p^* having minimal total cost among all possible warping paths. The *DTW distance* $\mathrm{DTW}(X, Y)$ between X and Y is then defined as the total cost of p^*:

$$\begin{aligned} \mathrm{DTW}(X, Y) &:= c_{p^*}(X, Y) \\ &= \min\{c_p(X, Y) \mid p \text{ is an } (N, M)\text{-warping path}\} \end{aligned} \tag{4.3}$$

We continue with several remarks about the DTW distance. First, note that the DTW distance is well-defined even though there may be several warping paths of minimal total cost. Second, it is easy to see that the DTW distance is symmetric in case that the local cost measure c is symmetric. However,

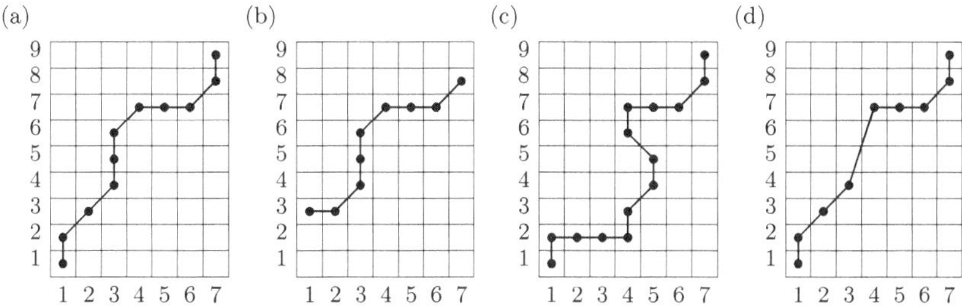

Fig. 4.3. Illustration of paths of index pairs for some sequence X of length $N = 9$ and some sequence Y of length $M = 7$. **(a)** Admissible warping path satisfying the conditions (i), (ii), and (iii) of Definition 4.1. **(b)** Boundary condition (i) is violated. **(c)** Monotonicity condition (ii) is violated. **(d)** Step size condition (iii) is violated

the DTW distance is in general not positive definite even if this holds for c. For example, one obtains $\mathrm{DTW}(X,Y) = 0$ for the sequences $X := (x_1, x_2)$ and $Y := (x_1, x_1, x_2, x_2, x_2)$ in case $c(x_1, x_1) = c(x_2, x_2) = 0$. Furthermore, the DTW distance generally does not satisfy the triangle inequality even in case c is a metric. This fact is illustrated by the following example.

Example 4.2. Let $\mathcal{F} := \{\alpha, \beta, \gamma\}$ be a feature space consisting of three features. We define a cost measure $c : \mathcal{F} \times \mathcal{F} \to \{0,1\}$ by setting $c(x,y) := 0$ if $x = y$ and $c(x,y) := 1$ if $x \neq y$ for $x, y \in \mathcal{F}$. Note that c defines a metric on $\mathcal{F}$ and particularly satisfies the triangle inequality. Now consider $X := (\alpha, \beta, \gamma)$, $Y := (\alpha, \beta, \beta, \gamma)$, and $Z := (\alpha, \gamma, \gamma)$. Then one easily checks that $\mathrm{DTW}(X,Y) = 0$, $\mathrm{DTW}(X,Z) = 1$, and $\mathrm{DTW}(Y,Z) = 2$. Therefore, $\mathrm{DTW}(Y,Z) > \mathrm{DTW}(X,Y) + \mathrm{DTW}(X,Z)$, i. e., the DTW distance does not satisfy the triangle inequality. Finally, note that the paths $p^1 = ((1,1), (2,2), (3,2), (4,3))$, $p^2 = ((1,1), (2,1), (3,2), (4,3))$, and $p^3 = ((1,1), (2,2), (3,3), (4,3))$ are different optimal warping paths between Y and Z of total cost two. This shows that an optimal warping path is generally not unique.

To determine an optimal path p^*, one could test every possible warping path between X and Y. Such a procedure, however, would lead to a computational complexity that is exponential in the lengths N and M. We will now introduce an $O(NM)$ algorithm that is based on *dynamic programming*. To this end, we define the prefix sequences $X(1\!:\!n) := (x_1, \ldots x_n)$ for $n \in [1:N]$ and $Y(1\!:\!m) := (y_1, \ldots y_m)$ for $m \in [1:M]$ and set

$$D(n,m) := \mathrm{DTW}(X(1\!:\!n), Y(1\!:\!m)). \tag{4.4}$$

The values $D(n,m)$ define an $N \times M$ matrix D, which is also referred to as the *accumulated cost matrix*. Obviously, one has $D(N,M) = \mathrm{DTW}(X,Y)$. In the following, a tuple (n,m) representing a matrix entry of the cost matrix C or of the accumulated cost matrix D will be referred to as a *cell*. The next theorem shows how D can be computed efficiently.

Theorem 4.3. *The accumulated cost matrix D satisfies the following identities: $D(n,1) = \sum_{k=1}^{n} c(x_k, y_1)$ for $n \in [1:N]$, $D(1,m) = \sum_{k=1}^{m} c(x_1, y_k)$ for $m \in [1:M]$, and*

$$D(n,m) = \min\{D(n-1,m-1), D(n-1,m), D(n,m-1)\} + c(x_n, y_m) \tag{4.5}$$

for $1 < n \leq N$ and $1 < m \leq M$. In particular, $\mathrm{DTW}(X,Y) = D(N,M)$ can be computed with $O(NM)$ operations.

Proof. Let $m = 1$ and $n \in [1:N]$. Then there is only one possible warping path between $Y(1:1)$ and $X(1:n)$ having a total cost of $\sum_{k=1}^{n} c(x_k, y_1)$. This proves the formula for $D(n,1)$. Similarly, one obtains the formula for $D(1,m)$. Now, let $n > 1$ and $m > 1$ and let $q = (q_1, \ldots, q_{L-1}, q_L)$ be an optimal warping path for $X(1\!:\!n)$ and $Y(1\!:\!m)$. Then the boundary condition implies $q_L = (n,m)$. Setting $(a,b) := q_{L-1}$, the step size condition implies $(a,b) \in \{(n-1,m-1),(n-1,m),(n,m-1)\}$. Furthermore, it follows that $(q_1,\ldots,q_{L-1})$ must be an optimal warping path for $X(1:a)$ and $Y(1:b)$ (otherwise, q would not be optimal for $X(1\!:\!n)$ and $Y(1\!:\!m)$). Since $D(n,m) = c_{(q_1,\ldots,q_{L-1})}(X(1:a),Y(1:b)) + c(x_n,y_m)$, the optimality of q implies the assertion of (4.5). □

Theorem 4.3 facilitates a recursive computation of the matrix D. The initialization can be simplified by extending the matrix D with an additional row and column and formally setting $D(n,0) := \infty$ for $n \in [1:N]$, $D(0,m) := \infty$ for $m \in [1:M]$, and $D(0,0) := 0$. Then the recursion of (4.5) holds for $n \in [1:N]$ and $m \in [1:M]$. Furthermore, note that D can be computed in a column-wise fashion, where the computation of the m-th column only requires the values of the $(m-1)$-th column. This implies that if one is only interested in the value $\mathrm{DTW}(X,Y) = D(N,M)$, the storage requirement is $O(N)$. Similarly, one can proceed in a row-wise fashion, leading to $O(M)$. However, note that the running time is $O(NM)$ in either case. Furthermore, to compute an optimal warping path p^*, the entire $(N \times M)$-matrix D is needed. It is left as an exercise to show that the following algorithm OPTIMALWARPINGPATH fulfills this task.

Algorithm: OPTIMALWARPINGPATH

Input: Accumulated cost matrix D.
Output: Optimal warping path p^*.

Procedure: The optimal path $p^* = (p_1,\ldots,p_L)$ is computed in reverse order of the indices starting with $p_L = (N,M)$. Suppose $p_\ell = (n,m)$ has been computed. In case $(n,m) = (1,1)$, one must have $\ell = 1$ and we are finished. Otherwise,

$$p_{\ell-1} := \begin{cases} (1,m-1), & \text{if } n = 1 \\ (n-1,1), & \text{if } m = 1 \\ \operatorname{argmin}\{D(n-1,m-1), & \\ \qquad D(n-1,m), D(n,m-1)\}, & \text{otherwise,} \end{cases} \tag{4.6}$$

where we take the lexicographically smallest pair in case "argmin" is not unique.

Figure 4.4 shows the optimal warping path p^* (white line) for the sequences of Fig. 4.2. Note that p^* covers only cells of C that exhibit low costs (cf. Fig. 4.4a). The resulting accumulated cost matrix D is shown in Fig. 4.4b.

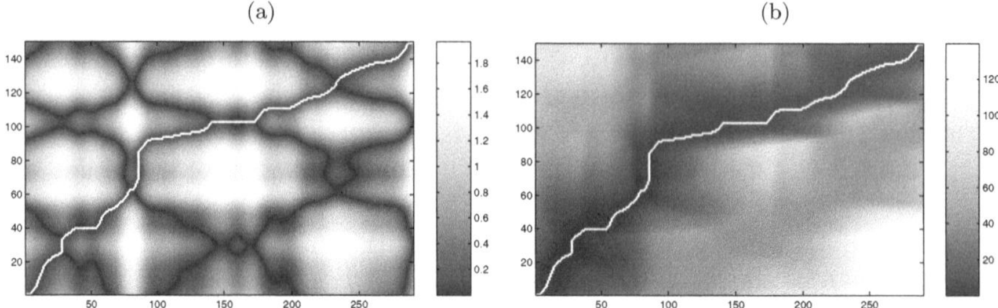

Fig. 4.4. **(a)** Cost matrix C as in Fig. 4.2 and **(b)** accumulated cost matrix D with optimal warping path p^* (*white line*)

4.2 Variations of DTW

Various modifications have been proposed to speed up DTW computations as well as to better control the possible routes of the warping paths. In this section, we discuss some of these variations and refer to [170] for further details.

4.2.1 Step Size Condition

Recall that the step size condition (iii) of Definition 4.1 represents a kind of local continuity condition, which ensures that each element of the sequence $X = (x_1, x_2, \ldots, x_N)$ is assigned to an element of $Y = (y_1, y_2, \ldots, y_M)$ and vice versa. However, one drawback of this condition is that a single element of one sequence may be assigned to many consecutive elements of the other sequence, leading to vertical and horizontal segments in the warping path, see Fig. 4.6a. Intuitively, the warping path can get stuck at some position with respect to one sequence, corresponding to a local deceleration by a large factor (or, conversely, to a local acceleration by a large factor regarding the second sequence).

To avoid such degenerations, one can modify the step size condition to constrain the slope of the admissible warping paths. As a first example, we replace the step size condition (iii) of Definition 4.1 by the condition $p_{\ell+1} - p_\ell \in \{(2,1), (1,2), (1,1)\}$ for $\ell \in [1 : L]$, see Fig. 4.5b. This leads to warping paths having a local slope within the bounds $\frac{1}{2}$ and 2. The resulting accumulated cost matrix D can then be computed by the recursion

$$D(n,m) = \min\{D(n-1,m-1), D(n-2,m-1), D(n-1,m-2)\} + c(x_n, y_m) \tag{4.7}$$

for $n \in [2 : N]$ and $m \in [2 : N]$. As initial values, we set $D(0,0) := 0$, $D(1,1) := c(x_1, y_1)$, $D(n,0) := \infty$ for $n \in [1 : N]$, $D(n,1) := \infty$ for $n \in [2 : N]$, $D(0,m) := \infty$ for $m \in [1 : M]$, and $D(1,m) := \infty$ for $m \in [2 : M]$. Note that, with respect to the modified step size condition, there is a warping

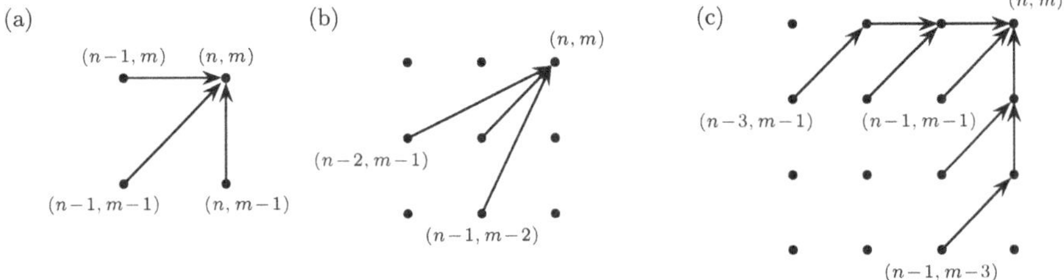

Fig. 4.5. Illustration of three different step size conditions, which express different local constraints on the admissible warping paths. (**a**) corresponds to the step size condition (iii) of Definition 4.1

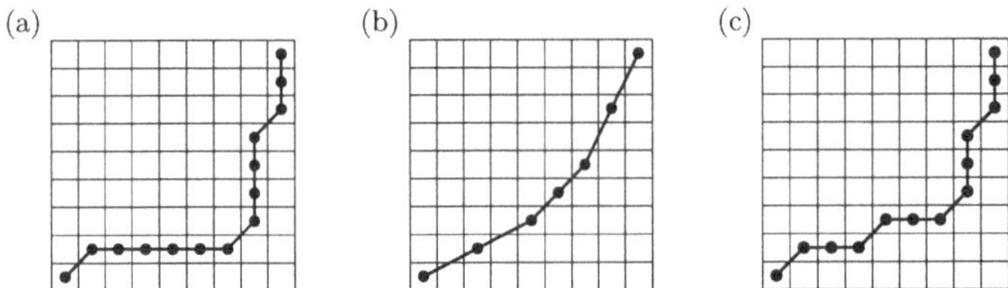

Fig. 4.6. Three warping paths with respect to the different step size conditions indicated by Fig. 4.5. **(a)** Step size condition of Fig. 4.5a may result in degenerations of the warping path. **(b)** Step size condition of Fig. 4.5b may result in the omission of elements in the alignment of X and Y. **(c)** Warping path with respect to the step size condition of Fig. 4.5c

path between two sequences X and Y if and only if the lengths N and M differ at most by a factor of two. Furthermore, note that not all elements of X need to be assigned to some element of Y and vice versa. This is illustrated by Fig. 4.6b: here, x_1 is assigned to y_1, x_3 is assigned to y_2, but x_2 is not assigned to any element of Y (i.e., x_2 is omitted and does not cause any cost at all).

Figure 4.5c gives a second example for a step size condition, which avoids such omission while imposing constraints on the slope of the warping path. The recursion of the resulting accumulated cost matrix D is given by

$$D(n,m) = \min \begin{cases} D(n-1,m-1) + c(x_n, y_m) \\ D(n-2,m-1) + c(x_{n-1}, y_m) + c(x_n, y_m) \\ D(n-1,m-2) + c(x_n, y_{m-1}) + c(x_n, y_m) \\ D(n-3,m-1) + c(x_{n-2}, y_m) + c(x_{n-1}, y_m) + c(x_n, y_m) \\ D(n-1,m-3) + c(x_n, y_{m-2}) + c(x_n, y_{m-1}) + c(x_n, y_m) \end{cases} \tag{4.8}$$

for $(n,m) \in [1:N] \times [1:M] \setminus \{(1,1)\}$. Here, the initial values are set to $D(1,1) := c(x_1,y_1)$, $D(n,-2) := D(n,-1) := D(n,0) := \infty$ for $n \in [-2:N]$, and $D(-2,m) := D(-1,m) := D(0,m) := \infty$ for $m \in [-2:M]$. The slopes of the resulting warping paths lie between the values $\frac{1}{3}$ and 3. Note that this step size conditions enforce that all elements of X are aligned to some element of Y and vice versa. In other words, in the recursion (4.8) all elements of X and Y generate some cost in the accumulated cost matrix D – opposed to the recursion (4.7). Figure 4.6 illustrates the differences of the resulting optimal warping paths computed with respect to different step size conditions.

4.2.2 Local Weights

To favor the vertical, horizontal, or diagonal direction in the alignment, one can introduce an additional weight vector $(w_d, w_h, w_v) \in \mathbb{R}^3$, yielding the recursion

$$D(n,m) = \min \begin{cases} D(n-1,m-1) + w_d \cdot c(x_n, y_m) \\ D(n-1,m) + w_h \cdot c(x_n, y_m) \\ D(n,m-1) + w_v \cdot c(x_n, y_m) \end{cases} \tag{4.9}$$

for $n \in [2:N]$ and $m \in [2:M]$. Furthermore, $D(n,1) := \sum_{k=1}^{n} w_h \cdot c(x_k, y_1)$ for $n > 1$, $D(1,m) = \sum_{k=1}^{m} w_v \cdot c(x_1, y_k)$ for $m > 1$, and $D(1,1) := c(x_1, y_1)$. The equally weighted case $(w_d, w_h, w_v) = (1,1,1)$ reduces to classical DTW, see (4.5). Note that for $(w_d, w_h, w_v) = (1,1,1)$ one has a preference of the diagonal alignment direction, since one diagonal step (cost of one cell) corresponds to the combination of one horizontal and one vertical step (cost of two cells). To counterbalance this preference, one often chooses $(w_d, w_h, w_v) = (2,1,1)$. Similarly, one can introduce weights for other step size conditions.

4.2.3 Global Constraints

One common DTW variant is to impose global constraint conditions on the admissible warping paths. Such constraints do not only speed up DTW computations but also prevent pathological alignments by globally controlling the route of a warping path. More precisely, let $R \subseteq [1:N] \times [1:M]$ be a subset referred to as global *constraint region*. Then a *warping path relative to* R is a warping path that entirely runs within the region R. The *optimal warping path relative to* R, denoted by p_R^*, is the cost-minimizing warping path among all warping paths relative to R.

Two well-know global constraint regions are the *Sakoe-Chiba band* and the *Itakura parallelogram*, as indicated by Fig. 4.7. Alignments of cells can be selected only from the respective shaded region. The *Sakoe-Chiba band* runs along the main diagonal and has a fixed (horizontal and vertical) width $T \in \mathbb{N}$. This constraint implies that an element x_n can be aligned only to one of the elements y_m with $m \in \left[\frac{M-T}{N-T} \cdot (n-T), \frac{M-T}{N-T} \cdot n + T\right] \cap [1:M]$, see Fig. 4.7a. The *Itakura parallelogram* describes a region that constrains the slope of a warping path. More precisely, for a fixed $S \in \mathbb{R}_{>1}$, the Itakura parallelogram consists of all cells that are traversed by some warping path having a slope

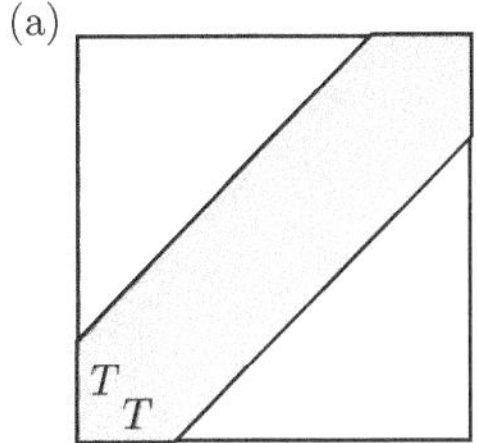

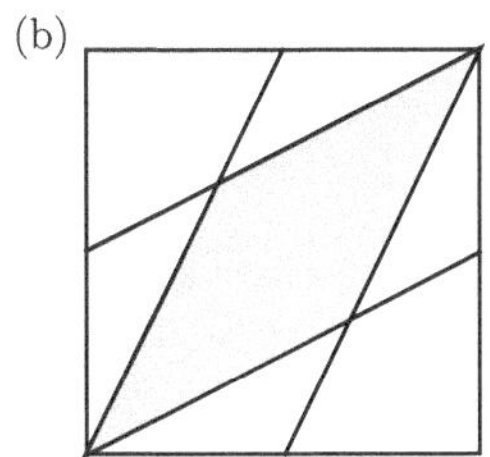

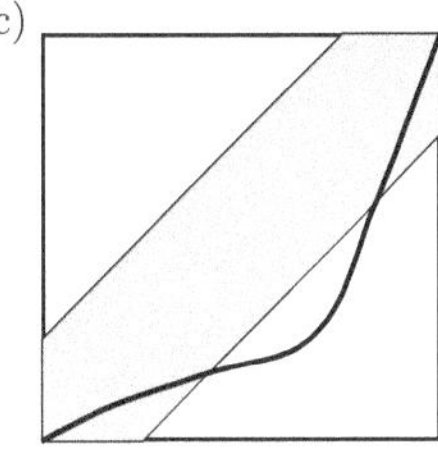

Fig. 4.7. **(a)** Sakoe-Chiba band of (horizontal and vertical) width T. **(b)** Itakura parallelogram with $S = 2$. **(c)** Optimal warping path p^* (*black line*) which does not run within the constraint region R

between the values $1/S$ and S, see Fig. 4.7b. Note that the local step size condition introduced in Sect. 4.2.1 may also imply some global constraint. For example, the step size condition $p_{\ell+1} - p_\ell \in \{(2,1),(1,2),(1,1)\}$ for $\ell \in [1:L]$, as indicated by Fig. 4.5b, implies a global constraint region in form of an Itakura parallelogram with $S = 2$.

For a general constraint region R, the path p_R^* can be computed similar to the unconstrained case by formally setting $c(x_n, y_m) := \infty$ for all $(n,m) \in [1:N] \times [1:M] \setminus R$. Therefore, in the computation of p_R^* only the cells that lie in R need to be evaluated. This may significantly speed up the DTW computation. For example, in case of a Sakoe-Chiba band of a fixed width T, only $O(T \cdot \max(N,M))$ computations need to be performed instead of $O(NM)$ as required in classical DTW. Here, note that one typically has $T \ll M$ and $T \ll N$.

However, the usage of global constraint regions is also problematic, since the optimal warping path may traverse cells outside the specified constraint region. In other words, the resulting optimal (constrained) warping path p_R^* generally does not coincide with the optimal (unconstrained) warping path p^*, see Fig. 4.7c. This fact may lead to undesirable or even completely useless alignment results.

4.2.4 Approximations

An effective strategy to speed up DTW computations is based on the idea to perform the alignment on coarsened versions of the sequences X and Y, thus reducing the lengths N and M of the two sequences. Such a strategy is also known as dimensionality reduction or data abstraction. For example, to reduce the data rate, one could process the sequences by some suitable low-pass filter followed by downsampling. Another strategy is to approximate the sequences by some piecewise linear (or any other kind of) function and then to perform the warping at the approximation level. For further strategies and an overview, we refer to [102].

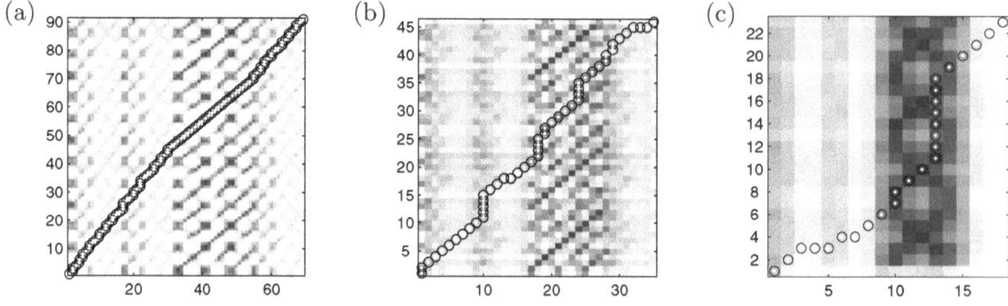

Fig. 4.8. (a) Cost matrix and optimal warping path (*dotted path*) between two feature sequences X of length $N = 91$ and Y of length $M = 68$. **(b)** Cost matrix obtained after low-pass filtering and downsampling (by a factor of two) the feature sequences. The resulting optimal warping path at the lower resolution level does not accord well to the optimal warping path of (a). **(c)** Further decreasing the feature resolution destroys the structure of the cost matrix and leads to a completely useless alignment

One important limitation of this approach, however, is that the user must carefully specify the approximation levels used in the alignment. If the user chooses too fine of an approximation, the gains in speed are negligible. Conversely, if the user chooses too coarse of an approximation, e. g., by decreasing the sampling rate of the feature sequences X and Y, the resulting optimal warping path may become inaccurate or even completely useless, see [102]. This fact is also illustrated by Fig. 4.8.

4.3 Multiscale DTW

To obtain an efficient as well as robust algorithm to compute DTW-based alignments, one can combine the strategies described in Sects. 4.2.3 and 4.2.4 in some iterative fashion to generate data-dependent constraint regions. The general strategy is to recursively project an optimal warping path computed at a coarse resolution level to the next higher level and then to refine the projected path. In this section, we summarize the main ideas of this approach, which will be referred to as multiscale DTW (MsDTW). For details we refer to [184]. A similar approach has been applied, e. g., to melody alignment [1] and to audio alignment [142].

Let $X_1 := X$ and $Y_1 := Y$ be the sequences to be synchronized, having lengths $N_1 := N$ and $M_1 := M$, respectively. It is the objective to compute an optimal warping path p^* between X_1 and Y_1. The highest resolution level will also be referred to as Level 1. By reducing the feature sampling rate by a factor of $f_2 \in \mathbb{N}$, one obtains sequences X_2 of length $N_2 := N_1/f_2$ and Y_2 of length $M_2 := M_1/f_2$. (Here, we assume that f_2 divides N_1 and M_1, which can be achieved by suitably padding X_1 and Y_1.) Next, one computes an optimal warping path p_2^* between X_2 and Y_2 on the resulting resolution

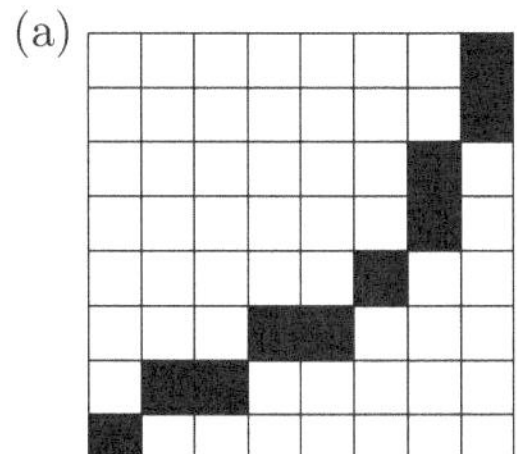

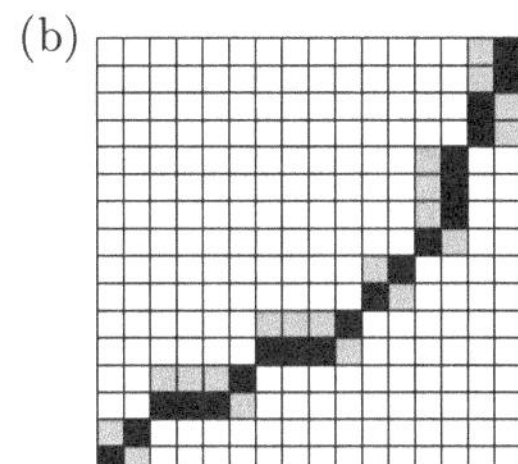

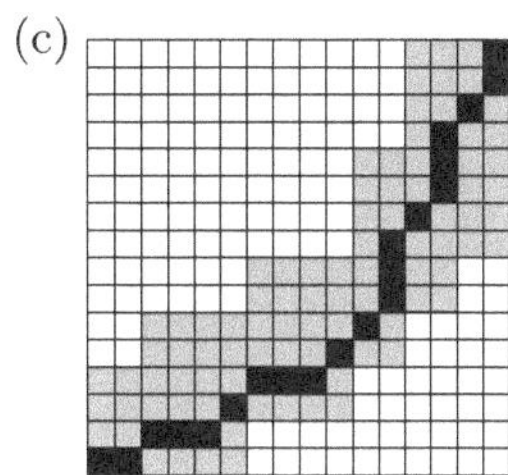

Fig. 4.9. **(a)** Optimal warping path p_2^* on Level 2. **(b)** Optimal warping path p_R^* with respect to the constraint region R obtained by projecting path p_2^* to Level 1. (Here, p_R^* does not coincide with the (unconstrained) optimal warping path p^*.) **(c)** Optimal warping path $p_{R^\delta}^*$ using an increased constraint region $R^\delta \supset R$ with $\delta = 2$. Here, $p_{R^\delta}^* = p^*$

level (Level 2). This path is projected onto Level 1 and there defines a constraint region R. Note that R consists of $L_2 \times f_2^2$ cells, where L_2 denotes the length of p_2^*. Finally, an optimal warping path p_R^* relative to R is computed. We say that this procedure is *successful* if $p^* = p_R^*$. The overall number of cells to be computed in this procedure is $N_2M_2 + L_2f_2^2$, which is generally much smaller than the total number N_1M_1 of cells on Level 1. In an obvious fashion, this procedure can be recursively applied by introducing further levels of decreasing resolution. For a complexity analysis, we refer to [184].

The constrained path p_R^* may not coincide with the optimal path p^*. To alleviate this problem, one can increase the constraint region R – at the expense of efficiency – by adding δ cells to the left, right, top, and bottom of every cell in R for some parameter $\delta \in \mathbb{N}$. The resulting region R^δ will be referred to as δ-neighborhood of R, see Fig. 4.9.

4.4 Subsequence DTW

In many applications, the sequences to be compared exhibit a significant difference in length. Instead of aligning these sequences globally, one often has the objective to find a subsequence within the longer sequence that optimally fits the shorter sequence, see Fig. 4.10. For example, assuming that the longer sequence represents a given database and the shorter sequence a query, a typical goal is to identify the fragment within the database that is most similar to the query. The problem of finding optimal subsequences can be solved by a variant of dynamic time warping, as will be described in this section.

Let $X := (x_1, x_2, \ldots, x_N)$ and $Y := (y_1, y_2, \ldots, y_M)$ be feature sequences, where we assume that the length M is much larger than the length N. In the following, we fix a local cost function c. It is the goal to find a subsequence $Y(a^* : b^*) := (y_{a^*}, y_{a^*+1}, \ldots, y_{b^*})$ with $1 \leq a^* \leq b^* \leq M$ that

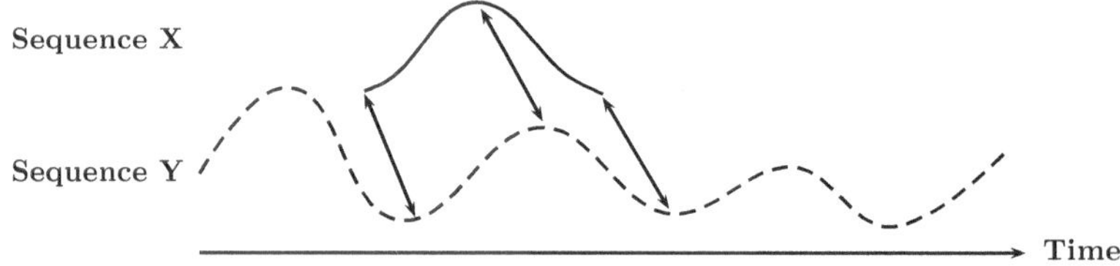

Fig. 4.10. Optimal time alignment of the sequence X with a subsequence of Y. Aligned points are indicated by the *arrows*

minimizes the DTW distance to X over all possible subsequences of Y. In other words,

$$(a^*, b^*) := \underset{(a,b)\,:\,1\leq a\leq b\leq M}{\operatorname{argmin}} \Big(\mathrm{DTW}\big(X\,,\,Y(a\!:\!b)\big)\Big). \tag{4.10}$$

The indices a^* and b^* as well as an optimal alignment between X and the subsequence $Y(a^* : b^*)$ can be computed by a small modification in the initialization of the DTW algorithm as described in Theorem 4.3. The basic idea is not to penalize the omissions in the alignment between X and Y that appear at the beginning and at the end of Y. More precisely, we modify the definition of the accumulated cost matrix D by setting $D(n,1) := \sum_{k=1}^{n} c(x_k, y_1)$ for $n \in [1 : N]$ and $D(1,m) := c(x_1, y_m)$ (opposed to $D(1,m) := \sum_{k=1}^{m} c(x_1, y_k)$) for $m \in [1 : M]$). The remaining values of D are defined recursively as in (4.5) for $n \in [2 : N]$ and $m \in [2 : N]$. Similar to Sect. 4.1, one can also define an extended accumulated cost matrix by setting $D(n,0) := \infty$ for $n \in [0 : N]$ and $D(0,m) := 0$ (opposed to $D(0,m) := \infty$) for $m \in [0 : M]$. The index b^* can be determined from D by

$$b^* := \underset{b\in[1:M]}{\operatorname{argmin}}\, D(N,b). \tag{4.11}$$

To determine a^* as well as the optimal warping path between X and the subsequence $Y(a^* : b^*)$, we apply Algorithm OptimalWarpingPath from Sect. 4.1, this time, however, starting with $p_L = (N, b^*)$. Let $p^* = (p_1, \ldots, p_L)$ denote the resulting path. Then $a^* \in [1 : M]$ is the maximal index such that $p_\ell = (a^*, 1)$ for some $\ell \in [1 : L]$. In other words, all elements of Y to the left of y_{a^*} and to the right of y_{b^*} are left unconsidered in the alignment and do not cause any additional costs. It is left as an exercise to show that the subsequence $Y(a^* : b^*)$ indeed has minimal total cost in the alignment with X among all possible subsequences of Y. The optimal warping path between X and $Y(a^* : b^*)$ is given by $(p_\ell, \ldots, p_L)$. Obviously, the computational complexity of the subsequence DTW algorithm is $O(NM)$. Furthermore, note that the optimal subsequence $Y(a^* : b^*)$ is in general not uniquely defined – there may be several choices for b^* in (4.11), and in the construction of a^* we have used a maximality criterion.

We finally describe how the accumulated cost matrix D can be used to derive an entire list of subsequences of Y that are close to X with respect to the DTW distance. To this end, we define a distance function

$$\Delta : [1:M] \to \mathbb{R}, \qquad \Delta(b) := D(N, b), \tag{4.12}$$

which assigns to each index $b \in [1:M]$ the minimal DTW distance $\Delta(b)$ that can be achieved between X and a subsequence $Y(a\!:\!b)$ of Y ending in y_b. For each $b \in [1:M]$, the DTW-minimizing $a \in [1:M]$ can be computed analogously to a^* as described above using Algorithm OPTIMALWARPINGPATH and starting with $p_L = (N, b)$. Note that if $\Delta(b)$ is small for some $b \in [1:M]$ and if $a \in [1:M]$ denotes the corresponding DTW-minimizing index, then the subsequence $Y(a\!:\!b)$ is close to X. This observation suggests the following algorithm to compute all (up to some specified overlap) subsequences of Y similar to X:

Algorithm: COMPUTESIMILARSUBSEQUENCES

Input:	$X = (x_1, \ldots, x_N)$	query sequence
	$Y = (y_1, \ldots, y_M)$	database sequence
	$\tau \in \mathbb{R}$	cost threshold
Output:	Ranked list of all (essential distinct) subsequences of Y that have a DTW distance to X below the threshold τ.	

(0) Initialize the ranked list to be the empty list.
(1) Compute the accumulated cost matrix D w.r.t. X and Y,
(2) Determine the distance function Δ as in (4.12).
(3) Determine the minimum $b^* \in [1:M]$ of Δ.
(4) If $\Delta(b^*) > \tau$ then terminate the procedure.
(5) Compute the corresponding DTW-minimizing index $a^* \in [1:M]$.
(6) Extend the ranked list by the subsequence $Y(a^*\!:\!b^*)$.
(7) Set $\Delta(b) := \infty$ for all b within a suitable neighborhood of b^*.
(8) Continue with Step (3).

Note that Step (7) is introduced to exclude an entire neighborhood of b^* from further consideration, thus avoiding that the ranked output list contains many subsequences that only differ by a slight shift. (For example, if $Y(a:b)$ is in the list, one can prevent that $Y(a:b+1)$ is in the list as well.) Depending on the application, one may choose a fixed size of the neighborhood around b^* or one may adaptively adjust the size according to the local property of Δ around b^*. For example, one may require that b^* is a local minimum of Δ and then determines the neighborhood confined by the closest local maxima to the left and to the right of b^*.

We illustrate the procedure of Algorithm COMPUTESIMILARSUBSEQUENCES by the example shown in Fig. 4.11a. The input consists of the query sequences X and the database sequence Y as well as the cost threshold $\tau = 5$. We iteratively look for all indices $b \in [1 : M]$ that are local minima of the distance function Δ with $\Delta(b) \leq \tau$, see Fig. 4.11b. In the first iteration, we obtain the local minimum $b^* = 291$ with $\Delta(b^*) = 2.13$, which also constitutes the global minimum of Δ. Based on Algorithm OPTIMALWARPINGPATH, one obtains $a^* = 163$ as well as the optimal warping path that aligns X with the subsequence $Y(a^* : b^*)$, see Figs. 4.11c, d. We next determine the closest local maximum $b^*_\ell = 265$ to the left and closest local maximum $b^*_r = 313$ to the right of b^* and exclude the neighborhood $[b^*_\ell : b^*_r]$ for further consideration by setting $\Delta(b) := \infty$ for $b \in [b^*_\ell : b^*_r]$. We then proceed with the modified Δ in the same fashion, obtaining the local minima $b^*_2 = 454$ with $\Delta(b^*_2) = 2.66$ and $a^*_2 = 367$. In a third iteration, one obtains $b^*_3 = 137$ with $\Delta(b^*_3) = 3.50$ and $a^*_2 = 44$. The three resulting "matches," i. e., subsequences of Y close to X, are shown in Fig. 4.11d.

4.5 Further Notes

Originating from speech processing, dynamic time warping (DTW) has become a well-established method to account for temporal variations in the comparison of related time series. A classical and comprehensive account on DTW and related pattern recognition techniques is given by Rabiner and Juang [170] in the context of speech recognition. In their book, one also finds a detailed introduction to Hidden Markov Models (HMMs) – a statistical method that extends the concept of DTW-based pattern recognition.

In addition to speech recognition, dynamic time warping has found numerous applications in a wide range of fields including data mining, information retrieval, bioinformatics, chemical engineering, signal processing, robotics, or computer graphics, see, e. g., [100] and the references therein. Basically any data that can be transformed into a (linear) sequence of features can be analyzed with DTW, which includes data types such as text, video, audio, or general time series. In the field of music information retrieval, DTW plays in important role for synchronizing music data streams [57,94,141,142,196,202]. DTW has also been used in the field of computer animation to analyze and align motion data [22, 75, 93, 106, 143, 217, 220]. In Chaps. 5 and 10, some of these reference will be discussed in more detail.

Extensive research has been performed on how to accelerate DTW computations, in particular for one-dimensional (or low-dimensional) real-valued sequences, often referred to as time series, see [102, 184] and the references therein. The problem of indexing large time series databases has also attracted great interest in the database community, see Last et al. [117] for an overview. Keogh [100] describes an indexing method based on lower bounding techniques that makes retrieval of time-warped time series feasible even for large

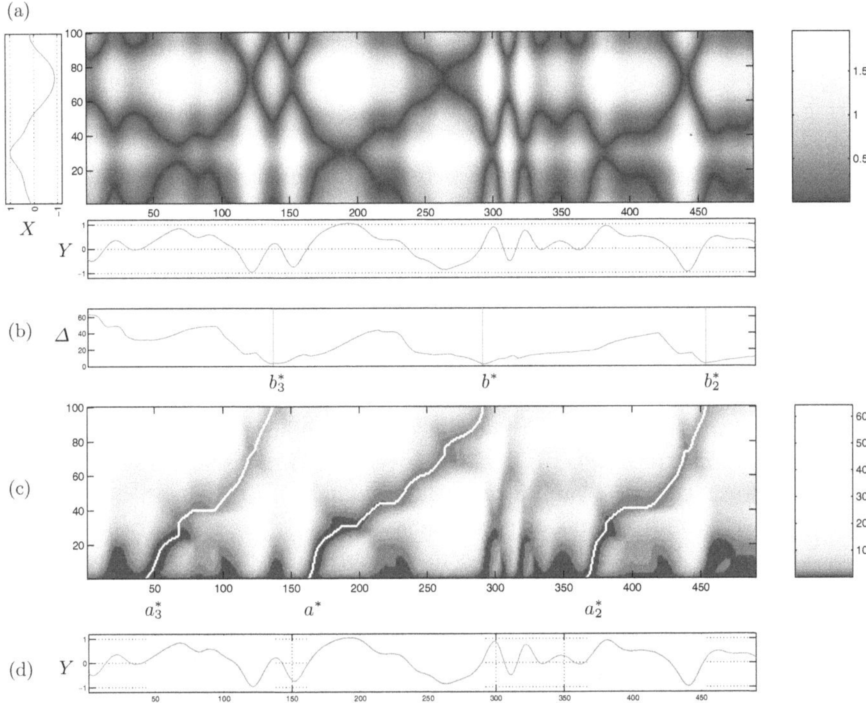

Fig. 4.11. (**a**) Cost matrix between the query sequences X (vertical axis) and the database sequence Y (horizontal axis) using the absolute value as local cost measure c. (**b**) Distance function Δ corresponding to the top row of the matrix D. The *red vertical lines* indicate the three local minima b^*, b_2^*, and b_3^* having a Δ-value below the cost threshold $\tau = 5$. (**c**) Accumulated cost matrix D for subsequence DTW and the three optimal warping paths (*white lines*) corresponding to the minima of (b). (**d**) Resulting subsequences $Y(a_3^* : b_3^*)$, $Y(a^* : b^*)$, and $Y(a_2^* : b_2^*)$ of Y (indicated by *grey regions*)

datasets. An early account on efficient indexing for subsequence matching of one-dimensional time series is given by Faloutsos et al. [62]. Most of these approaches follow the general strategy to first extract a coarse approximation from each time series of the database. Such approximations may be based on taking the first few coefficients of a Fourier [2] or wavelet transform [39], or the average values of adjacent analysis windows [101]. Next, based on a suitable distance measure on the approximations, one computes lower bounds for the distances between the corresponding time series. To view of efficient retrieval, the approximations can then be stored by means of multidimensional indexing methods such as an R-tree [90].

Closely related to DTW is the *edit distance*, which is sometimes also referred to as *Levenshtein distance* [120]. The edit distance is used to compute a distance between strings, i. e., one-dimensional sequences consisting of discrete symbols (rather than sequences consisting of continuous features). It is defined as the minimum number of operations needed to transform one string into the other, where an operation is an insertion, a deletion, or a substitution of a single symbol. The edit distance is used in fields such as text retrieval (spell checkers, plagiarism detection) or molecular biology (to compute a distance between DNA sequences, i. e., strings over $\{A,C,G,T\}$). For a detailed account on the edit distance with its applications to bioinformatics, we refer to [18].

Vlachos et al. [209] introduced similarity measures for multidimensional time series (having values in $\mathbb{R}^d$) based on the concept of longest common subsequences (LCS) – a variant of the edit distance that is more robust to noise and outliers. For low dimensions ($d \leq 3$), they also describe some efficient approximation algorithms that compute these similarity measures.

The computation of DTW, edit, as well as LCS distances can be done efficiently by means of *dynamic programming*, which is a general method for reducing the running time of algorithms exhibiting the properties of overlapping subproblems and optimal substructure. For a general introduction to dynamic programming, we refer the reader to the standard text book by Cormen et al. [47].

5

Music Synchronization

Modern digital music libraries contain textual, visual, and audio data. Recall that musical information is represented in diverse data formats which, depending upon the respective application, differ fundamentally in their structure and content. In this chapter, we introduce various synchronization tasks to automatically link different data streams given various formats (score, MIDI, audio) that represent the same piece of music (Sect. 5.1). Particularly, two different synchronization procedures are described in detail. First, we present an efficient and robust multiscale DTW approach for time-aligning two different CD recordings of the same piece (Sect. 5.2). Using chroma-based audio features, our algorithm yields good synchronization results for harmony-based music at a reasonable resolution level that is sufficient in view of music retrieval and navigation applications. Second, we discuss an algorithm for score–audio synchronization, which aligns the musical onset times given by a score with their physical occurrences a CD recording of the same piece (Sect. 5.3). Using semantically meaningful onset features, this algorithm works particularly well for piano music and yields alignments at a high temporal resolution. In Sect. 5.4, we describe possible research directions, give further references to the literature, and discuss some problems related to music synchronization. The first three sections of this chapter closely follow [5], [142], and [141], respectively.

5.1 Synchronization Tasks

Before we specify various synchronization tasks, we summarize some characteristics of music data as discussed in Sect. 2.1. Recall that the musical score gives an explicit specification of notes in terms of pitch and duration. Additionally, textual notations such as *allegro* or *ritardando* are used to indicate the global tempo or the local tempo variations. Similarly, loudness and dynamics are described by terms such as *piano*, *forte*, *crescendo*, or *diminuendo*. Hence, the score is a description of the piece of music that leaves a lot of room for various interpretations in some actual performance. These may concern not only the tempo or dynamics of individual passages but also the multinote execution implied by a single signifier such as a trill, arpeggio, or grace-note.

From a physical point of view, the musician, by means of his voice or instrument, generates an audio signal (Sect. 2.1.2). This audio signal has the form of a sound-wave emerging at its source and spreading through the air. Graphically, such an audio signal may be represented by its waveform, which depicts the amplitude of the air pressure over time. The PCM (pulse code modulation) format [164], as used for CD-recordings, is a discretized and encoded version of such a waveform. In the following, we do not distinguish between the waveform and its PCM representation both being referred to as *audio format.*

As was discussed in Sect. 2.1.3, the MIDI format may be thought of as a hybrid of the score and waveform-based audio format: it can encode all relevant content-based information in the notes of the score as well as agogic and dynamic niceties of a specific interpretation. However, MIDI is quite limited especially in modeling the timbre of a sound. MIDI data streams are frequently visualized by the piano-roll representation, see Figs. 2.8 and 5.1.

A musical work is, in the digital context, far from simple or singular, since it can exist both in multiple formats and in multiple realizations in each of them. This heterogeneity makes content-based browsing and retrieval in

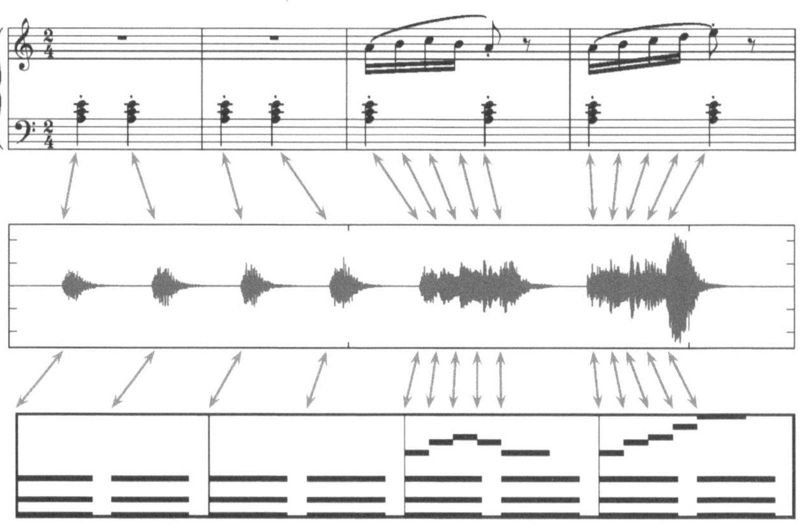

Fig. 5.1. First four measures of Op. 100, No. 2 by Friedrich Burgmüller in a score, audio, and MIDI (piano roll) representation. The red arrows, which constitute a typical synchronization result, indicate the aligned time positions of corresponding note events in the different representations

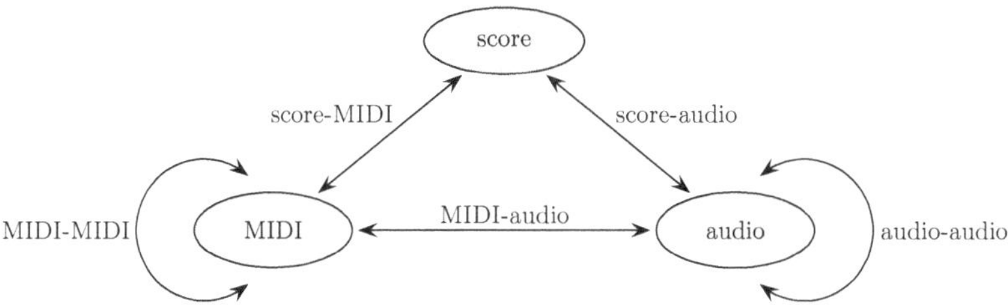

Fig. 5.2. Various synchronization tasks

digital music libraries a challenging task. Thus the value of synchronization algorithms automatically linking the various data streams lies in their promise of interrelating the multiple information sets related to a single piece of music. In the music framework, *synchronization* is taken to mean a procedure which, for a given position in one representation of a piece of music, determines the corresponding position within another representation, see Fig. 5.1. As a possible application, the synchronization results can be used to facilitate convenient access to audio recordings. Automatic annotation in different data formats can be exploited in the context of content-based retrieval. Moreover, alignments can ease the investigation of tempo studies. Finally, temporal links between score and audio data are useful for automatic tracking of the score position in a performance. Considering the three representative data formats – symbolic score data, audio data, and semi-symbolic MIDI data – one obtains the following five meaningful types of synchronization tasks:

(1) Score–audio synchronization
(2) Score–MIDI synchronization
(3) MIDI–audio synchronization
(4) MIDI–MIDI synchronization (simply referred to as MIDI synchronization)
(5) Audio–audio synchronization (simply referred to as audio synchronization)

Figure 5.2 presents a schematic view of all the possible relations between score, MIDI, and audio data. In Sect. 5.4 we discuss further synchronization tasks involving, e.g., scanned images of sheet music.

5.2 A Multiscale Approach to Audio Synchronization

For one and the same piece of music, there often exists a large number of CD recordings representing different interpretations by various musicians. In particular for Western classical music, these interpretations may exhibit considerable deviations in tempo, note realizations, dynamics, and instrumentation. For example, in case of Beethoven's Fifth Symphony, a music database may contain interpretations by Karajan and Bernstein, historical recordings by Furtwängler and Toscanini, Liszt's piano transcription of Beethoven's Fifth played by Sherbakov and Glenn Gould, and some synthesized version of corresponding MIDI files. The goal of *audio synchronization* is to time-align corresponding note events in two different interpretations. These alignments can be used to jump freely between different audio recordings, thus affording efficient and convenient audio browsing.

In the last few years, several music synchronization strategies have been proposed, see, e.g., [4, 5, 50, 57, 94, 141, 142, 173, 196, 202] and the references therein. Most of these approaches rely on some variant of dynamic time warping (DTW): first, a cost matrix is computed based on suitable features sequences extracted from the two audio recordings to be synchronized. Then, the cost-minimizing alignment path is determined from this matrix via dynamic programming. However, because of the quadratic time and space complexity, DTW-based strategies become infeasible for long pieces.

To obtain an efficient audio synchronization algorithm, we make use of the multiscale DTW (MsDTW) approach as described in Sect. 4.3. Recall that the general idea was to recursively project an alignment path computed at a coarse feature resolution level to the next higher level and then to refine the projected path. One hazard with this approach is that an incorrect alignment on a low resolution level propagates to higher levels resulting in erroneous alignment results. This hazard is fostered by the fact that coarsening the features can lead to heavily deteriorated cost matrices as illustrated by Fig. 5.4a–c. In view of finding a good tradeoff between robustness, efficiency, and practicability of the resulting synchronization procedure, we discuss several crucial issues, including the usage of scalable audio features and adaptation strategies for the feature sampling rate (Sect. 5.2.1), the specification of suitable local cost measures (Sect. 5.2.2), the determination of MsDTW resolution levels and constraint regions (Sect. 5.2.3), as well as enhancement strategies of DTW cost matrices (Sect. 5.2.5). We report on our extensive experiments based on a wide range of classical music demonstrating the practicability of our algorithm (Sect. 5.2.4). The synchronization results have been integrated in an advanced audio player (as described in Chap. 8) and sonified for a qualitative evaluation.

5.2.1 Audio Features

The first step in audio synchronization consists of transforming the audio recordings into sequences of suitable acoustic features. Since audio recordings of different interpretations may exhibit significant variations with respect to dynamics, timbre, or articulation, normalized chroma-based audio features, as was discussed in Sect. 3.3, have proven to be a valuable tool for audio synchronization, see Hu et al. [94]. These features show a large degree of robustness to variations in the above parameters, while keeping sufficient information to characterize harmony-based music. We therefore convert an audio signal into a sequence $X = (x_1, x_2, \ldots, x_N)$ of normalized (this time with respect to the Euclidean norm) 12-dimensional feature vectors $x_n \in [0,1]^{12}$, $1 \leq n \leq N$, which express the local energy distribution in the 12 chroma classes. Here, we use a feature sampling rate of 10 Hz, where each feature vector x_n covers 200 ms of audio with an overlap of 100 ms. This rate, which will constitute the finest resolution level, turns out to be sufficient in view of our intended applications. For details on the computation of chroma features we refer to Sect. 3.3.

For our multiscale approach, we need a flexible and computationally inexpensive way to adjust the feature resolution. Instead of simply modifying the analysis window in the chroma computation, we use the CENS features as described in Sect. 3.3. Recall that we introduced a second, much larger statistics window (covering $\mathbf{w}$ consecutive chroma vectors) and considered short-time statistics of the chroma energy distribution over this window. This again resulted in a sequence of 12-dimensional vectors, which was then downsampled by a factor of $\mathbf{d}$ and renormalized with respect to the Euclidean norm. For example, $w = 41$ and $q = 10$ yields a feature sampling rate of 1 Hz, where each feature vector represents information of the audio signal within a window of roughly 4 100 ms. The resulting feature sequence was referred to as $\mathrm{CENS}^{\mathbf{w}}_{\mathbf{d}}$-sequence. Note that by modifying the parameters $\mathbf{w}$ and $\mathbf{d}$, one can adjust the feature granularity and sampling rate without repeating the cost-intensive chroma computations. For examples, we refer to Figs. 3.9 and 3.10.

5.2.2 Local Cost Measure

The normalized chroma and CENS features are elements in the set $\mathcal{F} = \{v \in [0,1]^{12} \mid \|v\|_2 = 1\}$ as defined in (3.4). To compare two features $x, y \in \mathcal{F}$, we use the cost measure $c_\alpha : \mathcal{F} \times \mathcal{F} \to [0,1] + \alpha$ defined by $c_\alpha(x,y) := 1 - \langle x, y \rangle + \alpha$ for some offset $\alpha \in \mathbb{R}_{\geq 0}$. Note that $\langle x, y \rangle$ is the cosine of the angle between x and y since the features are normalized. The offset α is introduced for the following reason. Audio recordings often contain long segments of little variance such as pauses or sustained chords. This leads to rectangular regions within the cost matrix, where all cells are of low cost. Such a region will be referred to as *low-cost plain*. If $\alpha = 0$, all cells within a low-cost plain reveal some cost close to zero. Being close to zero, there may be large, more or less random *relative* differences among the costs of these cells (e.g., one cell has cost 0.01 and another one 0.001). As a result, one obtains an uncontrollable run of the optimal warping path within a low-cost plain. By increasing α, the relative differences decrease, whereas the absolute differences are retained unchanged (e.g., for $\alpha = 1$, one cell has cost 1.01 and another one 1.001). As a consequence, the effect of the weight vector (w_d, w_h, w_v) introduced in Sect. 4.2.2 becomes more dominant. For the parameters $(w_d, w_h, w_v) = (2, 1.5, 1.5)$ and $\alpha = 1$, which have turned out to be suitable in our experiments, the diagonal direction receives a slight but stable preference in low-cost plains, see Fig. 5.3. In the following, we set $c := c_1$.

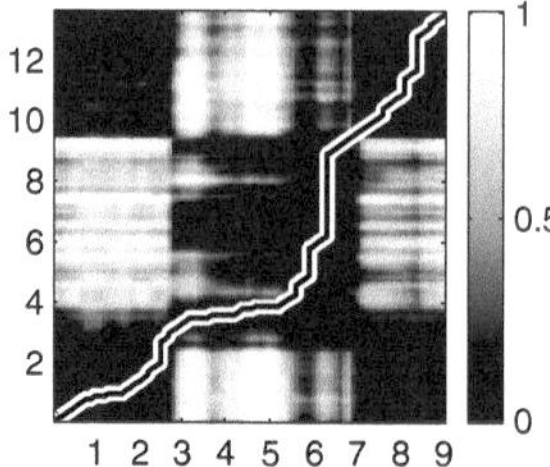

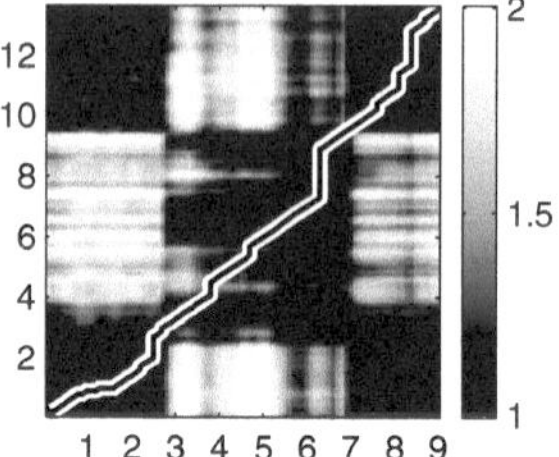

Fig. 5.3. Optimal warping path based on the cost measure c_α with $\alpha = 0$ (*left*) and $\alpha = 1$ (*right*)

Table 5.1. Specification of the features used in the MsDTW-based audio synchronization algorithm

Level	Feature	Resolution (Hz)	Factor
1	Chroma	10	–
2	CENS^{41}_{10}	1	10
3	CENS^{121}_{30}	1/3	3
4	CENS^{271}_{90}	1/9	3

5.2.3 Resolution Levels and δ-Neighborhood

As described in Sect. 5.2.1, the chroma features at a time resolution of 10 Hz constitute the basic resolution (Level 1) of our audio synchronization. The CENS features, which can be efficiently derived from the chroma features, are used at the lower resolution levels. To determine a suitable number of levels as well as the resolutions at each level used in the MsDTW, we conducted comprehensive experiments. We report on some experiments that are based on the features indicated by Table 5.1. For our tests, we used 363 pairs of CD recordings, where each pair corresponds to two different interpretations of the same piece. The recordings have a duration of 3–20 min and cover a wide range of classical music, see Table 5.4 for examples. In a preprocessing step, we computed and stored the chroma features of all recordings. For each test series, we performed an audio synchronization for all 363 pairs. The algorithms have been implemented in C/C++ and tests were run on an Intel Pentium M, 1.7 GHz, 1 GByte RAM, under Windows XP.

In one test series, we used classical DTW on Level 1 and MsDTW based on the first two, three, and four levels. In all these cases we used $\delta = 30$, where δ denotes the neighborhood parameter used for the constraint region (Sect. 4.3). This choice of δ is discussed later. The resulting running times are shown in Table 5.2. The DTW-based strategy required 1 434.0 s to synchronize all of the 363 pairs. In contrast, the two-level MsDTW-based strategy required $t_{\mathrm{CENS}} = 16.5$ s to compute the CENS^{41}_{10}-features used for Level 2,

Table 5.2. Total running time (for 363 synchronization pairs) against the number of levels used in the MsDTW algorithm based on the features of Table 5.1

Levels	t_{Cells} (s)	t_{CENS} (s)	t_{MsDTW} (absolute) (s)	t_{MsDTW} (relative) (%)
1	1434.0	0.0	1434.0	100
1–2	78.5	16.5	95.0	6.62
1–3	67.6	24.7	92.3	6.44
1–4	67.2	29.8	97.0	6.76

and $t_{\text{Cells}} = 78.5$ s to compute all required cells on the two levels, amounting to a total running time of $t_{\text{MsDTW}} = 95.0$ s – only 6.62% of the running time of classical DTW. Similarly, it took 92.3 and 97.0 s when using the three-level and four-level MsDTW-based strategy, respectively. In particular, we obtained the lowest total running time for three levels. Using a fourth level indeed further decreased the total number of cells to be evaluated, but the computational overhead due to the additional CENS features needed at Level 4 deteriorated the overall result. Also introducing additional intermediate levels had only a marginal effect on the overall running time. We therefore use three multiscale levels as default setting in our audio synchronization system. Note that the sampling rate of 1/3 Hz of Level 3 affords an efficient computation for basically every relevant music recording. For example, recordings of a duration of 30 min (pieces of music such as a movement rarely exceed this duration) lead to a feature sequence of length 600. The resulting DTW would then require the evaluation of $360,000$ cells, which can be efficiently handled by a computer.

As was mentioned in Sect. 4.3, the MsDTW-based strategy may not be successful in the sense that the resulting (constrained) optimal warping path may differ from the (unconstrained) optimal path obtained from classical DTW. For our evaluation, we use the DTW-alignment as ground truth and check whether the MsDTW-alignment entirely coincides with the DTW-alignment (then MsDTW is called successful) or not (then MsDTW produces an error). In another test series, we evaluated different δ-neighborhoods for a three-level MsDTW. Table 5.3 shows the performance for $\delta = 0, 10, 20, 30$. For $\delta = 0$, in 92 of the 363 cases the MsDTW-based strategy was not successful (corresponding to an error rate of 25.34%). Increasing δ leads to an increase of the running time and a decrease of the error rate. For $\delta = 30$, all 363 pairs have been successfully aligned by the MsDTW strategy. The running time to evaluate the cells has increased from $t_{\text{Cells}} = 29.3$ s ($\delta = 0$) to $t_{\text{Cells}} = 67.6$ s ($\delta = 30$), which is a moderate increase with regard to the total running time. For our audio synchronization system, we therefore use $\delta = 30$ as default.

Finally, we also employed an adaptive strategy to locally adjust the size of the neighborhood. The basic ideas is as follows: starting from R^{20} (the δ-neighborhood of R with $\delta = 20$, see Sect. 4.3), we added up to 20 additional cells in lower and upper direction at certain critical cells of R^{20}. Intuitively, a cells of R^{20} is considered as critical if it belongs to a low-cost plain or if it

Table 5.3. Running time, absolute and relative error against the size of the δ-neighborhood and adaptive neighborhood based on a three-level MsDTW

δ	t(cells) (s)	t(CENS) (s)	t(MsDTW) (abs.) (s)	t(MsDTW) (rel.) (s)	Error (absolute)(%)	Error (relative)(%)
0	29.3	25.0	54.3	3.79	92	25.34
10	41.6	24.7	66.3	4.62	27	7.44
20	54.6	24.6	79.2	5.52	6	1.65
30	67.6	24.7	92.3	6.44	0	0
Adap.	60.5	24.8	85.3	5.95	3	0.83

belongs to some "edge" of such a plain. Mathematically, critical points can be characterized by expectation and variance values along verticals of R^{20}. Since only marginal improvements could be achieved, see Table 5.3, this strategy was not used in the following experiments.

5.2.4 Experimental Results

In this section, we discuss some representative experimental results based on the three-level MsDTW with $\delta = 30$. For these parameters, as was reported above, the audio synchronization was successful for all of the 363 pairs of audio recordings. Table 5.4 shows a selection of these recordings including complex orchestral pieces having a duration of 3–20 min. Some of the interpretations significantly differ in tempo, instrumentation, and articulation. For example, Sacchi's interpretation of Schubert's Unfinished is much faster (817 s) than Solti's interpretation (951 s). Or, there is an orchestral as well as a piano version (piano transcription) of Beethoven's Fifth and Wagner's Prelude, respectively. Furthermore, the Mae interpretation of Vivaldi's "Spring" includes many additional ornamentations, which cannot be found in the Zukerman interpretation. In view of such significant variations, the CENS^{41}_{10}-features as used on Level 2 constitute a good compromise between reasonable feature resolution, computational efficiency, and robustness of the alignment result. (Actually, in situations where one has significant differences in note realizations – e.g., one interpreter plays a sustained chord whereas another interpreter plays an ornamentation – the alignment on a finer resolution level may not even be semantically meaningful.)

The increase in performance of our MsDTW-based audio synchronization in comparison to a classical DTW-based approach is illustrated by Table 5.5. For example, to synchronize "Beet9Bern" and "Beet9Kar" only 1.75% of the cells on Level 1 have to be evaluated by MsDTW, decreasing the memory requirements roughly by a factor of 57. Here, note that the overall memory requirement is proportional to the maximal number of cells needed to be evaluated at some level. The number of additional cells needed to be computed at Levels 2 and 3 is comparatively small. The overall running time (including the

Table 5.4. Some audio recordings (with identifier) contained in our test database comprising 33 h of audio and 363 synchronization pairs

Identifier	Composer	Piece	Interpreter	L. (s)
Beet5Bern	Beethoven	Symphony No. 5, Op. 67, 1st mov.	Bernstein	519.0
Beet5Kar	Beethoven	Symphony No. 5, Op. 67, 1st mov.	Karajan	443.9
BeLi5Sher	Beeth. (Liszt)	Symphony 5, Op. 67, 1st mov. (piano)	Sherbakov	444.1
Beet9Bern	Beethoven	Symphony No. 9, Op. 125, 4th mov.	Bernstein	1144.9
Beet9Kar	Beethoven	Symphony No. 9, Op. 125, 4th mov.	Karajan	1054.8
Dvo9Franc	Dvorak	Symphony No. 9, Op. 95, 1st mov.	Francis	710.7
Dvo9Maaz	Dvorak	Symphony No. 9, Op. 95, 1st mov.	Maazel	704.2
GriegMorBee	Grieg	Op. 46, No. 1, 4th mov. (morning)	Beecham	190.1
GriegMorGun	Grieg	Op. 46, No. 1, 4th mov. (morning)	Gunzenhauser	222.1
ElgEnigDel	Elgar	Op. 36, Andante (Enigma)	Del Mar	91.8
ElgEnigSino	Elgar	Op. 36, Andante (Enigma)	Sinopoli	93.1
RavBolAbb	Ravel	Bolero	Abbado	862.5
RavBolOza	Ravel	Bolero	Ozawa	901.0
Schub8Sac	Schubert	Symphony No. 8, D759, 1st mov.	Sacchi	817.3
Schub8Sol	Schubert	Symphony No. 8, D759, 1st mov.	Solti	950.8
SchosJazzCha	Shostakovich	Jazz Suite No. 2, 6th mov. (Waltz)	Chailly	223.6
SchosJazzYab	Shostakovich	Jazz Suite No. 2, 6th mov. (Waltz)	Yablonsky	193.6
VivSpringMae	Vivaldi	RV 269 (Spring), 1st mov.	Mae	192.5
VivSpringZuk	Vivaldi	RV 269 (Spring), 1st mov.	Zukerman	218.9
WagPreArm	Wagner	Meistersinger Prelude	Armstrong	595.0
WagPreGould	Wagner	Meistersinger Prelude (piano)	Gould	576.9

The last column indicates the length of the respective recording (in seconds)

Table 5.5. Performance of the implementation of our MsDTW-based audio synchronization algorithm for a representative selection of recordings using three-levels and $\delta = 30$

Synchronization		Number of cells evaluated by DTW and MsDTW							Total run time (s)		
Recordings		Level 1			Level 2			Level 3	Levels 1–3		
Id 1 Id 2	L. (s)	DTW	Ms DTW	(%)	DTW	Ms DTW	(%)	DTW	DTW	Ms DTW	(%)
Beet9Bern Beet9Kar	1144.9 1054.8	120808050	2117929	1.75	1209030	17657	1.46	134464	31.18	1.08	3.46
RavBolAbb RavBolOza	862.5 901.0	77737897	1694610	2.18	778426	14121	1.81	86688	20.04	0.80	3.99
Schub8Sac Schub8Sol	817.3 950.8	77736075	1704150	2.19	777918	14322	1.84	86541	20.01	0.80	3.99
Dvo9Maaz Dvo9Franc	704.2 710.7	50068752	1356308	2.71	501255	11295	2.25	55695	12.85	0.60	4.67
WagPreArm WagPreGould	595.0 576.9	34337270	1124029	3.27	343892	9387	2.73	38407	8.85	0.48	5.42
Beet5Bern BeLi5Sher	519.0 444.1	23068056	923444	4.00	231400	7721	3.34	25926	5.84	0.37	6.34
Beet5Bern Beet5Kar	519.0 443.9	23052480	923028	4.00	230880	7716	3.34	25752	5.87	0.38	6.47
Beet5Kar BeLi5Sher	443.9 444.1	19726920	848640	4.3	197580	7042	3.56	22052	5.02	0.32	6.37
GriegMorGun GriegMorBee	222.1 190.1	5561666	448024	8.06	55973	3772	6.74	6300	1.44	0.17	11.81
SchosJazzCha SchosJazzYab	223.6 193.6	4335306	394259	9.09	43456	3291	7.57	4875	1.11	0.15	13.51
VivSpringMae VivSpringZuk	192.5 218.9	4217940	389034	9.22	42267	3261	7.72	4745	1.07	0.13	12.15
ElgEnigDel ElgEnigSino	91.8 93.1	856508	167699	19.58	8648	1404	16.23	992	0.22	0.06	27.27

For each level, the total number of cells (DTW) as well as the number of cells to be evaluated by MsDTW are indicated. The last three columns show a comparison of the DTW and MsDTW running times. In all examples, the MsDTW-based approach was successful.

CENS feature computation) to synchronize the two recordings (each version having a duration of almost 20 min) was 1.08 s – nearly thirty times faster than classical DTW. Obviously, the relative savings increase with the lengths of the pieces.

To evaluate the absolute alignment quality achieved by our system, we conducted the following experiment. Based on the alignment result of two recordings, we time-warped the second recording to run synchronously to the first recording. For the warping, we used an overlap-add technique based on waveform similarity (WSOLA) as described in Verhelst and Roelands [207]. We then produced a stereo audio file containing the mono version of the original first recordings in one channel and a mono version of the time-warped second recording in the other channel. Listening to this stereo audio file exhibits even small temporal deviations of less than 100 ms between corresponding note events.

5.2.5 Enhancing Cost Matrices

We have also conducted experiments with manually distorted and highly repetitive audio material, which constitutes an extreme scenario for MsDTW.

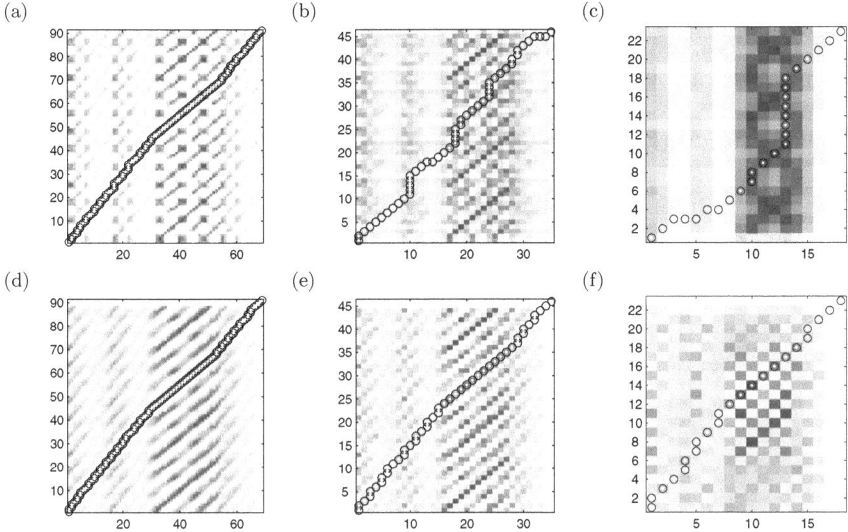

Fig. 5.4. Optimal warping path between "Bach12" and "Bach12Warp" using **(a)** CENS^{41}_{10}, **(b)** CENS^{81}_{20}, and **(c)** CENS^{161}_{40}. The paths are semantically incorrect for the two lower resolution levels. Locally filtering the cost matrix based on CENS^{41}_{10} (using eight different gradients) and downsampling by **(d)** a factor 1, **(e)** a factor 2, and **(f)** a factor 4 leads to correct optimal warping paths even on the lower resolution levels

As one example, we used the first 22 s of an Cabrera interpretation of Bach's Toccata BWV 565, where the theme is repeated three times at three different octaves. We generated an audio file, referred to as "Bach12," by concatenating four copies of this segment, resulting in 12 repetitions of the theme. We then generated a time-warped version of "Bach12," referred to as "Bach12Warp," by locally increasing and decreasing the tempo up to 50%. The cost matrix using CENS^{41}_{10} and the resulting optimal warping between "Bach12" and "Bach12Warp" is shown in Fig. 5.4a. Now, reducing the feature sampling rate leads to a heavily deteriorated cost matrix, where the repetitions cannot be resolved any longer. This, in turn, results in absurd optimal warping paths on the lower resolution levels, see Fig. 5.4b, c.

To alleviate this problem, one can improve the structural properties of the cost matrix by incorporating contextual information into the local cost measure. For a detailed description of such a strategy, we refer to Sect. 7.2. Intuitively, the idea is to enhance the diagonal path structure of the 2D cost matrix by applying a local 1D low-pass filter along the diagonals (having gradient $(1, 1)$). Note that this process only works if corresponding segments of the two audio recordings reveal the same tempo progression. To account for tempo differences in the two interpretations, the idea is to simultaneously filter along different directions (in our implementation we used eight different gradients in a neighborhood of $(1,1)$, which cover tempo variations of roughly -30% to $+40\%$) and then to take the minimum over the filter outputs. The technical details of this approach are described in Müller and Kurth [137]. The effect of this enhancement strategy, which allows to reduce the feature sampling rate without completely destroying the structural properties of the cost matrix, is illustrated by Fig. 5.4d–f. Even at the lowest resolution level (obtained by filtering with an averaging filter of length 8 and downsampling by a factor of 4), the optimal warping path leads to the "correct" alignment. In practice, one has to assess the tradeoff between increased computational complexity caused by the additional filtering step and the boost of robustness and confidence due to the structural enhancement.

5.3 Onset-Based Score–Audio Synchronization

In this section, we describe an algorithm for score–audio synchronization, where one data stream represents the score of a piece of music and the other data stream represents a recorded performance of the same piece. The *score–audio synchronization* then amounts to associating the note events given by the score data stream with their physical occurrences in the audio file. As was discussed in Sect. 2.1, score and audio data fundamentally differ in their respective structure and content. Therefore, the first step in any score–audio synchronization procedure consists in extracting suitable parameters from the score and audio data streams to make them comparable. Here, different strategies are possible. First, one can convert the score data into an audio data stream using a synthesizer, thus translating the score–audio into some audio–audio synchronization task, which can then be solved as described in Sect. 5.2. Such a strategy has been used, e.g., in Turetsky and Ellis [202]. As a second strategy, as suggested in Hu et al. [94], one can directly convert the score data into some chroma-like representation, which can then be compared with a chroma representation of the audio data stream. A third strategy is to extract event-related data such as note onsets and pitches from the audio data, which can then be compared with the symbolic note parameters given by the score. Each of these strategies has its assets and drawbacks and the performance of the respective strategy very much depends on the type of music, the instrumentation used in the audio recording, as well as musical deviations in the two data streams to be aligned. In particular, the combination of various strategies, as discussed in Sect. 5.4, may yield efficient, robust as well as accurate synchronization procedures for large classes of music.

Following the third strategy, we now discuss a score–audio synchronization procedure based on a sparse set of highly expressive audio features, which encode note onset candidates within different pitch subbands (Sect. 5.3.1). Such onset features work particularly well for instruments such as a piano or a guitar, which exhibit a characteristic energy increase (attack phase) at the beginning of a note realization. Because of its expressiveness, this feature set allows for an accurate synchronization at a high temporal resolution (around 20 ms). Because of its sparseness, it facilitates a time and memory efficient alignment procedure (based on 0–20 features per second depending on the respective segment of the piece of music). In Sect. 5.3.2, we present a local similarity function that directly relates the audio features to the note parameters of the score data. Similar to DTW, we then use dynamic programming (DP) to compute the actual score–audio alignment (Sect. 5.3.3). Here, we are led by the following simple but important principles: first, because of possible deviations in the data stream to be aligned, we prefer to have missing matches over having bad or wrong matches. Hence, we do not force the alignment of all score notes but rather allow note objects to remain unmatched. Second, the score data will guide us in what to look for in the audio data stream. In other words, all information contained in the extracted audio features, which is not reflected by the score data, remains unconsidered. As for classical DTW, the running time as well as the memory requirements in the DP matching procedure are proportional to the product of the lengths of the two sequences to be aligned. The synchronization algorithm can be accelerated considerably if one knows matches prior to the actual DP computation. To account for such kind of prior knowledge, we introduce the notion of *anchor configurations*, which may be thought of as note objects having some salient dynamic or spectral properties, e.g., some isolated fortissimo chord with some salient harmonic structure or some long pause. The counterparts of such note objects in the audio data streams can be determined by a linear-time linear-space algorithm, which efficiently provides us with so-called *anchor matches*. The remaining matches can then be computed by much shorter, local DP computations between these anchor matches (Sect. 5.3.4). The synchronization results as well as the running time behavior of our algorithm tested on complex polyphonic piano pieces including Chopin's Etudes Op. 10 are presented in Sect. 5.3.5.

5.3.1 Audio Features

In the following, we consider the class of polyphonic piano music of any genre and complexity. This allows us to exploit certain characteristics of the piano sound in the extraction of audio features. However, dealing with piano music is still a difficult task due to the following facts (see, e.g., [16, 66] for more details):

- Striking a single piano key already generates a complex sound consisting not only of the fundamental pitch and several harmonics but also comprising inharmonicities caused by the keystroke (mechanical noise) as well as transient and resonance effects.
- Especially due to the usage of the right (sustaining) pedal, the note lengths in piano performances may differ considerably from the note lengths specified by the score. This results in complex sounds in polyphonic music which are not reflected by the score. Furthermore, pedaling also has a great effect on the timbre (sound spectrum) of a piano sound.

- The piano has a large pitch range as well as dynamic range. The respective sound spectra are not just translated, scaled, or amplified versions of each other but differ fundamentally in their respective structure depending on pitch and velocity.

Because of the complexity of the data, the extraction of exact note parameters from an audio recording is difficult and often infeasible for polyphonic music. However, in view of our score–audio synchronization, exact information may not be required as long as the extracted audio feature contain sufficient information on the physical onset times of the note realizations. As basis for our score–audio synchronization, we transform the audio signal into the onset representation as described in Sect. 3.2. Recall that an onset feature reflects a locally maximal energy increase in a certain pitch subband at a certain physical time. Each onset feature is encoded by a triple (p, t, s), where p denotes the pitch, t the time position in seconds within the audio file, and s the size of the peak indicating the note velocity. Typically, an onset feature (p, t, s) indicates a physical onset of a piano note having some harmonics corresponding to pitch p, see also Fig. 3.6 for an example.

In deriving our onset representation, we use a feature sampling rate of 88.2 Hz to compute the necessary STMSPs for the 88 pitch subbands, see Sect. 3.2. Hence, the resulting onset features possess a time resolution of 11.3 ms.

5.3.2 Local Similarity Measure

To align the score data with the audio data, we introduce a local similarity measure[1] that allows us to assign a *similarity value* to a pair of sets, one consisting of suitable score parameters and the other consisting of onset features. We divide the notes of the score into *score bins*, where each score bin consists of a set of notes that have the same musical onset time. Each note of a score bin is specified by its corresponding MIDI pitch p. For example, the first score bin of the score shown in Fig. 3.3 is given by $S_1 := \{48, 52, 55\}$ corresponding to the first three notes, and so on. Similarly, we divide up the onset representation into *peak bins*. To this end, we evenly split up the time axis into segments of length 50 ms. Then, we define peak bins by assigning each peak to the segment corresponding to its time position. Finally, we discard all empty peak bins. Altogether, we obtain a list $S = (S_1, S_2, \ldots, S_N)$ of score bins and a list $P = (P_1, P_2, \ldots, P_M)$ of peak bins, where N and M denote the respective number of bins. The division into peak bins seems to introduce a time resolution of 50 ms. As we see in Sect. 5.3.3, this is not the case since we further process the individual notes after the bin matching procedure.

[1] Similar to a local cost or distance measure as used in DTW, one often uses a similarity measure to compare the parameters in question. Contrary to a cost measure, a similarity measure assumes large values for similar and small values for dissimilar parameters.

We now introduce a local similarity measure d that measures the similarity between a note bin S_n and a peak bin P_m. Recall that a note bin S_n consists of a set of MIDI pitches p, whereas a peak bin P_m consists of a set of triples (q, t, s), where q denotes the MIDI pitch, t the time position, and s the size of the peak. Then, we define the local similarity $d(n, m)$ by

$$d(n,m) := d(S_n, P_m) := \sum_{p \in S_n} \sum_{(q,t,s) \in P_m} (\delta_{p,q} + \delta_{p+12,q} + \delta_{p+19,q}) \cdot s, \quad (5.1)$$

where $\delta_{a,b}$ equals one if $a = b$ and zero if $a \neq b$ for any two integers a and b. Note that the sum $\delta_{p,q} + \delta_{p+12,q} + \delta_{p+19,q}$ is either one or zero. It is one if and only if the peak (q, t, s) appears in a subband pertaining to either one of the first three harmonics of the note p. In this case, the peak (q, t, s) contributes to the similarity value $d(n, m)$ according to its size s. In other words, the local score $d(n, m)$ is high if there are many significant peaks in P_n pertaining to notes (or harmonics of the notes) in S_m. Note that the peaks not corresponding to score notes or to their harmonics are left unconsidered by $d(n, m)$, i.e., the score data indicates which kind of information to look for in the audio signal. This principle makes the similarity function robust against additional or erroneous notes in the performance as well as "bad" peaks. Since the note and peak bins typically contain only very few (around 1–10) elements, $d(n, m)$ can be computed efficiently.

Finally, we want to indicate how to modify the definition (5.1) to obtain other local similarity functions. In an obvious way, one can account for a larger number of harmonics by introducing further terms $\delta_{p+24,q}$ (fourth harmonic), $\delta_{p+28,q}$ (fifth harmonic), or $\delta_{p+31,q}$ (sixth harmonic). Moreover, one can introduce note-dependent weights to favor certain harmonics over others. For example, the fundamental pitch dominates the piano sound spectrum over most of its range except for the lower two octaves, where most of the energy is in the second or even third harmonic. This suggests to favor the fundamental pitch for the upper notes and the second or third harmonic for the lower ones. Furthermore, omitting the factor s in the above definition of $d(n, m)$ leads to a local similarity measure that, intuitively spoken, is invariant under dynamics, i.e., strongly played notes and softly played notes are treated equally.

5.3.3 Matching Model and Alignment

The next step of the synchronization algorithm is to match the sequences S of score bins and the sequence P of peak bins. Before doing so, we have to specify a suitable matching model. Because of note ambiguities in the score such as trills or arpeggios as well as due to missing and wrong notes in the performance, not every note object of the score needs to have a realization in the audio recording. There also may be "bad" peaks extracted from the audio file. Hence, we do not want to force every note bin to be matched with a peak bin and vice versa. Furthermore, each note of the score should be aligned with at most one time position in the audio data stream. Finally, notes with the different musical onset times should be assigned to different physical onset times. These requirements lead to the following formal notion of a match:

Definition 5.1. *A* match *between the sequences* $S = (S_1, S_2, \ldots, S_N)$ *and* $P = (P_1, P_2, \ldots, P_M)$ *is a partial map* $\mu : [1 : N] \to [1 : M]$ *that is strictly monotonously increasing.*

This definition needs some explanations. The fact that objects in S or P may not have a counterpart in the other data stream is modeled by defining μ as a partial function and not as a total one. The monotony of μ reflects the requirement of faithful timing: if a note bin in S precedes a second one this should also hold for the μ-images of these bins. μ being a function and strictness of μ ensures that each note bin is assigned to at most one peak bin and vice versa.

Based on the local similarity measure d, the *global similarity value* of a match μ between S and P is given by the sum

$$\sum_{(n,m):m=\mu(n)} d(n,m). \tag{5.2}$$

Similarly to DTW, we use dynamic programming (DP) to compute an optimal match between S and P that maximizes the global similarity value. To this end, we recursively define the accumulated similarity matrix $D = (D_{n,m})$ by

$$D_{n,m} := \max\{D_{n,m-1}, D_{n-1,m}, D_{n-1,m-1} + d(n,m)\} \tag{5.3}$$

and $D_{0,0} := D_{n,0} := D_{0,m} := 0$ for $1 \leq n \leq N$ and $1 \leq m \leq M$. Then an optimal match μ^* can be constructed from D by the following algorithm (compare with Algorithm OptimalWarpingPath of Sect. 4.1).

Algorithm: OptimalMatch

Input: Accumulated similarity matrix D.
Output: Optimal match μ^*.

Procedure: $n := N$, $m := M$, μ^* defined on $\emptyset$
while $(n > 0)$ and $(m > 0)$ do
if $D(n,m) = D(n,m-1)$ then $m := m - 1$
    else if $D(n,m) = D(n-1,m)$ then $n := n - 1$
    else $\mu^*(n) := m$, $n := n - 1$, $m := m - 1$
return μ^*

After matching the note bins with the peak bins, we individually align the notes of S_n to time positions in the audio file, improving the time resolution of 50 ms imposed by the peak bins. To this end, for a note $p \in S_n$, we consider the subset of all peaks $(q, t, s) \in P_{\mu(n)}$ with $q = p$, $q = p + 12$, or $q = p + 19$.

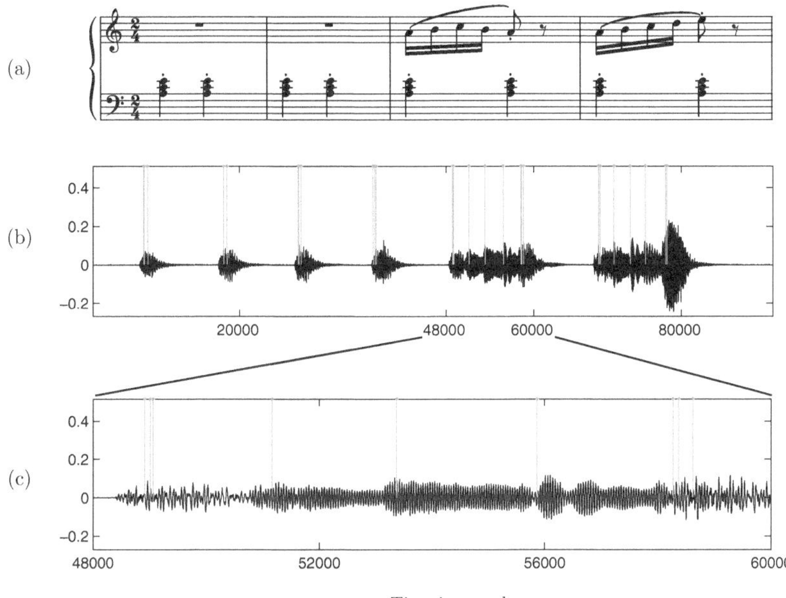

Fig. 5.5. **(a)** Score representation of the first four measures of Op. 100, No. 2 by Friedrich Burgmüller. **(b)** Audio representation of a corresponding piano recording (roughly 4.2 s) sampled at 22 050 Hz. The matched notes obtained by the score–audio synchronization are indicated by the vertical lines. **(c)** Enlargement of a segment of (b)

If this subset is empty, the note p is left unmatched. Otherwise, we assign the note $p \in S_n$ to the time position t belonging to the peak (q, t, s) of maximal size s within this subset. In other words, among all peaks in $P_{\mu(n)}$ that are in accordance with p the one with the highest size determines the point of time to be assigned to the note represented by $p \in S_n$. This strategy emphasizes the significance of the peaks. Alternative strategies may emphasize other aspects such as the fundamental pitch or relative onset times. The final assignment of the individual notes constitutes the synchronization result.

As an example, Fig. 5.5 illustrates the synchronization result of our Burgmüller example introduced in Sect. 3.2. Observe that notes with the same musical onset time may be aligned to distinct physical onset times. (This takes into account that a pianist may play the notes of a chord not exactly at the same time.) Finally, we want to point out that the assigned time positions generally tend to be slightly delayed. The reason is that it takes some time to build up a sound after a keystroke and that we actually measure the maximal increase of energy. In general, this delay is larger for lower pitches than for higher pitches.

5.3.4 Efficiency and Anchor Matches

The running time as well as memory requirements of DP are proportional to the *product* of the number of score and peak bins to be aligned, which makes DP inefficient for an increasing number of bins. The best possible complexity for a synchronization algorithm is proportional to the *sum* of the number of score and peak bins. This may be achieved by using online techniques as discussed in Sect. 5.4. Such techniques, however, are extremely sensible towards wrong or missing notes, local time deviations, or erroneously extracted features, which may result in very poor or completely useless synchronization results.

The quality of the computed alignment and the robustness of the synchronization algorithm are of great importance. In other words, increasing the efficiency of the algorithm should not degrade the synchronization result. To substantially increase the efficiency, we suggest the following simple but powerful procedure: first, identify in the score certain configurations of notes, also referred to as *anchor configurations* which possess salient dynamic and spectral properties. Such a configuration may be some isolated fortissimo chord, a note or chord played after or before some long pause, or a note with a salient fundamental pitch. Because of their special characteristics, anchor configurations can be efficiently detected in the corresponding audio file using a linear-time linear-space algorithm. From this, one computes score–audio matches referred to as *anchor matches*. Then, one aligns the remaining notes by *locally* applying the DP-based synchronization algorithm on the segments defined by two adjacent anchor matches. The acceleration of the overall procedure will depend on the number and distribution of the anchor matches. The best overall improvements are obtained with evenly distributed anchor matches. For example, $(n-1)$ anchor matches dividing the piece into equally long segments speeds up the total running time needed to perform all local DP computations by a factor of n. The memory requirements are even cut down by a factor of n^2 since the similarity matrix of only one local DP computation has to be stored at a time.

Of course, finding suitable anchor configurations is a difficult research problem by itself. As a first step, we use a semiautomatic ad-hoc approach in which the user has to specify a small number of suitable anchor configurations for a given piece of music. We have implemented several independent detection algorithms for different types of anchor configurations which are applied concurrently in order to decrease the detection error rate. Pauses in the audio data, as well as isolated fortissimo chords are detected by suitably thresholding the ratio between the short-time and the long-time signal's energy computed with a sliding window. Additionally, since pauses as well as long isolated chords correspond to segments with a small number of note onsets, such events can be detected in our onset representation of the audio signal by means of a suitable sparseness criterion. Notes of salient fundamental pitch, i.e., notes whose fundamental pitch does not clash with harmonics of other notes within a large time interval, may be detected by scanning through the corresponding subband using an energy-based measure. To further enhance detection reliability, we also investigate the neighborhoods of the detected candidate anchor matches comparing notes before and after the anchor configuration to the corresponding subband peak information. Finally, we discard candidate anchor matches exhibiting a certain likelihood of confusion with the surrounding note objects or onset events. The resulting anchor matches may be presented to the user for manual verification prior to the local DP matching stage.

5.3.5 Experimental Results

A prototype of our synchronization algorithm has been implemented in MATLAB. For the evaluation we used MIDI files representing the score data and corresponding CD recording. Our test material consists mainly of classical polyphonic piano pieces of various lengths ranging from several seconds up to ten minutes. In particular, it contains complex pieces such as Chopin's Etudes Op. 10 and Beethoven's piano sonatas (Table 5.6).

The evaluation of synchronization results is problematic and constitutes a research problem for itself. First, one has to specify the granularity of the alignment, which very much depends on the particular application. For example, if one is interested in a system that simultaneously highlights the current measure of the score while playing a corresponding interpretation (as a reading aid for the listener), an alignment deviation of a note or even several notes might be tolerable. However, for musical studies or when used as training data for statistical methods a synchronization at note level or even onset level might be required. Intuitive objective measures of synchronization quality are the percentage of note events correctly matched, the percentage of mismatched notes, or the deviation between the computed and optimal tempo curve. (The output of a synchronization algorithm may be regarded as a tempo deviation or *tempo curve* between the two input data streams.) However, such a measure will fail if the musical and physical note events to be aligned do not exactly correspond (such as for trills, arpeggios, or wrong notes). In this case, the

Table 5.6. Some piano pieces and corresponding CD recordings (with identifier) contained in our test database

Identifier	Composer	Piece	Interpreter
Scale	–	C-major scale played four times with increasing tempo	–
Burg2	Burgmüller	Etude No. 2, Op. 100	David
Chop3	Chopin	Etude No. 3, Op. 10 (Tristesse)	Varsi
Chop12	Chopin	Etude No. 12, Op. 10 (Revolution)	Varsi
Beet1	Beethoven	Sonata Op. 2, No. 1, 1st movement	Barenboim
Beet4	Beethoven	Sonata Op. 2, No. 1, 4st movement	Barenboim

measure might give a low grade, which is not due to a bad performance of the algorithm but due to the nature of the input streams. A semantically meaningful evaluation requires manual interaction, which makes such a procedure unfeasible for large-scale examinations. Similarly, the measurement of tempo curves requires some ground truth about the desired outcome of the synchronization procedure. Moreover, if a synchronization algorithm is intended to handle even changes in the global structure of the musical piece such as an additional chorus or a missing bridge [202], modified notions of tempo curves have to be considered.

Similar to Sect. 5.2.4, we evaluated our synchronization results via *sonification*. Recall that our score–audio synchronization algorithm aligns the musical onset times given by the score (MIDI file) with the corresponding physical onset times extracted from the audio file. According to this alignment, we modified the MIDI file such that the musical onset times correspond to the physical onset times. In doing so we only considered those notes of the score that are actually matched and disregard the unmatched notes. Then we converted the modified MIDI file into an audio file by means of a synthesizer (in our experiments, we used a simple sine generator). Finally, we produced a stereo audio file containing in one channel a mono version of the original CD recording and in the other channel a mono version of the synthesized audio file. Listening to this stereo audio file exhibits, due to the sensibility of the human auditory system, even smallest temporal deviations of less than 100 ms between note onsets in the two version. As the evaluations showed, our synchronization algorithm generally computes accurate global alignments even for complex piano pieces such as Chopin's Etudes or Beethoven's piano sonatas. These alignments are more than sufficient for applications such as the retrieval scenario, the reading aid scenario, or for musical studies. Moreover, most onsets of individual notes are matched with high accuracy – even for passages with short notes in fast succession being blurred due to extensive usage of the sustain pedal. Furthermore, aligning sudden tempo changes such as ritardandi, accelerandi, or pauses generally poses no problem for our algorithm.

However, in some specific situations our current algorithm may produce some local mismatches or may not be able to find any suitable match. For example, pianissimo passages are problematic since softly played notes do not generate significant energy increases in the respective subbands. Therefore, such onsets may be missed by our extraction algorithm. Furthermore, a repetition of the same chord first played in forte and then played in piano may be problematic. Here, the forte chord may cause "bad" peaks (see Sect. 3.2), which can then be mixed up with the peaks corresponding to the softly played piano chord. Such problematic situations may be handled by using a combination of different features and synchronization strategies as discussed in Sect. 5.4.

We now give some examples to illustrate the running time behavior and the memory requirements of our MATLAB implementation. Tests were run on an

Table 5.7. Running time in seconds of our score–audio synchronization algorithm for the piano pieces and corresponding CD recordings given by Table 5.6

Piece	#(notes) in score	Length (s) of audio	#(note bins)	#(peak bins)	t(peak) (s)	t(DP) (s)
Scale	32	20	32	65	3	0.4
Burg2	480	45	244	615	22	37
Chop3	1 877	173	618	1 800	114	423
Chop12	2 082	172	1 318	2 664	116	714
Beet1	2 173	226	1 322	2 722	149	716
Beet4	4 200	302	2 472	3 877	201	2 087

Table 5.8. Accumulated running time t(DP) in seconds and memory requirements (MR) in megabyte needed for the local DP computations between anchor matches

Piece	Length (s)	List of anchor matches (positions are given in (s))					t(DP) (s)	MR [MB]
Chop3	173	–	–	–	–	–	423	8.90
Chop3	173	–	–	98.5	–	–	222	3.20
Chop3	173	42.5	–	98.5	–	146.2	142	1.45
Chop3	173	42.5	74.7	98.5	125.3	146.2	87	0.44
Beet1	226	–	–	–	–	–	716	28.79
Beet1	226	–	106.5	–	–	–	363	8.24
Beet1	226	53.1	106.5	146.2	168.8	198.5	129	1.54
Beet4	302	–	–	–	–	–	2 087	76.67
Beet4	302	–	–	125.8	–	–	1 042	20.4
Beet4	302	55.9	118.8	–	196.3	249.5	433	5.09

Intel Pentium IV, 3 GHz with 1 GByte RAM under Windows 2000. Table 5.7 shows the running times for several piano pieces, where the pieces are specified by the first column. The second column shows the number of notes in the score of the respective piece and the third column the length in seconds of some performance of that piece. In the fourth and fifth columns one finds the number of note bins and peak bins (Sect. 5.3.2). The next column shows that the running time for the peak extraction, denoted by t(peak), is about linear in the length of the performance. Finally, the last column illustrates that the actual running time t(DP) of the DP algorithm is, as expected, roughly proportional to the product of the number of note bins and peak bins. The running time of the overall synchronization algorithm is essentially the sum of t(peak) and t(DP). The sonifications of the corresponding synchronization results can be found on our web page mentioned above.

Table 5.8 shows how running time and memory requirements of the DP computations decrease significantly when using suitable anchor configurations. In column three to seven one finds the anchor matches that were used in the first matching stage. Here, an anchor match is indicated by its assigned time position within the audio data stream. The computation time of these anchor matches is negligible relative to the overall running time. The eighth column shows the accumulated running time for all local DP computations. As can be seen, this running time depends heavily on the distribution of the anchor matches. For example, in the "Chop3" piece, one anchor match located in the middle of the pieces roughly accelerates the DP computation by a factor of two. Also the memory requirements (MR), which are dominated by the largest local DP computation, decrease drastically, see the last column of Table 5.8.

5.4 Further Notes

In this chapter, we introduced various music synchronization tasks and presented algorithms for audio–audio and score–audio synchronization. In the audio–audio synchronization algorithm, we used chroma-based audio features and an efficient MsDTW-based matching procedure yielding stable alignments even in the presence of significant variations. The second score–audio synchronization algorithm, which has been particularly designed for the class of piano music, is based on a sparse but expressive set of onset features and produced alignments at a high temporal resolution. Because of the complexity and diversity of music data, a universal synchronization algorithm yielding optimal solutions for all kinds of music seems to be unrealistic. Recall that one generally has to account for various aspects such as the data format (e.g., score, MIDI, audio), the genre (e.g., pop music, classical music, jazz), the instrumentation (e.g., orchestra, piano, drums, voice), and many other parameters (e.g., dynamics, tempo, timbre). Therefore, for music analysis tasks it seems promising to combine different, competing strategies instead of relying on one single strategy in order to cope with the richness and variety of music.

Exemplarily, we discuss two strategies for combining the synchronization algorithms from Sects. 5.2 and 5.3 to improve score–audio synchronization. The first strategy consists of two steps, where one first uses the chroma-based MsDTW-algorithm to efficiently align the score with the audio data stream at a relatively low resolution level. Here, the score data can be directly transformed into a chroma-like representation as suggested in Hu et al. [94]. In a second step, the alignment can then be refined by using the onset-based synchronization algorithm. Here, similar to the MsDTW strategy, one can use a suitable neighborhood of the alignment path obtained from the first step as constraint region for the DP computation in the second step. The overall synchronization algorithm inherits the efficiency and robustness from the first step, while yielding accurate alignments at a high temporal resolution as produced by the second step. As a second strategy, one can construct a combined similarity measure that incorporates spectral (chroma) information as well as onset information. A similar approach using a combined sustain and attack model has been suggested in Soulez et al. [196].

As was discussed before, one main application of music synchronization is to support content-based retrieval and browsing in complex and inhomogeneous music collections. In Chap. 8, we describe an advanced audio player, the SyncPlayer [114], which utilizes the synchronization results to facilitate additional retrieval and browsing functionality. For example, during playback of a CD recording, the SyncPlayer allows the user to directly jump to the corresponding position of any other interpretation of the same piece. Beyond the scenarios discussed in Sect. 5.1, there are many other synchronization tasks of practical relevance. One interesting yet unsolved problem is the automatic synchronization of scanned sheet music with a corresponding CD recording, where the pixels of the scanned digital images are to be aligned with time positions of the audio file. Even though the output of current OMR software (Sect. 2.1.1) is often erroneous and contains significant recognition errors, the music parameters extracted from the digital images still may by sufficient to derive a reasonable alignment with the audio data stream. Such a linking structure, as is also mentioned in Dunn et al. [61], could be used to highlight the current position in the scanned score or to automatically turn pages during playback of a CD recording. As a further application, synchronized audio data can be used to continuously blend from one interpretation to another one. In this context, an interesting research problem would be to employ synchronization techniques for mixing and morphing different interpretations.

So far, we have considered synchronization scenarios, where the two data streams to be aligned are entirely known prior to the actual synchronization. This assumption is exploited by our DP-based synchronization procedures, which computes an optimal global match between the two complete data streams. Opposed to such an *offline* scenario, one often has to deal with scenarios where the data streams are to be processed *online*. One such prominent online scenario is known as *score following*, which can be regarded as score–audio synchronization problem. While a musician is performing a piece of music according to a given musical score, the goal of score following is to identify the musical events depicted in the score with high accuracy and low latency [52]. Note that such an online synchronization procedure inherently has a linear running time. As a main disadvantage, however, an online strategy is very sensitive to local tempo variations and deviations from the score – once the procedure is out of sync, it is very hard to recover and return to the right track. Similar to score following, Dixon et al. [57] describe a linear-time DTW approach to audio synchronization based on forward path estimation. Even though the proposed algorithm is very efficient, the risk of missing the optimal alignment path is still relatively high.

A further synchronization problem, which involves score following, is known as *automatic accompaniment*. Here, one typically has a solo part played by a musician which is to be accompanied in real time by a computer system. The problem of real-time music accompaniment has first been studied by Dannenberg et al. [48,51]. Vercoe [206] developed a system for the automatic accompaniment of a transverse flute. Raphael [172] describes an accompaniment system based on Hidden Markov Models.

Finally, we want to mention the problem of *music transcription*, which has the goal of automatically converting an audio recording into a score notation. A general approach to the transcription problem is to first extract note parameters (onset times, pitches, and note durations) from the audio file and then to reassess the extracted data by suitable quantization and normalization methods to obtain "pure" score parameters. (Other approaches to music transcription are discussed, e.g., in [34, 99, 174].) In some way, the problem of synchronization can be considered to be complementary to the problem of transcription. While in music transcription there is only one data stream from which the score parameters are to be determined, in synchronization one starts with two data streams. The goal of transcription is to obtain "pure" score-data by adjusting the deviations in the extracted data resulting from the specific underlying interpretation. On the contrary, the goal of synchronization (particularly of score–audio synchronization) is to catch exactly those local time deviations in the interpretation required for linking the audio recording with the uninterpreted score data stream. For an overview of further related problems and pointers to the literature we refer to Arifi et al. [5].

6 Audio Matching

In the context of music retrieval, the *query-by-example* paradigm has attracted a large amount of attention: given a query in the form of a music excerpt, the task is to automatically retrieve all excerpts from the database containing parts or aspects which are somehow similar to the query. This problem is particularly difficult for digital waveform-based audio data such as CD recordings. Because of the complexity of such data, the notion of *similarity* used to compare different audio clips is a delicate issue and largely depends on the respective application as well as on the user requirements.

In this chapter, we consider the problem of *audio matching*. Here the goal is to retrieve all audio clips from the database that in some sense represent the same musical content as the query clip. This is typically the case when the same piece of music is available in several interpretations and arrangements. For example, given a 20-s excerpt of Bernstein's interpretation of the theme of Beethoven's Fifth Symphony, the goal is to find all other corresponding audio clips in the database; this includes the repetition in the exposition or in the recapitulation within the same interpretation as well as the corresponding excerpts in all recordings of the same piece conducted, e.g., by Karajan or Sawallisch. Even more challenging is to also include arrangements such as Liszt's piano transcription of Beethoven's Fifth or a synthesized version of a corresponding MIDI file. Obviously, the degree of difficulty increases with the degree of variations one wants to permit in the audio matching.

In view of different interpretations and instrumentations, the audio matching algorithm should be able to deal with variations regarding timbre, dynamics, articulation, and tempo. Here, the usage of chroma-based audio features (Sect. 3.3) allows for absorbing most of these variations while retaining enough information to distinguish musically unrelated audio clips. The audio matching can then be performed at the feature level by basically comparing the query feature sequence to any subsequence (of the same length) of the database feature sequence. Such a procedure has been proposed in Müller et al. [140] and will be described in Sect. 6.1. We evaluated our matching procedure on the basis of an audio collection consisting of more than one thousand audio recordings that represent a wide range of classical music and include complex orchestral and vocal works. In Sect. 6.2, we will report on our experimental results.

We then extend and improve this matching procedure by introducing an index-based matching algorithm, which significantly speeds up the retrieval process while obtaining competitive retrieval results. To this end, we describe two methods for deriving semantically meaningful feature codebooks, which are used for quantizing the chroma-based audio features (Sect. 6.3). According to the assigned codebook vector, the features can then be stored in some inverted file index – a well-known index structure as used in standard text retrieval. In Sect. 6.4, we then describe our efficient index-based audio matching procedure. In particular, we introduce various concepts of fault tolerance including fuzzy search, mismatch search, and the strategy of using multiple queries. Combining these concepts, one can handle significant variations in articulation, note realizations, tempo, and musical key. Furthermore, we discuss a two-stage ranking strategy, which allows us to suppress false positive matches in the first stage and to produce a ranked list of high-quality audio matches in the second stage. Our experiments demonstrates the practicability and efficiency of our novel matching algorithm (Sect. 6.5). One of the experimental results shows that our proposed index-based matching approach reduces the retrieval time (in comparison to the previous matching technique) by a factor of more than 15. In our current MATLAB implementation running on a standard PC it takes less then a second to retrieve a ranked list of matches from a 112 h database. Thus in a more global context of a large scale real-world music collection, our matching procedure constitutes an important building block to facilitate high-quality audio matching. In Sect. 6.6, we discuss related problems, give references to the literature, and indicate possible research directions.

6.1 Diagonal Audio Matching

In this section, we quickly describe the audio features used for the audio matching while fixing the notation (Sect. 6.1.1), introduce the basic idea of our audio matching procedure (Sect. 6.1.2), and then explain how to handle global variations in tempo and musical key (Sect. 6.1.3). It turns out that the matching algorithm can be interpreted as convolution, which can be efficiently computed via FFT-based algorithms (Sect. 6.1.4).

6.1.1 Audio Features

In the audio matching problem, the notion of similarity used to compare different recordings is of crucial importance. In our scenario, we are interested in identifying audio clips irrespective of certain details concerning the interpretation or instrumentation. Therefore, the idea is to use very coarse audio features that strongly correlate to the harmonic progression of the audio signal while showing a high degree of robustness to variations in parameters such as timbre, dynamics, articulation, and local tempo deviations. To this end, we will use the chroma-based $\mathrm{CENS}^{\mathbf{w}}_{\mathbf{d}}$-features as introduced in Sect. 6.3, which are elements of the set $\mathcal{F} = \left\{v \in [0,1]^{12} \mid \|v\|_2 = 1\right\}$ as defined in (3.4). In other words, $\mathcal{F}$ consists of all points on the unit sphere S^{11} that have nonnegative entries. For later use, we define a projection operator $\pi^{\mathcal{F}} : \mathbb{R}^{12} \to S^{11}$ by

$$\pi^{\mathcal{F}}(v) := \begin{cases} v/\|v\|_2 & \text{if } \|v\|_2 \neq 0, \\ (1,1,\ldots,1)/\sqrt{12} & \text{if } \|v\|_2 = 0. \end{cases} \tag{6.1}$$

Recall that the main idea of CENS features is to take statistics over relatively large windows (consisting of $\mathbf{w}$ consecutive vectors). This not only smoothes out local time deviations as may occur for articulatory reasons but also compensates for different realizations of note groups such as trills or arpeggios. For an example, we refer to Fig. 3.10. Finally, recall that one obtains a flexible and computationally inexpensive procedure to adjust the feature resolution by simply modifying the parameters $\mathbf{w}$ and $\mathbf{d}$ in the CENS computation. This will be exploited in Sect. 6.1.3 to efficiently simulate tempo changes.

6.1.2 Basic Matching Procedure

The audio database consists of a collection of CD recordings, typically containing various interpretations for one and the same piece of music. To simplify things, we may assume that this collection consists of one large document D by concatenating the individual recordings and keeping track of the boundaries in a supplemental data structure. In a preprocessing step, D is transformed into a sequence of CENS^{41}_{10}-features denoted by $\mathrm{CENS}^{41}_{10}[D] = (w_0, w_1, \ldots, w_{M-1})$.

In our settings, a typical query Q is an audio clip of a duration of 10–30 s. In the basic matching procedure, the query Q is also transformed into a CENS feature sequence $\mathrm{CENS}^{41}_{10}[Q] = (v_0, v_1, \ldots, v_{N-1})$. This query sequence is compared to any subsequence $(w_i, w_{i+1}, \ldots, w_{i+N-1})$ consisting of N consecutive vectors of the database sequence and starting at position $i \in [0 : M-N]$. Then, a *match* of the query Q with respect to the database is defined to be a tuple (k, N) consisting of the *match position* $k \in [0 : M-1]$ and the *match length* $N \in [1 : M-1]$. We then also say that the query Q matches the audio clip corresponding to the feature subsequence $(w_k, \ldots, w_{k+N-1})$. For the comparison, we use the distance measure

$$\Delta(i) := 1 - \frac{1}{N} \sum_{n=0}^{N-1} \langle w_{i+n}, v_n \rangle. \tag{6.2}$$

Note that since all CENS vectors are of unit length, the inner products $\langle w_{i+n}, v_n \rangle$ coincide with the cosine of the angle between w_{i+n} and v_n. Altogether, we obtain a distance function $\Delta : [0 : M-1] \to [0,1]$ by setting

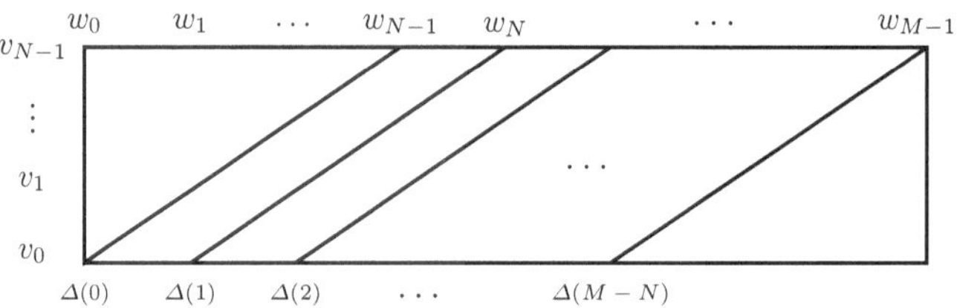

Fig. 6.1. The computation of the distance function Δ with respect to $\mathrm{CENS}_{10}^{41}[Q] = (v_0, v_1, \ldots, v_{N-1})$ and $\mathrm{CENS}_{10}^{41}[D] = (w_0, w_1, \ldots, w_{M-1})$

$\Delta(i) := \infty$ for $i \in [M - N + 1 : M - 1]$. As the measure (6.2) is computed basically by summing up diagonals of a matrix $(\langle w_m, v_n \rangle)_{mn}$, the proposed matching technique will be referred to as *diagonal matching*, see Fig. 6.1.

To determine the best match between Q and D, we simply look for the index $i_0 \in [0 : M - 1]$ minimizing Δ. Then the best match is given by the tuple (i_0, N), which encodes the audio clip corresponding to the feature sequence $(w_{i_0}, w_{i_0+1} \ldots, w_{i_0+N-1})$. We then exclude a neighborhood of length M of the best match from further considerations by setting $\Delta(i) = \infty$ for $i \in [i_0 - \lceil N/2 \rceil : i_0 + \lceil N/2 \rceil] \cap [0 : M - 1]$, thus avoiding matches with a large overlap to the subsequent matches. To find subsequent matches, the latter procedure is repeated until a certain number of matches is obtained or a specified distance threshold is exceeded.

As an illustrating example, we consider a database D consisting of four pieces: one interpretation of Bach's Toccata BWV565, two interpretations (Bernstein, Sawallisch) of the first movement of Beethoven's Fifth Symphony Op. 67, and one interpretation of Shostakovich's Waltz 2 from his second Jazz Suite. The query Q again consists of the first 21 s (20 measures) of Bernstein's interpretation of Beethoven's Fifth Symphony (cf. Fig. 3.10b). The upper part of Fig. 6.2 shows the resulting distance function Δ. The lower part shows the feature sequences corresponding to the ten best matches sorted from left to right according to their distance. Here, the best match (coinciding with the query) is shown on the leftmost side, where the matching rank and the respective Δ-distance (1/0.011) are indicated above the feature sequence and the position (0-21, measured in seconds) within the audio file is indicated below the feature sequence. Corresponding parameters for the other nine matches are given in the same fashion.

Note that the distance 0.011 for the best match is not exactly zero, since the interpretation in D starts with a small segment of silence, which has been removed from the query Q. Furthermore, note that the first 20 measures of Beethoven's Fifth, corresponding to Q, appear again in the repetition of the exposition and once more with some slight modifications in the recapitulation. Matches 1, 2, and 5 correspond to these excerpts in Bernstein's interpretation,

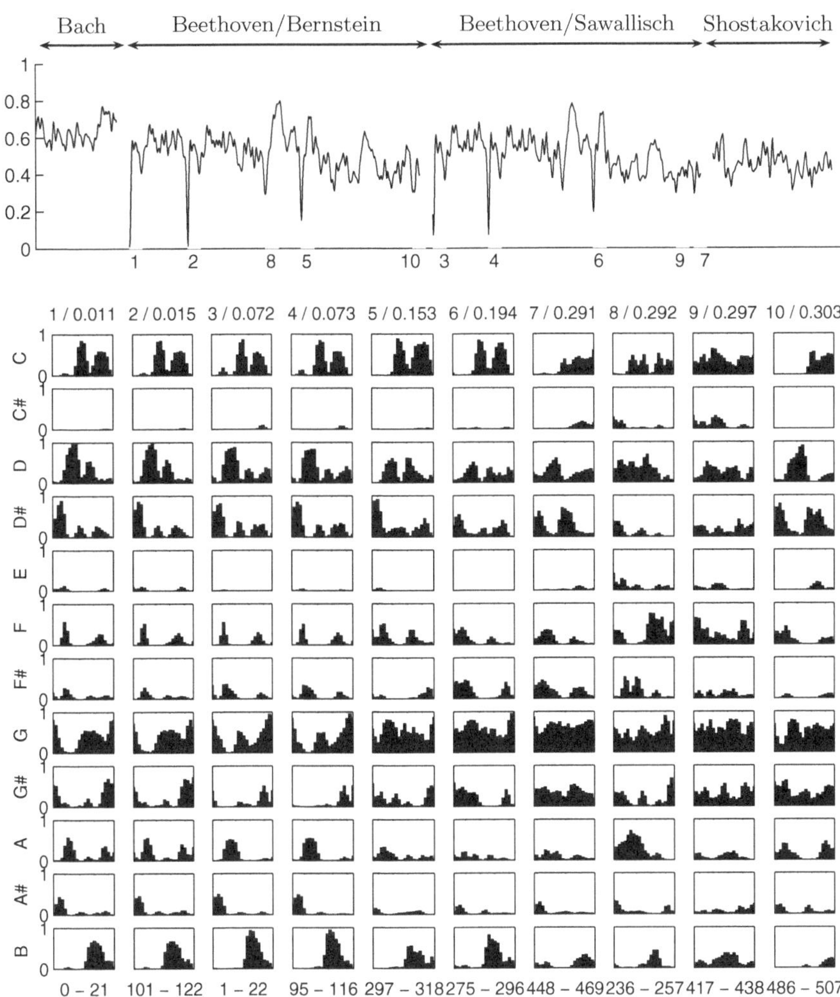

Fig. 6.2. Distance function Δ (*top*) and CENS feature sequences of the first 10 matches for a dataset D consisting of four pieces and a query Q corresponding to Fig. 3.10b

whereas matches 3, 4, and 6 to those in Sawallisch's interpretation. In Sect. 6.2, we continue this discussion and give additional examples.

6.1.3 Global Variations in Tempo and Key

Because of the temporal blurring of the CENS features, this basic matching procedure works well even in the presence of local tempo variations. However,

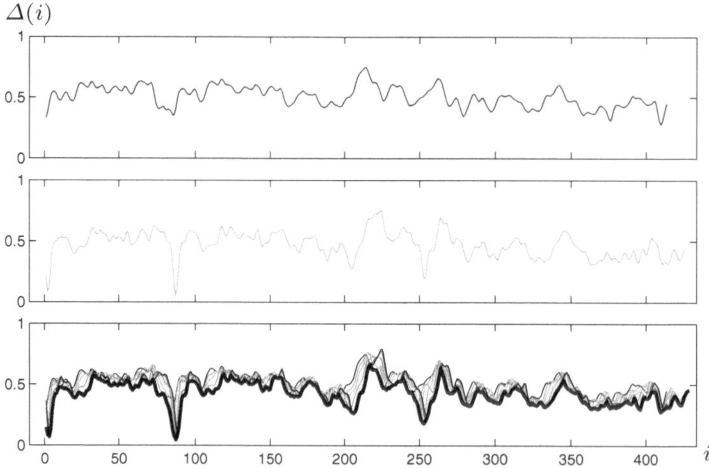

Fig. 6.3. The query Q consists of the first 21 s of a Bernstein interpretation and the database D of a Karajan interpretation (443 s) of the first movement of Beethoven's Fifth. The tempo in the Bernstein interpretation is much slower (roughly 80%) than the one in the Karajan interpretation. *Top*: Distance function Δ_4 for $\mathrm{CENS}_{10}^{41}[Q]$. *Middle*: Distance function Δ_7 for $\mathrm{CENS}_{13}^{53}[Q]$. *Bottom*: Distance functions Δ_j for $j \in [1:8]$ and joint distance function $\Delta_{\min}$ (*bold line*)

in the presence of global tempo differences this matching procedure does not yet work well, see Fig. 6.3. For example, Bernstein's interpretation of the first movement of Beethoven's Fifth has a much slower tempo (roughly 80%) than Karajan's interpretation. While there are 23 CENS feature vectors for the first 20 measures computed from Bernstein's interpretation, there are only 19 in Karajan's case. To account for such global tempo variations, we create several versions of the query audio clip corresponding to different tempi. These tempo changes are simulated on the CENS feature level by suitably modifying the parameters $\mathbf{w}$ and $\mathbf{d}$. More precisely, instead of only considering $\mathrm{CENS}_{10}^{41}[Q]$, we calculate feature sequences $\mathrm{CENS}_{\mathbf{d}_j}^{\mathbf{w}_j}[Q]$ of length N_j for eight different pairs $(\mathbf{w}_j, \mathbf{d}_j)$ with $\mathbf{d}_j = j + 6$ and $\mathbf{w}_j = 4\mathbf{d}_j + 1$, $j \in [1:8]$. Each pair effectively simulates a tempo change. For example, using $\mathbf{d}_7 = 13$ (instead of $\mathbf{d}_4 = 10$) and $\mathbf{w}_7 = 53$ (instead of $\mathbf{w}_4 = 41$) results in a scaled version of the CENS features simulating a tempo change of $10/13 \approx 0.77$. The eight pairs cover global tempo variations of roughly -30% to $+40\%$, see Table 6.1. For each of the resulting query sequences, we separately calculate a distance function Δ_j, $j \in [1:8]$. Finally, the joint distance function $\Delta_{\min} : [0:M-1] \to [0,1]$ is defined by $\Delta_{\min}(i) := \min(\Delta_1(i), \ldots, \Delta_8(i))$. Then the parameter $i_0 \in [0:M-1]$ minimizing $\Delta_{\min}$ yields the best match between the query and a database subsequence starting with w_{i_0}. To obtain the length of this subsequence, let $j_0 \in [1:8]$ denote the index that minimizes

Table 6.1. Tempo changes (tc) simulated by changing the statistics window size $\mathbf{w}$ and the downsampling factor $\mathbf{d}$

j	$\mathbf{w}_j$	$\mathbf{d}_j$	tc
1	29	7	1.43
2	33	8	1.25
3	37	9	1.1
4	**41**	**10**	**1.0**
5	45	11	0.9
6	49	12	0.83
7	53	13	0.77
8	57	14	0.7

$\{\Delta_j(i_0) \mid j \in [1:8]\}$. Then the best match is given by (i_0, N_{j_0}) corresponding to the subsequence $(w_{i_0}, \ldots, w_{i_0+N_{j_0}-1})$. To obtain further matches, one can proceed as described earlier.

A similar strategy can be employed to account for a global difference in the musical key between the query and the database matches. The basic idea is to simulate all possible 12 transpositions by cyclically shifting the 12-dimensional CENS vectors of the query sequence, see [81]. The resulting 12 query versions are then processed as in the case of the global tempo simulation, resulting in a transposition-invariant joint distance function. For details we refer to Sect. 7.5.3, where we will apply the same strategy in the context of the audio structure problem.

6.1.4 Efficient Implementation

At this point, we want to mention that the distance function Δ defined by (6.2) can be computed efficiently. Here, one has to note that each of the 12 components of the vector $\sum_{n=0}^{N-1}\langle w_{i+n}, v_n\rangle$ can be expressed as a convolution, which can then be evaluated efficiently using FFT-based convolution algorithms. By this technique, Δ can be calculated with $O(DM \log N)$ operations, where $D = 12$ denotes the dimension of the vectors. In other words, the query length N only contributes a logarithmic factor to the total arithmetic complexity. Thus, even long queries may be processed efficiently. We next describe the experimental setting and the running time to process a typical query.

6.2 Experimental Results for Diagonal Audio Matching

For our experiments, we set up a database (DB112) containing 112 h of uncompressed audio material (mono, 22 050 Hz), which requires 16.5 GB of disk space. The database comprises 1 167 audio files reflecting a wide range of classical music, including, among others, pieces by Bach, Bartok, Bernstein, Beethoven, Chopin, Dvorak, Elgar, Mozart, Orff, Ravel, Schubert, Shostakovich, Vivaldi, and Wagner. In particular, it contains all Beethoven symphonies, all Beethoven piano sonatas, all Mozart piano concertos, several Schubert and Dvorak symphonies – many of the pieces in several versions. Some of the orchestral pieces are also included as piano arrangements or synthesized MIDI-versions. In a preprocessing step, we computed the CENS features for all audio files of the database, resulting in a single sequence $\mathrm{CENS}^{41}_{10}[D]$ as described in Sect. 6.1.2. Storing the features $\mathrm{CENS}^{41}_{10}[D]$ requires only 40.3 MB (opposed to 16.5 GB for the original data), which amounts in a data reduction of a factor of more than 400. Note that the feature sequence $\mathrm{CENS}^{41}_{10}[D]$ is all we need during the matching procedure. We implemented our audio matching procedure in MATLAB. The tests were run on an Intel Pentium IV, 3 GHz with 1 GByte RAM under Windows 2000. Processing a query of 10–30 s of duration takes roughly 1 s w.r.t. Δ and about 7–10 s w.r.t. Δ_{min}. We will see in Sect. 6.4, how the processing time can be reduced significantly by employing suitable indexing methods.

6.2.1 Representative Matching Results

We now discuss in detail some representative matching results obtained from our procedure, using the query clips shown in Table 6.2. For each query clip, the columns contain from left to right an identifier, the composer, the specification of the piece of music, the measures corresponding to the clip, and the interpreter.

Table 6.2. Query audio clips used in the experiments

Identifier	Composer	Piece	Measures	Interpreter
BachAn	Bach	BWV 988, Goldberg Variations, Aria	1–n	MIDI
BeetF	Beethoven	Symphony No. 5, Op. 67, 1st mov.	1–20	Bernstein
BeLiF	Beethoven	Symphony No. 5, Op. 67, 1st mov. (piano)	129–170	Scherbakov
BeetT	Beethoven	Sonata Op. 31, No. 2, 1st mov. (Tempest)	41–62	Barenboim
OrffCB	Orff	Carmina Burana	1–4	Jochum
SchuU	Schubert	Symphony No. 8, D759, 1st mov. (Unfinished)	9–21	Abbado
ShoWn	Shostakovich	Jazz Suite No. 2, 6th mov. (Waltz)	1–n	Chailly
VivaS	Vivaldi	RV269 No.1 (Spring)	44–55	MIDI

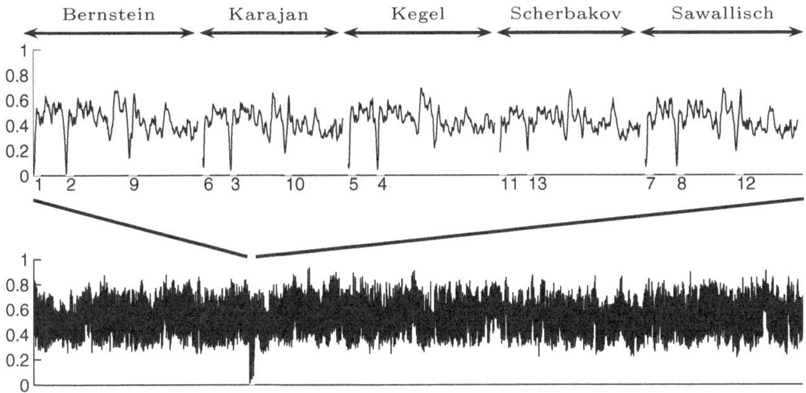

Fig. 6.4. *Bottom*: Distance function Δ_{min} for the entire database w.r.t. the query "BeetF." *Top*: Enlargement showing the five interpretations of the first movement of Beethoven's Fifth containing all of the 13 matches with Δ_{min}-distance to the query below 0.2

We continue our Beethoven example. Recall that the query, in the following referred to as "BeetF" (see Table 6.2), corresponds to the first 20 measures, which appear once more in the repetition of the exposition and with some slight modifications in the recapitulation. Since our database contains Beethoven's Fifth in five different versions – four orchestral version conducted by Bernstein, Karajan, Kegel, and Sawallisch, respectively, and Liszt's piano transcription played by Scherbakov – there are altogether 15 occurrences in our database similar to the query "BeetF." Using our matching procedure, we automatically determined the best 15 matches in the entire database w.r.t. Δ_{min}. Those 15 matches contained 14 of the 15 "correct" occurences – only the 14th match (distance 0.217) corresponding to some excerpt of Schumann's third symphony was "wrong." Furthermore, it turned out that the first 13 matches are exactly the ones having a Δ_{min}-distance below 0.2 from the query, see also Fig. 6.4 and Table 6.3. The 15th match (excerpt in the recapitulation by Kegel) already has a distance of 0.220. Note that even the occurrences in the exposition of Scherbakov's piano version were correctly identified as 11th and 13th match, even though differing significantly in timbre and articulation from the orchestral query. Only the occurrence in the recapitulation of the piano version was not among the top matches.

As a second example, we queried the piano version "BeLiF" of about 26 s of duration (see Table 6.2), which corresponds to the first part of the development of Beethoven's Fifth. The Δ_{min}-distances of the best 20 matches are shown in Table 6.3. The first six of these matches contain all five "correct" occurrences in the five interpretations corresponding to the query excerpt, see also Fig. 6.5. Only the 4th match comes from the first movement (measures 200–214) of Mozart's symphony No. 40, KV 550. Even though seemingly

Table 6.3. The columns show the respective $\Delta_{\min}$-distances of the 20 best matches for the queries indicated by Table 6.2

No.	BachA8	BeetF	BeLiF	OrffCB	ShoW20	SchuU	VivaS	BeetT	BeetT (cyc)
1	0.005	0.011	0.010	0.005	0.017	0.024	0.095	0.022	0.022
2	0.020	0.015	0.139	0.037	0.051	0.052	0.139	0.029	0.029
3	0.090	0.044	0.142	0.065	0.098	0.061	0.154	0.059	0.059
4	0.093	0.051	0.168	0.138	0.104	0.070	0.155	0.064	0.064
5	0.093	0.058	0.168	0.148	0.109	0.071	0.172	0.088	0.088
6	0.095	0.069	0.172	0.150	0.140	0.072	0.210	0.095	0.088
7	0.098	0.072	0.200	0.152	0.148	0.073	0.221	0.180	0.090
8	0.102	0.073	0.203	0.155	0.163	0.091	0.238	0.187	0.095
9	0.104	0.143	0.204	0.158	0.167	0.097	0.241	0.191	0.121
10	0.107	0.180	0.214	0.165	0.173	0.100	0.244	0.195	0.165
11	0.107	0.183	0.221	0.166	0.186	0.101	0.248	0.196	0.178
12	0.108	0.195	0.221	0.166	0.187	0.103	0.257	0.197	0.178
13	0.110	0.197	0.225	0.167	0.188	0.107	0.262	0.197	0.180
14	0.110	0.217	0.229	0.179	0.192	0.108	0.267	0.197	0.183
15	0.112	0.220	0.230	0.179	0.193	0.122	0.268	0.200	0.184
16	0.114	0.224	0.231	0.172	0.194	0.151	0.271	0.206	0.186
17	0.117	0.225	0.232	0.173	0.197	0.158	0.273	0.212	0.187
18	0.120	0.229	0.234	0.174	0.198	0.205	0.275	0.213	0.187
19	0.122	0.237	0.235	0.174	0.199	0.207	0.276	0.214	0.188
20	0.122	0.238	0.236	0.176	0.199	0.214	0.279	0.218	0.188

The last column shows the transposition-invariant joint distance function for the query "BeeT," see Sect. 6.1.3 and Fig. 6.7 (b).

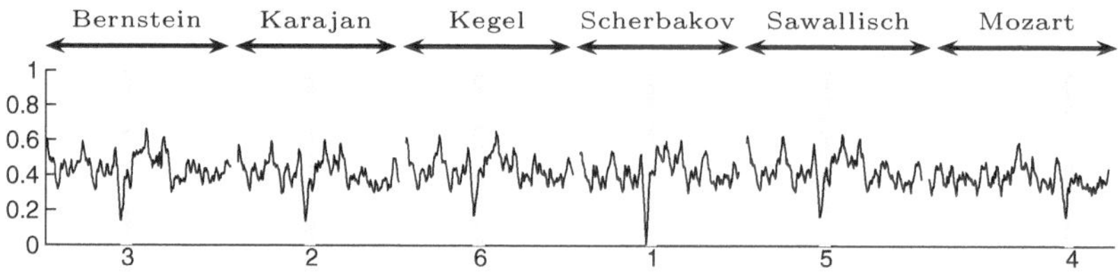

Fig. 6.5. Section consisting of the five interpretations of the first movement of Beethoven's Fifth and the first movement of Mozart's symphony No. 40, KV 550. The five occurences in the Beethoven interpretations are among the best six matches, all having $\Delta_{\min}$-distance to the query "BeLiF" below 0.2

unrelated to the query, the harmonic progression of Mozart's piece exhibits a strong correlation to the Beethoven query at these measures. As a general tendency, it has turned out in our experiments that for queries of about 20 s of duration the "correct" matches have a distance to the query below 0.2. In general, only few "false" matches have a $\Delta_{\min}$-distance to the query below this distance threshold.

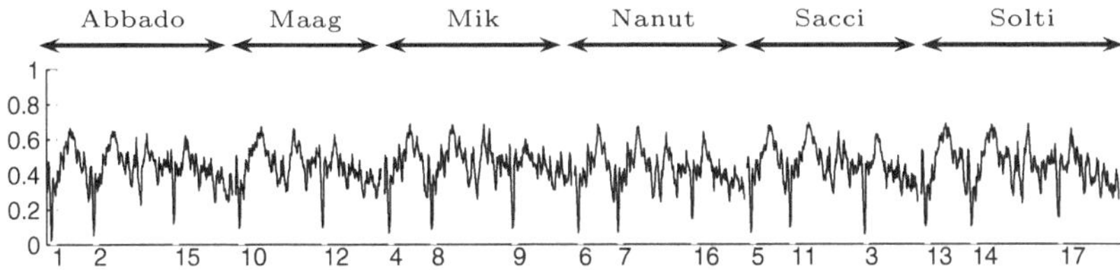

Fig. 6.6. Section consisting of the five interpretations of the first movement of Schubert's Unfinished. The 17 occurences exactly correspond to the 17 matches with Δ_{min}-distance to the query "SchuU" below 0.2

A similar result was obtained when querying "SchuU" corresponding to measures 9–21 of the first theme of Schubert's "Unfinished" conducted by Abbado. Our database contains the "Unfinished" in six different interpretations (Abbado, Maag, Mik, Nanut, Sacci, Solti), the theme appearing once more in the repetition of the exposition and in the recapitulation. Only in the Maag interpretation the exposition is not repeated, leading to a total number of 17 occurrences similar to the query. The best 17 matches retrieved by our algorithm exactly correspond to these 17 occurences, all of those matches having a Δ_{min}-distance well below 0.2, see Table 6.3 and Fig. 6.6. The 18th match, corresponding to some excerpt of Chopin's Scherzo Op. 20, already had a Δ_{min}-distance of 0.205.

Our database also contains two interpretations (Jochum, Ormandy) of the Carmina Burana by Carl Orff, a piece consisting of 25 short episodes. Here, the first episode "O Fortuna" appears again at the end of the piece as 25th episode. The query "OrffCB" corresponds to the first four measures of "O Fortuna" in the Jochum interpretation (22 s of duration), employing the full orchestra, percussion, and chorus. Again, the best four matches exactly correspond to the first four measures in the first and 25th episodes of the two interpretations. The fifth match is then an excerpt from the third movement of Schumann's Symphony No. 4, Op. 120. When asking for all matches having a Δ_{min}-distance to the query below 0.2, our matching procedure retrieved 75 matches from the database. The reason for the relatively large number of matches within a small distance to the query is the relatively unspecific, unvaried progression in the CENS feature sequence of the query, which is shared by many other pieces as well. In Sect. 6.2.2, we will discuss a similar example ("BachAn") in more detail. It is interesting to note that among the 75 matches, there are 22 matches from various episodes of the Carmina Burana, which are variations of the original theme.

To test the robustness of our matching procedure to the respective instrumentation and articulation, we also used queries synthesized from uninterpreted MIDI versions. For example, the query "VivaS" (see Table 6.2) consists of a synthesized version of the measures 44–55 of Vivaldi's "Spring"

RV269, No. 1. This piece is contained in our database in seven different interpretations. The best seven matches were exactly the "correct" excerpts, where the first five of these matches had a $\Delta_{\min}$-distance to the query below 0.2 (see also Table 6.3). The robustness to different instrumentations is also shown by the Shostakovich example in the next section.

The next example shows how one can deal with global difference in the musical key between the query and the database matches. The query "BeetT" consists of the second theme of the first movement of Beethoven's Tempest Sonata played by Barenboim. This theme appears three times: two times in the exposition and its repetition, respectively, and a third time in the recapitulation, where the theme is transposed by five semitones upwards and, after some measures, and then moves to a lower octave. The piece is contained in our database three times played by Barenboim, Gilels, and Pollini. Using the audio matching procedure as before, the appearances of the theme in the expositions appear as the top six matches, all within a $\Delta_{\min}$-distance to the query below 0.1 (see also Table 6.3). However, the transposed theme in the three recapitulations were not identifed, see Fig. 6.7a. Using the transposition-invariant joint distance function (Sect. 6.1.3), one obtains all expected nine matches as the top nine matches, see Fig. 6.7b. For example, the ninth match is the transposed theme in Pollini's interpretation having a transposition-invariant distance of 0.121, see Table 6.3. Note that the overall transposition-invariant distance function is much closer to zero than before. The correct matches, however, are still clearly separated.

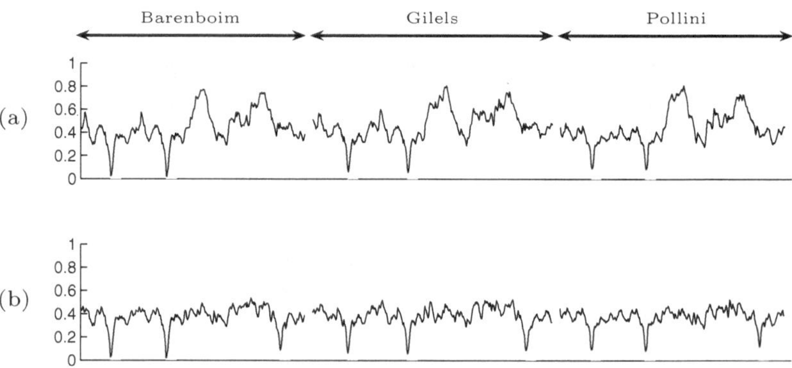

Fig. 6.7. **(a)** $\Delta_{\min}$-distance function to the query "BeetT" for the three interpretations of the first movement of Beehoven's Tempest Sonata. Only the six matches in the expositions appear as top matches. **(b)** Corresponding transposition-invariant distance function. Here, also the transposed themes in the recapitulations appear as top matches

6.2.2 Dependence on Query Length

Not surprisingly, the quality of the matching results depends on the length of the query: queries of short duration will generally lead to a large number of matches in a close neighborhood of the query. Enlarging the query length will generally reduce the number of such matches. We illustrate this principle by means of the second Waltz of Shostakovich's Jazz Suite No. 2. This piece is of the musical form $A_1A_2BA_3A_4$, where the first theme consists of 38 measures and appears four times (parts A_1, A_2, A_3, A_4), each time in a different instrumentation. In part A_1 the melody is played by strings, then in A_2 by a clarinet and wood instruments, in A_3 by a trombone and brass, and finally in A_4 in a tutti version. The Waltz is contained in our database in two different interpretations (Chailly,Yablonsky), leading to a total number of eight occurrences of the theme.

The query "ShoWn" (see Table 6.2) consists of the first n measures of the theme in the Chailly interpretation. Table 6.4 compares the total number of matches to the query duration. For example, the query clip "ShoW12" (duration of 13 s) leads to 590 matches with a Δ_{min}-distance below 0.2. Among these matches the four occurrences A_1, A_2, A_3, and A_4 in the Chailly interpretation could be found at position 1 (the query itself), 2, 6, and 10, respectively. Similarly, the four occurrences in the Yablonsky interpretation could be found at the positions 119/59/103/138. Enlarging the query to 20 measures (22 s) led to a much smaller number of 23 matches with a Δ_{min}-distance below 0.2. Only the trombone theme in the Yablonsky version (36th match with Δ_{min}-distance of 0.207) was not among the first 23 matches. Finally,

Table 6.4. Total number of matches with Δ_{min}-distance below 0.2 for queries of different durations

Query	Length (s)	#(matches) with $\Delta_{\mathrm{min}} \leq 0.2$	Chailly	Yablonsky
ShoW12	13	590	1	119
			2	59
			6	103
			10	138
ShoW20	22	23	1	4
			2	5
			7	36
			3	6
ShoW27	29	8	1	3
			2	5
			7	8
			4	6

The third and forth column indicate the rank of the four "correct" matches contained in the Chailly and Yablonsky recording, respectively.

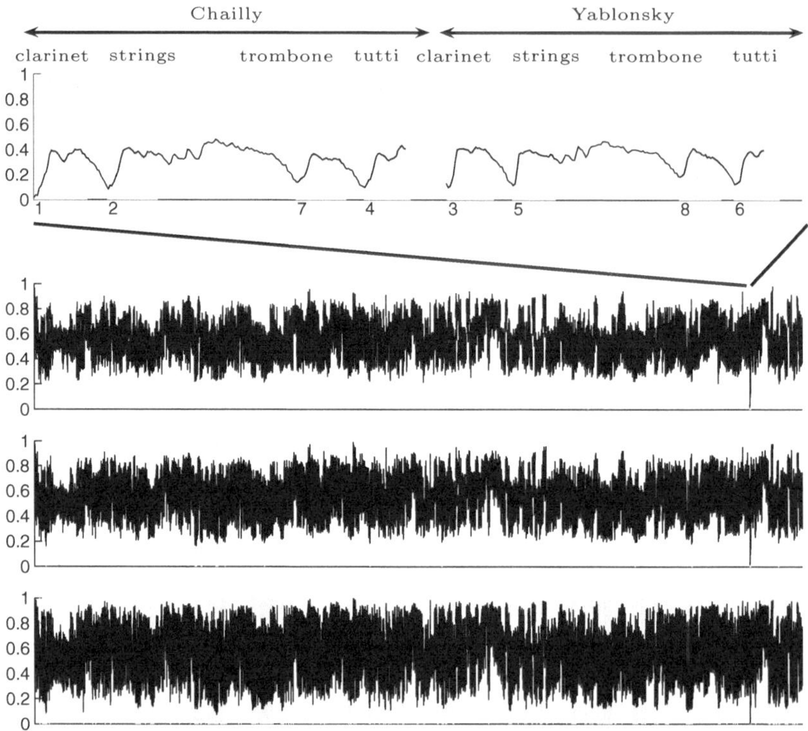

Fig. 6.8. *Second to fourth row*: $\Delta_{\min}$-distance functions for the entire database w.r.t. the queries "ShoW27," "ShoW20," and "ShoW12," respectively. The green bars indicate the matching regions. *First row*: Enlargement for the query "ShoW27" showing the two interpretations of the Waltz. Note that the theme appears in each interpretation in four different instrumentations

querying "ShoW27" (29 s) led to 8 matches with a $\Delta_{\min}$-distance below 0.2, exactly corresponding to the eight "correct" occurrences, see Fig. 6.8. Among these matches, the two trombone versions have the largest $\Delta_{\min}$-distances. This is caused by the fact that the spectra of low-pitched instruments such as the trombone generally exhibit phenomena such as oscillations and smearing effects resulting in degraded CENS features.

As a final example, we consider the Goldberg Variations by J.S. Bach, BWV 988. This piece consists of an Aria, 30 variations, and a repetition of the Aria at the end of the piece. The interesting fact is that the variations are on the Aria's bass line, which closely correlates with the harmonic progression of the piece. Since the sequence of CENS features also closely correlates with this progression, a large number of matches is to be expected when querying the theme of the Aria. The query "BachAn" consists of the first n measures of the Aria synthesized from some uninterpreted MIDI, see Table 6.2.

Querying "BachA4" (10 s of duration) led to 576 matches with $\Delta_{\min}$-distance below 0.2. Among these matches, 214 correspond to some excerpt originating from a variation of one of the four Goldberg interpretations contained in our database. Increasing the duration of the query, we obtained 307 such matches for "BachA8" (20 s), 195 of them corresponding to some Goldberg excerpt. Similarly, one obtained 144 such matches for "BachA12" (30 s), 127 of them corresponding to some Goldberg excerpt.

In conclusion, our experimental results suggest that a query duration of roughly 20 s seems to be sufficient for a good characterization of most audio excerpts. Enlarging the duration generally makes the matching process even more stable and reduces the number of false positives.

6.3 Codebook Selection for CENS Features

The matching procedure described so far has a running time that linearly depends on the length of the database. To speed up the matching procedure, we will introduce an index-based approach using an inverted file index (see Sect. 6.4). This kind of indexing depends on a suitable *codebook* that consists of a finite set of characteristic CENS vectors. In this section, we discuss various strategies for constructing such codebooks.

Recall that the set $\mathcal{F}$ of possible CENS features consists of all points on the unit sphere S^{11} with nonnegative entries, see (3.4). A codebook w.r.t. $\mathcal{F}$ is given by a finite set $\mathcal{C}_R = \{c_1, \ldots, c_R\} \subset \mathcal{F}$. We will now assign a set of features to each index $r \in [1 : R]$ by a nearest neighbor criterion partitioning the set $\mathcal{F}$ into pairwise disjoint feature classes. More precisely, an arbitrary feature vector $v \in \mathcal{F}$ is assigned to the class label $\mathcal{Q}[v] \in [1 : R]$ defined by

$$\mathcal{Q}[v] := \mathrm{argmin}_{r\in[1:R]}\arccos(\langle v, c_r\rangle). \qquad (6.3)$$

(In the case that the minimizing index is not unique, we randomly choose one of these indices.) In other words, the vector $c_{\mathcal{Q}[v]}$ minimizes the angle to v among all elements in the $\mathcal{C}_R$. The function $\mathcal{Q} : \mathcal{F} \to [1 : R]$ is also referred to as *quantization function* associated to the codebook $\mathcal{C}_R$. In the remainder of this section, we describe two fundamentally different approaches for selecting the codebook $\mathcal{C}_R$.

6.3.1 Codebook Selection by Unsupervised Learning

A common strategy of selecting a codebook without exploiting any further knowledge about the underlying data is based on unsupervised clustering. In the following, we show how the well-known LBG (Linde–Buzo–Gray) algorithm [55] for vector quantization in the Euclidean space can be adapted to perform clustering in the spherical feature space $\mathcal{F} \subset S^{11}$.

Let $W = (w_0, \ldots, w_{M-1})$ denote the database feature sequence, which will be used for constructing a codebook $\mathcal{C}_R = \{c_1, \ldots, c_R\} \subset \mathcal{F}$. The LBG algorithm consists of two steps. In an *initialization* step, an initial codebook $\mathcal{C}_R^0 = \{c_1^0, \ldots, c_R^0\} \subset \mathcal{F}$ is chosen. In this work, we randomly chose this initial codebook from a uniform distribution on $\mathcal{F}$ (it may also be randomly chosen from V). The parameter R is chosen to yield a suitable number of inverted lists, see Sect. 6.4. Using the nearest neighbor assignment on the sphere, the codebook induces a quantization function $\mathcal{Q}^0$ analogous to (6.3). The mean angular *quantization error* is then defined by

$$\varepsilon^0 := \frac{1}{M} \sum_{m=0}^{M-1} \arccos(\langle w_m, c^0_{\mathcal{Q}^0[w_m]} \rangle). \tag{6.4}$$

In the subsequent *iterative* step, the following procedure of alternately selecting a new codebook and updating the quantization function is repeated for steps $\ell = 1, 2, \ldots$ until a suitable stop criterion is satisfied.

1. Let $P_r := (\mathcal{Q}^{\ell-1})^{-1}(r)$ denote the set of features which are assigned the class label $r \in [1:R]$ by the quantization function $\mathcal{Q}^{\ell-1}$. Define the new codebook $\mathcal{C}_R^\ell = \{c_1^\ell, \ldots, c_R^\ell\}$ by updating the codebook vectors according to
$$c_r^\ell := \pi^{\mathcal{F}} \left(\frac{1}{|P_r|} \sum_{v \in P_r} v \right) \tag{6.5}$$
for all $r \in [1:R]$.
2. Determine a new quantization function $\mathcal{Q}^\ell : \mathcal{F} \to \mathcal{C}_R^\ell$ by setting
$$\mathcal{Q}^\ell(w_m) := \mathrm{argmin}_{r \in [1:R]} \arccos(\langle w_m, c_r^\ell \rangle). \tag{6.6}$$
3. Calculate the new mean angular quantization error
$$\varepsilon^\ell := \frac{1}{M} \sum_{n=0}^{M-1} \arccos(\langle w_m, c^\ell_{\mathcal{Q}^\ell[w_m]} \rangle). \tag{6.7}$$

The iteration is terminated as soon as the difference $|\varepsilon^\ell - \varepsilon^{\ell-1}|$ falls below a suitable threshold. We then define $\mathcal{Q} := \mathcal{Q}^\ell$ and $\mathcal{C}_R := \mathcal{C}_R^\ell$.

Besides choosing a spherical distance (angles) on $\mathcal{F}$, the difference to the classical LBG-approach is the additional projection on the unit sphere expressed by the operator $\pi^{\mathcal{F}}$. This projection is necessary because the convex combination $|P_r|^{-1} \sum_{v \in P_r} v$ computed in (6.5) is generally not located on $\mathcal{F}$. Instead, one could use spherical averages to better account for the spherical geometry of $\mathcal{F}$, see Buss and Fillmore [23]. In practice, however, the effects on the final clustering result are only marginal. The convergence of the modified LBG-algorithm is somewhat slower than in the classical case. However, as the codebook is computed offline, our proposed algorithm turns out to be sufficiently fast for our purpose. As an example, the algorithm required $\ell = 127$ iterations to select a codebook of size $R = 200$ for a training set W consisting of $M = 186\,929$ vectors. (The set W was derived from a test database DB55, which is a subset of the database DB112, see Sect. 6.2.)

6.3.2 Codebook Selection Based on Musical Knowledge

One main advantage of the unsupervised learning procedure for the codebook selection is the freedom in choosing the number R of codebook vectors, which is an important parameter for the efficiency of index-based matching, see Sect. 6.4. However, the learning approach depends on the availability of a suitable set of training data to yield a codebook that generalizes to an arbitrary dataset. To overcome this limitation, we directly construct a codebook without resorting to any training data by exploiting musical knowledge.

The main idea is that a CENS vector contains some explicit information regarding the harmonic content of the underlying audio fragment. For example, a CENS vector close to

$$\frac{1}{\sqrt{3}}(1,0,0,0,1,0,0,1,0,0,0,0) \in \mathcal{F} \tag{6.8}$$

indicates the harmonic relation to C major. Since our database mainly consists of harmonic Western music based on the equal-tempered scale, one can expect clusters of CENS vectors that correspond to single notes, intervals (combination of two notes), or certain triads (combination of three notes) such as major, minor, augmented or diminished chords. An analysis of 55 h of audio recordings chosen from our test database shows that for more than 95% of all extracted CENS vectors, more than 50% of a vector's energy is contained in at most four of the 12 components. In other words, for most CENS vectors the energy is concentrated in only a few components, which is the motivation for the following procedure. Let $\delta_1, \ldots, \delta_{12} \in \mathcal{F}$ denote the 12 unit vectors, which correspond to pure chroma that contain the entire energy within one component. Let $C_i \subset \mathcal{F}$, $i \in [1:12]$, denote the set of vectors with i dominant components:

$$C_i := \{\pi^{\mathcal{F}}(\delta_{k_1} + \ldots + \delta_{k_i}) \mid 1 \leq k_1 < \ldots < k_i \leq 12\}. \tag{6.9}$$

As basic building blocks for our codebook, we will consider all vectors with up to four dominant components resulting in

$$|C_1| + |C_2| + |C_3| + |C_4| = \sum_{i=1}^{4} \binom{12}{i} = 793 \tag{6.10}$$

vectors. Note that some of these vectors such as $\pi^{\mathcal{F}}(1,1,1,1,0,0,0,0,0,0,0,0)$ may correspond to harmonically rare combinations of notes. However, because of the temporal blurring of the CENS vectors, such combinations may occur particularly in passages such as harmonic transitions, chromatic scales, or ornamentations. Since the number of resulting vectors is well under control, we do not further thin out the set of building blocks.

To account for the harmonics contained in the acoustic realization of a single musical note, we introduce an additional note model. As an example, let us consider the concert pitch A4 of fundamental frequency $f = 440\,\mathrm{Hz}$, which corresponds to the MIDI pitch $p = 69$ and has the chroma index $(p \bmod 12) + 1 = 10$. Then, an acoustic realization of A4 basically consists of a weighted mix of harmonics having the frequencies $h \cdot f$ with $h \in \mathbb{N}$. Generally, let p denote the MIDI pitch of some musical note with fundamental frequency $f = 2^{\frac{p-69}{12}} \cdot 440\,\mathrm{Hz}$ and let $\gamma(p,h)$ denote the chroma index of the hth harmonic of p with respect to the equal-tempered scale. It is easy to show that

$$\gamma(p,h) = \big((p + \mathrm{round}(12 \log_2 h)) \bmod 12\big) + 1. \tag{6.11}$$

For example, considering the first 8 harmonics, the chroma indices $\gamma(p,h)$ for $h = 1,2,\ldots,8$ with respect to the musical note C4 (MIDI pitch $p = 60$) are $1,1,8,1,5,8,11,1$, which correspond to the chroma C, C, G, C, E, G, B, C. In our note model, we consider 8 harmonics, where the harmonic components are weighted according to a weight vector $(\alpha_1,\ldots,\alpha_8)$. In our experiments, we set $\alpha_h = 1$ for $h = 1,2,3$ and $\alpha_h := \frac{1}{h-2}$ for $h = 4,\ldots 8$. (Actually, as our experiments showed, the particular number of harmonics has only a marginal effect on the overall matching result. The reason we chose 8 harmonics is that these harmonics still can be reasonably assigned to pitches in the equal-tempered scale.) As a next step, the unit vector δ_k, which represents all pitches p of chroma index $k = (p \bmod 12) + 1$, is replaced by the linear combination

$$\tilde{\delta}_k := \pi^{\mathcal{F}}\left(\textstyle\sum_{h=1}^{8} \alpha_h \cdot \delta_{\gamma(p,h)}\right) \tag{6.12}$$

for a p with chroma index k. Note that $\tilde{\delta}_k$ can be obtained from $\tilde{\delta}_1$ by cyclically shifting the components of $\tilde{\delta}_1$ by $k-1$ positions.

Having incorporated harmonics, we now replace the sets C_i in (6.9) by modified versions

$$\tilde{C}_i := \{\pi^{\mathcal{F}}(\tilde{\delta}_{k_1} + \ldots + \tilde{\delta}_{k_i}) \mid 1 \le k_1 < \ldots < k_i \le 12\}. \tag{6.13}$$

The final codebook is then defined by $\mathcal{C}_{793} := \tilde{C}_1 \cup \tilde{C}_2 \cup \tilde{C}_3 \cup \tilde{C}_4 \subset \mathcal{F}$.

We conclude this section by noting that in our experiments both of the above approaches to codebook selection turn out to possess individual benefits as will be discussed in Sect. 6.5.

6.4 Index-Based Audio Matching

We now describe how our audio matching procedure of Sect. 6.1 can be supplemented by a much faster index-based matching procedure employing an inverted file index [42, 218]. The main idea is to classify (quantize) the database CENS features with respect to a selected codebook $\mathcal{C}_R$ and then to index the features according to their class labels $r \in [1:R]$. This results in an *inverted list* (also referred to as *inverted file*) for each label. Exact audio matching can then be performed efficiently by intersecting suitable inverted lists (Sect. 6.4.1). To account for musical variations, we soften the notion of exact matching by introducing various fault tolerance and ranking mechanisms (Sect. 6.4.2). To further improve the ranking, we finally combine the concepts of index-based and diagonal matching, which yields a ranked list of audio matches to the original query (Sect. 6.4.3).

6.4.1 Exact Matches

In this section, we introduce the concept of exact matches (with respect to a given quantization function) and show how these matches can be computed efficiently. Let $W = (w_0, \ldots, w_{M-1})$ be the database CENS sequence of the database D. We fix a codebook $\mathcal{C}_R = \{c_1, \ldots, c_R\} \subset \mathcal{F}$ and denote its associated quantization function by $\mathcal{Q}$, see (6.3). Furthermore, let $r_m := \mathcal{Q}[w_m]$, $m \in [0 : M-1]$, denote the quantized feature vector and

$$\mathcal{Q}[W] := (r_0, r_1, \ldots, r_{M-1}) \tag{6.14}$$

the quantized database sequence. For each class label $r \in [1 : R]$, we then create an inverted list $L(r)$, which consists of all index positions m such that the vector w_m is quantized to class r:

$$L(r) := \{m \in [0 : M-1] \mid r_m = r\}. \tag{6.15}$$

The *inverted file index* of the database, which can be computed in a preprocessing step, consists of all inverted lists $(L(r))_{r \in [1:R]}$.

To process a query Q, it is converted into a query CENS sequence $V = (v_0, v_1, \ldots, v_{N-1})$ (with respect to suitable parameters $\mathbf{w}$ and $\mathbf{d}$). Next, the query sequence is quantized to yield

$$\mathcal{Q}[V] := (s_0, s_1, \ldots, s_{N-1}), \tag{6.16}$$

where we set $s_n := \mathcal{Q}[v_n]$ for $n \in [0 : N-1]$. Then, an *exact match* is defined to be a tuple (k, N) such that $\mathcal{Q}[V]$ is a subsequence of consecutive feature vectors in $\mathcal{Q}[W]$ starting from index k. In other words, writing in this case $\mathcal{Q}[V] \sqsubset_k \mathcal{Q}[W]$, one obtains

$$\mathcal{Q}[V] \sqsubset_k \mathcal{Q}[W] \quad :\Leftrightarrow \quad \forall n \in [0 : N-1] : \; s_n = r_{k+n}. \tag{6.17}$$

The set of all match positions yielding exact matches of length N is given by

$$H(\mathcal{Q}[V]) := \{k \in [0 : M-1] | \; \mathcal{Q}[V] \sqsubset_k \mathcal{Q}[W]\}. \tag{6.18}$$

Now, this set can be calculated by intersecting suitably shifted inverted lists:

$$H(\mathcal{Q}[V]) = \bigcap_{n \in [0:N-1]} (L(s_n) - n), \tag{6.19}$$

where the difference of an inverted list and a natural number is defined elementwise. To prove (6.19), not that there is a match at position k if and only if each of the $n \in [0 : N-1]$ query features s_n occurs at position $k+n$ in the database sequence. Hence, $k+n \in L(s_n)$ must hold. The latter is equivalent to $k \in (L(s_n)-n)$ for all n, which proves the intersection formula. As each inverted list can be stored as a *sorted* list of integers, the intersections in (6.19) can be performed very efficiently by linearly processing sorted integer lists, see Clausen and Kurth [42].

6.4.2 Fault Tolerance Mechanisms

Even though the usage of CENS features and the quantization step introduce a high degree of robustness to variations in dynamics, instrumentation, and articulation, the requirement of exact matching on the quantized feature level is still too restrictive in order to obtain most of the relevant matches. In the following, we introduce several fault tolerance mechanisms to soften the strict requirements of exact matching.

Fuzzy Matches

The first general mechanism is known as *fuzzy search*, see Clausen and Kurth [42]. In the audio matching context, we use this mechanism to cope with the problem that arises when slight differences in the extracted CENS features lead to different assignments of codebook vectors in the quantization step (6.3). The idea is to admit at each position in the query sequence a whole set of possible, alternative class labels instead of a single one. To this end, we introduce the concept of a *fuzzy quantization function* $\mathcal{Q}_\lambda^\varrho$ for some $\lambda \in \mathbb{N}$ and $\varrho \in \mathbb{R}_{>0}$. Instead of using only the nearest codebook vector $\mathcal{Q}[v]$ to quantizing a CENS feature $v \in \mathcal{F}$, we now assign an entire set $\mathcal{Q}_\lambda^\varrho[v] \subset \mathcal{F}$ of alternative codebook vectors. By definition, this set contains at least the nearest neighbor $\mathcal{Q}[v]$ and all of those additional $\lambda - 1$ next nearest neighbors that have an angular distance to v below the threshold ϱ. Here, the threshold condition is introduced to avoid outliers in the assignment. In our experiments, $3 \leq \lambda \leq 7$ and $\varrho = 0.15\pi$ turn out to be reasonable parameter values. Now, the query sequence V is quantized with respect to $\mathcal{Q}_\lambda^\varrho$ to yield a sequence

$$\mathcal{Q}_\lambda^\varrho[V] := (S_0, S_1, \ldots, S_{N-1}), \tag{6.20}$$

where we set $S_n := \mathcal{Q}_\lambda^\varrho[v_n]$ for $n \in [0 : N-1]$. Extending the definition in (6.17), a *fuzzy match* is a tuple (k, N) such that $\mathcal{Q}_\lambda^\varrho[V] \sqsubset_k \mathcal{Q}[W]$, where

$$\mathcal{Q}_\lambda^\varrho[V] \sqsubset_k \mathcal{Q}[W] \quad :\Leftrightarrow \quad \forall n \in [0 : N-1] : \; S_n \ni r_{k+n}. \tag{6.21}$$

The set of all match positions yielding fuzzy matches of length N is given by

$$H(\mathcal{Q}_\lambda^\varrho[V]) := \{k \in [0 : M-1] |\; \mathcal{Q}_\lambda^\varrho[V] \sqsubset_k \mathcal{Q}[W]\}. \tag{6.22}$$

Finally, defining the lists $L(S_n) := \bigcup_{s \in S_n} L(s)$, one obtains

$$H(\mathcal{Q}_\lambda^\varrho[V]) = \bigcap_{n \in [0:N-1]} (L(S_n) - n), \tag{6.23}$$

thus yielding an efficient algorithm to compute all fuzzy matches from the inverted file index. Note that if $S_n = \{s_n\}$ for all $n \in [0 : N-1]$ one obtains the case of exact matches.

Fuzzy Matches with Mismatches

As a second general fault tolerance mechanism, we adopt the concept of *mismatch search*, see Clausen and Kurth [42]. Here, the idea is to admit a certain number of mismatches when matching the (fuzzy) query sequence with a database subsequence. More precisely, given the fuzzy query $\mathcal{Q}_\lambda^\varrho[V] = (S_0, S_1, \ldots, S_{N-1})$, we define a function $\mu : [0 : M-1] \to [0 : N]$ that counts the following multiplicities:

$$\mu(m) := \big| \{n \in [0 : N-1] \mid m \in (L(S_n) - n)\} \big|. \tag{6.24}$$

Obviously, $\mu(k) = N$ if and only if $k \in H(\mathcal{Q}_\lambda^\varrho[V])$, cf. (6.23). Generalizing the concept of a fuzzy match, we call a pair (k, N) a *fuzzy match with ν mismatches* for some mismatch parameter $\nu \in [0 : N]$ if $\mu(k) = N - \nu$. In this case, the condition $r_{k+n} \in S_n$ for $n \in [0 : N-1]$ as given in (6.21) is violated exactly at ν positions. The mismatch function can be computed efficiently either by dynamic programming [42] or by hashing techniques [219].

Multiple Queries

The third mechanism, which accounts for global tempo variations, proceeds in analogy to the one described in Sect. 6.1. Instead of using only a single query feature sequence $V = \mathrm{CENS}_{10}^{41}[Q]$, we use eight different query sequences $\mathrm{CENS}_{\mathbf{d}_j}^{\mathbf{w}_j}$, $1 \leq j \leq 8$, which are then processed separately. Similarly, one can account for global variations in the musical key by processing the query sequences obtained by cyclically shifting the components of the CENS vectors.

6.4.3 Retrieval Scenario and Ranking

Our index-based audio matching procedure basically consists of two stages, which are illustrated in Fig. 6.9. In the *preprocessing stage*, the music database D is converted into the feature sequence $\mathrm{CENS}_{10}^{41}[D]$, which is then quantized with respect to some fixed codebook $\mathcal{C}_R$ and indexed via inverted lists, see Fig. 6.9a. The resulting inverted file index is independent of the query, the fault tolerance settings, and the ranking strategies described at the end of this section. In the *query and retrieval stage*, the user supplies a query Q in

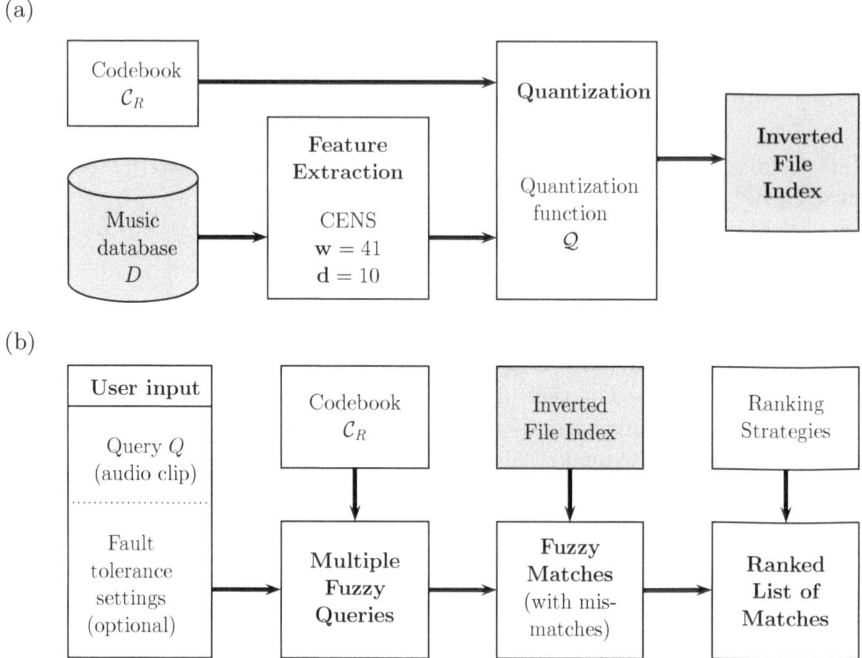

Fig. 6.9. Overview of our index-based audio matching procedure. **(a)** The preprocessing stage. **(b)** The query and retrieval stage

form of some audio clip, which typically has a duration between 10 and 30 s. Optionally, the user may also specify certain fault tolerance parameters that affect the fuzzyness of the query, the number of admissible mismatches, or the tolerance to global variations in tempo and musical key. Based on these parameters and the codebook $\mathcal{C}_R$, the query Q is transformed into multiple fuzzy queries, which are then processed by means of the inverted file index. Finally, after some suitable postprocessing, the system returns a ranked list of audio matches, see Fig. 6.9b.

In the remaining part of this section, we describe two different ranking strategies used in the postprocessing step. The first strategy ranks the matches according to the number of mismatches, which can be derived from the multiplicity function μ defined in (6.24). Let (k, N) denote a retrieved match with multiplicity $\mu(k) \in [0 : N]$. Note that because of the usage of different query sequences $\mathrm{CENS}^{\mathbf{w}_j}_{\mathbf{d}_j}$ (having a length of N_j, see Sect. 6.1.3), the length of the retrieved matches may vary. To compensate for the respective length, we define the ranking value $\Theta(k, N)$ of the match (k, N) to be the quotient

$$\Theta(k, N) := \frac{\mu(k)}{N} \in [0, 1]. \tag{6.25}$$

In case of an exact match or a fuzzy match without mismatches one obtains $\Theta(k, N) = 1$. The quality of matches decreases with a decreasing Θ-value. To avoid an excessive number of matches in the retrieval process, one can either introduce a fixed upper bound for the number of matches to be retrieved or one can a priori restrict the matches by introducing a lower bound for the ranking value (which can also be exploited to further speed up the computation of μ).

The ranking function Θ is relatively coarse and assumes only a small number of different values in particular for short queries. Furthermore, a codebook $\mathcal{C}_R$ with a relatively small number of elements as well as relatively large parameters λ and ϱ in the fuzzy quantization function $\mathcal{Q}_\lambda^\varrho$ may result in a large number of false positives and unexpected matches with a high or even maximal Θ-rank. To improve the ranking as well as to eliminate large overlaps in the retrieved matches, we further postprocess the top matches by the diagonal matching procedure described in Sect. 6.1. For each of those matches, we consider the corresponding subsequence in the database sequence W. To obtain a more flexible diagonal matching, each such subsequence is extended by a small number of vectors to the left and the right. Finally, diagonal matching is performed on each of the extended subsequences. The resulting matches are ranked according to the values assumed by the joint distance function Δ_{min}, see Sect. 6.1.3. To avoid multiple matches at very close positions within the same song, a match candidate is rejected if it overlaps a previously found (and hence higher ranking) match by more than 30%.

In a sense, the index-based matching can be seen as an efficient way to significantly reduce the search space (database) to a small fraction (top matches obtained by index-based audio matching), while retaining a superset of the desired matches. Then the more cost-intensive diagonal matching is performed only on the small database fraction.

6.5 Experimental Results for Index-Based Audio Matching

To evaluate our index-based audio matching procedure, we have conducted various experiments using the test database DB112 (Sect. 6.2). First, we qualitatively compare the results obtained from the pure diagonal matching procedure and the proposed index-based method (Sect. 6.5.1). A special focus is put on comparing the audio matching results with respect to various codebooks constructed by means of the learning-based and knowledge-based approaches (Sect. 6.5.2). Finally, we report on the speed-up factors obtained by our index-based matching approach (Sect. 6.5.3).

6.5.1 Diagonal vs. Index-Based Matching

We now summarize some of the results of our systematic experiments, which have been conducted for both the diagonal and the index-based audio matching procedure. In what follows, the index-based matching approach is always postprocessed using both of the ranking strategies introduced in Sect. 6.4.3. To qualitatively evaluate the proposed audio matching methods, we created a ground truth dataset of relevant query results for 36 queries (Q-GT36), which were manually selected from DB112. All of those queries constitute popular excerpts of classical pieces having durations between 10 and 40 s. Furthermore, for our quantitative tests, we generated three sets (Q-R10, Q-R15, Q-R20) each consisting of roughly 1 000 random queries having durations of 10, 15, and 20 s, respectively.

In a first experiment, we investigated the quality of the retrieval results for various combinations of audio matching parameters using queries from Q-GT36. A main goal of this experiment was to evaluate suitable parameter settings for the subsequent tests. In particular, we tested three different codebooks $\mathcal{C}_R$ with $R \in \{50, 100, 200\}$ obtained from the unsupervised learning procedure, five different parameters $\lambda \in \{3, 4, 5, 6, 7\}$ with a threshold of $\varrho = 0.15\pi$ used in the fuzzy quantization function $\mathcal{Q}_\lambda^\varrho$, as well as four different lower bounds $\theta \in \{0.2, 0.3, 0.4, 0.5\}$ for the Θ-ranking value.

As a first result, it turns out that the codebook $\mathcal{C}_{200}$ outperforms the smaller codebooks $\mathcal{C}_{50}$ and $\mathcal{C}_{100}$. A manual inspection of the retrieval results indicates that the number of vectors in the smaller codebooks is too low to sufficiently discriminate true and false positive matches. In particular, when using the codebook $\mathcal{C}_{100}$ the number of match candidates with a Θ-value greater than 0.3 (less than 70% mismatches) generally exceeds the corresponding number for $\mathcal{C}_{200}$ by almost one order of magnitude. As a further consequence of smaller codebooks, the inverted lists get longer which in turn results in increased running times ranging from 10–20% in the latter case (with fixed values of λ and θ). To find suitable parameter settings for λ and θ, we started with $\mathcal{C}_{200}$ and the query set Q-GT36, gradually increasing λ until the overall number of false negatives did not further increase. To compensate for the simultaneously increasing number of match candidates, the threshold θ of the required minimum rank was suitably adapted. In this process, the parameter combination $\lambda = 5$ (at most five alternatives in each fuzzy set) and $\theta = 0.4$ (at least 40% of the features have to match) turned out to yield the best retrieval results.

In a second experiment, we qualitatively compared the index-based matching approach ($R = 200$, $\lambda = 5$, $\theta = 0.4$) and the pure diagonal matching approach based on the queries from Q-GT36. Using the manually annotated ground truth, we obtain precision-recall (PR) values from the top 10 matches for each of the two methods. Figure 6.10 shows the corresponding PR-diagram obtained from averaging the PR values over all 36 queries. While diagonal matching appears to be slightly better in the range of the top 4–5 matches,

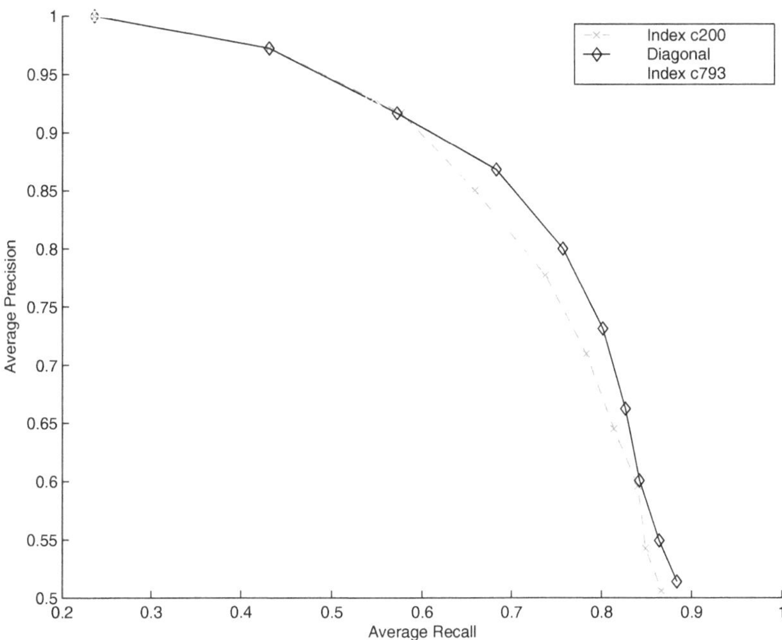

Fig. 6.10. Average precision-recall (PR) values on the query set Q-GT36 for the top ten retrieval results. *Diamond curve*: PR diagram for pure diagonal matching. *Cross curve*: PR diagram for index-based matching (learning-based codebook $\mathcal{C}_{200}$, $\lambda = 5$, $\theta = 0.4$). *Circle curve*: PR diagram for index-based matching (knowledge-based codebook $\mathcal{C}_{793}$, $\lambda = 7$, $\theta = 0.3$)

the PR values for the top 3 matches and the top 7–10 matches almost coincide. Note that because there are only about six correct matches on average for a query in Q-GT36, there is a certain bias in the PR-curves, which consider the top 10 matches. However, this bias does not affect the comparability of the curves relative to each other.

In a third experiment, we tested how the two audio matching procedures perform in the *audio identification* scenario, where the objective is to identify only the query within the audio database (rather than identifying all musically similar audio clips). To this end, we used the 3 000 queries from Q-R10, Q-R15, and Q-R20, which were randomly drawn from DB112, as an input. For both matching procedures, Table 6.5 shows the average recall values for the top match, the top two, and the top three matches, respectively. Considering only the top match, the percentage of correct identifications increases with the query length. Considering the top two matches, both procedures yield almost optimal identification results. Here, note that the identification task is not a trivial one: due to shifts in the analysis windows used in computation of the

Table 6.5. Comparison of audio identification capabilities of diagonal and index-based audio matching

Query set	Matching procedure	Average recall		
		@1	@2	@3
Q-R10	Diagonal	0.97	1	1
	Index	0.95	1	1
Q-R15	Diagonal	0.972	0.998	0.999
	Index	0.957	0.996	0.999
Q-R20	Diagonal	0.972	0.999	0.999
	Index	0.965	0.999	1

The columns indicate the average recall values for the top match, the top two, and the top three matches, respectively.

Table 6.6. Comparison of audio identification capabilities of diagonal and index-based audio matching for distorted query signals (test set Q-R15)

Distortion	Matching procedure	Average recall		
		@1	@2	@3
Lame MP3 compression @65 kbps	Diagonal	0.968	0.998	0.999
	Index	0.956	0.999	1
Time stretching (100–150% tempo)	Diagonal	0.962	0.998	0.999
	Index	0.944	0.998	1
Strong 3D echo chamber effect	Diagonal	0.925	0.979	0.985
	Index	0.881	0.954	0.966

chroma and CENS features (e.g., using CENS_{10}^{41}-features which have a temporal resolution of 1 Hz, these shifts can be up to half a second, see Sect. 3.3), the CENS feature sequence V of the query generally does not coincide with the corresponding CENS feature subsequence within the database sequence W. Most of the confusion in the audio identification are due to more or less identical repetitions of the query within the respective music piece.

Further experiments show that the audio identification still works with high precision even in the case that the queries are severely distorted. Table 6.6 shows the corresponding average recall values for three different types of transformations performed on all of the queries from Q-R15. The first transformation is lossy audio compression using the Lame MP3 audio encoder at the low bitrate of 65 kbps. As the compression follows perceptual criteria, the signals' harmonic content is retained, resulting in recall values that almost correspond to those of the original signals. More interesting are the relatively high recall values for the second transformation, being a nonlinear temporal deformation (time stretch) where the tempo of each query is linearly increased from the original tempo to the 1.5-fold tempo. To simulate an acoustic signal transmission, the third transformation consists of applying a strong 3D echo-effect

(*Ambient Metal Room* preset from the CoolEdit 2000 softwares' delay effects) to the query signals. Because of the echo effect, the harmonies are smeared over an interval of a few seconds, which significantly affects the signal's energy distribution in the chroma bands. However, even in this case the average recall for the top three matches is still high.

6.5.2 Comparison of Codebooks

In Sect. 6.3 we have introduced both a learning-based and a knowledge-based strategy for codebook selection. Various experiments have been conducted to investigate the effect of the respective codebook on the retrieval process. Similar to our second experiment, we have qualitatively compared the index-based matching approach for the codebook $\mathcal{C}_{200}$ (learning-based) with $\lambda = 5$ and $\theta = 0.4$ and the codebook $\mathcal{C}_{793}$ (knowledge-based) with $\lambda = 7$ and $\theta = 0.3$. Here, the fuzzy and ranking parameters were in each case chosen to yield optimum retrieval results within the above experimental setting. Note that due to the larger number of codebook vectors in $\mathcal{C}_{793}$, more relaxed parameters λ and θ are chosen. Figure 6.10 shows the average PR diagrams over the query set Q-GT36 using the two different codebooks. While the retrieval results are almost eqivalent, the manually constructed codebook yields slightly better results for the top three matches. In this case the index-based matching even outperforms diagonal matching.

We furthermore directly compared the retrieval results of the two index-based approaches using the retrieval results of pure diagonal matching as a reference. To this end, for increasing $K \in \mathbb{N}$, we compared each of the sets consisting of the top K matches obtained from the two index-based matching procedures with the corresponding set obtained from diagonal matching and determined the percentage of coinciding matches. Figure 6.11 shows the two resulting curves (coincidence in percentage over K) for the two index-based approaches based on $\mathcal{C}_{200}$ and $\mathcal{C}_{793}$. Both approaches yield comparable numbers of coincidences – for the top seven matches the matching with the manually constructed codebook $\mathcal{C}_{793}$ correlates slightly higher with pure diagonal matching. Furthermore, a manual inspection showed that the decrease in coincidences is mainly due to noncoinciding false positives. The overall high percentage of coincidences indicate that both of the index-based approaches yield nearly the same top matches as pure diagonal matching.

6.5.3 Running Times

In our final experiments, we investigate the speed-up in computation time obtained by the index-based matching in comparison to the diagonal matching procedure. All of the proposed algorithms were implemented in the MATLAB environment and the tests were run on an Athlon XP 2800+ PC with 1 GB of main memory under MS Windows 2000. Before the actual experiments, the audio clips contained in the query sets Q-R10, Q-R15, and Q-R20 were

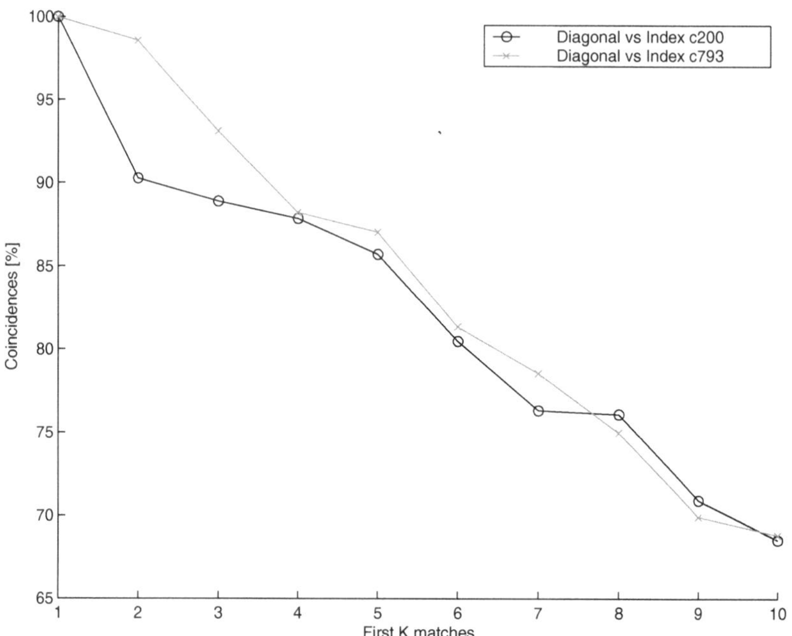

Fig. 6.11. Average coincidence of top K matches in percent: *Circle curve:* diagonal matching compared to index-based matching using the learning-based codebook $\mathcal{C}_{200}$ ($\lambda = 5$, $\theta = 0.4$). *Cross curve:* diagonal matching compared to index-based matching using the knowledge-based codebook c_{793} ($\lambda = 7$, $\theta = 0.3$)

transformed into sequences of chroma features at a feature resolution of 10 Hz (see Sect. 3.3) and stored. The running time for computing the chroma features linearly depend on the query length and will not be of interest for the following considerations. Table 6.7 shows the average running times (in seconds) required for audio matching performed on each of the three query sets. The running times correspond to the actual matching procedures, excluding the times for loading the database feature sequence (for diagonal matching) and for loading the inverted file index (for index-based matching), which have to be performed only once and are independent of the actual query. The running times for converting the chroma features into the eight different CENS sequences (for tempo change simulation, see Sect. 6.1.3) are included as this computation is part of the query process.

The running time of the pure diagonal matching procedure (first row of Table 6.7) is dominated by the computation of the distance function $\Delta_{\min}$ – the time to determine a desired number of matches ranked according to the $\Delta_{\min}$-value is then negligible. In our experiments, we always generated a ranked list of the top 20 matches in both the index-based and the pure diag-

Table 6.7. Average running times and speed-up factors of diagonal and index-based audio matching ($\mathcal{C}_{200}$, $\lambda = 5$, $\theta = 0.4$) for the query sets Q-R10, Q-R15, and Q-R20

	Average running time (s)		
	Q-R10	Q-R15	Q-R20
Diagonal matching	11.30	11.92	12.54
Index-based matching	0.57	0.67	0.8
Speed-up factor	19.66	17.79	15.67

onal matching approach. In the index-based approach we used the codebook $\mathcal{C}_{200}$ with $\lambda = 5$ and $\theta = 0.4$. Because of the practical considerations, index-based query processing is implemented as follows. For each of the eight different query sequences $\mathrm{CENS}^{\mathbf{w}_j}_{\mathbf{d}_j}$, $1 \leq j \leq 8$, we determine lists of the top 40 matches satisfying $\theta \geq 0.4$ (note that a list hence may contain less than 40 matches). Subsequently, diagonal matching-based postprocessing determines the best 20 matches from the resulting candidate set of $8 \cdot 40 = 320$ matches. It turns out that the running time (second row of Table 6.7) is dominated by the intersection and union operations on the inverted lists, whereas the running time of the postprocessing step (ranking and diagonal matching on top matches) is negligible.

The last row of Table 6.7 shows the substantial speed-up of the index-based approach as compared to pure diagonal matching, which ranges from factors of about 15–20 depending on the query length. We finally note that larger codebooks in principle lead to smaller inverted lists, typically resulting in faster query times. On the other hand, larger threshold and fuzzy parameters θ and λ increase the processing time. As both effects compensate each other in our parameter settings, the processing times for using the codebooks $\mathcal{C}_{200}$ and $\mathcal{C}_{793}$ in index-based matching are roughly the same.

6.6 Further Notes

In this chapter, we presented two audio matching procedures facilitating the retrieval of musically similar audio clips even in the presence of significant variations in articulation, tempo, and instrumentation. To the best of our knowledge, this kind of matching scenario has first been considered in Müller et al. [140]. The index-based procedure, which is described in Kurth and Müller [113], yields speed-up factors of 15–20 concerning the retrieval time as compared to the pure diagonal matching approach while providing competitive matching results concerning precision and recall. The matching procedure (implemented in MATLAB 6.5 running on a standard PC) requires less than a second to process a 20 s query with respect to a 100-h audio database. Ongoing work is concerned with further extending the index-based matching approach to handle even larger collections of several million pieces of music, e.g., by incorporating hashing techniques [92,211,219]. In a hierarchical context, our index-based audio matching algorithm provides a practical and efficient component for medium sized datasets while providing high-quality audio matches. As a further important contribution of this chapter, we discussed two strategies – a learning-based approach using a modified LBG algorithm and a knowledge-based approach – in order to reduce the set of the chroma-based CENS features to a small number of codebook vectors. The reduction of chroma-based features to a small codebook without loosing semantic expressiveness seems also very attractive in view of speeding up other MIR applications such as audio synchronization (Chap. 5) or audio structure analysis (Chap. 7). As an application, we may employ audio matching to substantially accelerate music synchronization. Here, the idea is to identify salient audio matches, which can then be used as anchor matches as suggested in Sect. 5.3.4. By considering other audio features related to rhythm and timbre, further ongoing work addresses the problem of audio matching for a broader class of music that does not necessarily exhibit a characteristic harmonic progression.

The problem of audio matching can be regarded as a generalization of the problem of *audio identification*, which is also sometimes referred to as *audio fingerprinting*. In the scenario of audio identification, the retrieval task is to locate a query clip within the original audio recording, which is assumed to be part of the audio database, and to deliver the title, composer, or artists of the underlying piece. Here, the notion of similarity is close to the identity. Recent identification algorithms show some degree of robustness towards noise, MP3 compression artifacts, or slight temporal distortions. Opposed to semantically more advanced audio matching problems, the problem of audio identification has been studied extensively and can be regarded as largely solved even for large scale music collections, see, e.g., [3, 28, 92, 112, 194, 211].

Related but somewhat different to the audio matching scenario is the recently studied problem of *cover song identification*. Here, the goal is to identify different versions of the same piece of music within a database, see, e.g., [33,59,79]. A cover song may differ from the original song with respect to the instrumentation, it may represent a different genre, or it may be a remix with a different musical structure. Further softening the notion of similarity, one arrives at problems known as *genre classification* and automatic *music recommendation*, see, e.g., [158,203]. The objective of these problems is to determine a single similarity value that expresses the relatedness of *entire* pieces or songs, which can then be used for music classification and recommendation tasks. On the contrary, the objective of our audio matching problem is to identify musically similar *fragments* of arbitrary lengths (typically 10–30 s of audio) within the database, which, for example, can then be used for efficient navigation within a single CD recording (intra-document navigation) or between different performances (inter-document navigation).

Finally, we give some references to techniques that were used in our audio matching procedure. The reduction of high dimensional features to small codebooks is used for a wide range of applications in the domain of speech and audio processing. For example, vector quantization of audio features based on K-means clustering has been used for audio identification [3] and audio analysis [32]. Inverted file indexing is a standard technique in text retrieval [218], which has also been applied in the music context, e.g, for music identification [42]. In our retrieval technique, the temporal order in which the extracted features occur plays a fundamental role in identifying related audio clips. The importance of using contextual information by considering feature sequences (rather than disregarding the temporal context of the features) in determining musical similarity has also been emphasized in Casey and Slaney [32] and Müller and Kurth [137].

7

Audio Structure Analysis

The alignments and crosslinks obtained from audio synchronization (Chap. 5) and audio matching (Chap. 6) can be used to conveniently switch between different versions of a piece of music (*interdocument navigation*). We will now address the problem of *audio structure analysis*, which lays the basis for *intradocument navigation*. One major goal of the structural analysis of an audio recording is to automatically extract the repetitive structure or, more generally, the musical form of the underlying piece of music. Recent approaches such as [14,37,46,49,81,127,129,139,161] work well for music where the repetitions largely agree with respect to instrumentation and tempo as it is typically the case for popular music. For other classes of music including Western classical music, however, musically similar audio segments may exhibit significant variations in parameters such as dynamics, timbre, execution of note groups, musical key, articulation, and tempo progression. In this chapter, we propose robust and efficient algorithms for structure analysis that identifies musically similar segments. To obtain a flexible and robust algorithm, the idea is to simultaneously account for possible variations at various stages and levels. At the feature level, we use coarse chroma-based audio features that absorb microvariations. To cope with local variations, we design an advanced cost measure by integrating contextual information (Sect. 7.2). Finally, we describe a new strategy for structure extraction that can cope with more global variations (Sects. 7.3 and 7.4). Our experimental results with classical and popular music show that our algorithm performs successfully even in the presence of significant musical variations (Sect. 7.5). In Sect. 7.1, we start by summarizing a general strategy for audio structure analysis and introduce some notation that is used throughout this chapter. Related work and future research directions will be discussed in Sect. 7.6. In this chapter, we closely follow [139]. The enhancement strategy of self-similarity matrices by introducing a contextual local cost measure has first been described in [137].

7.1 General Strategy and Notation

To extract the repetitive structure from audio signals, most of the existing approaches proceed in four steps. In the first step, a suitable high-level representation of the audio signal is computed. To this end, the audio signal is transformed into a sequence $V := (v_1, v_2, \ldots, v_N)$ of feature vectors $v_n \in \mathcal{F}$, $1 \leq n \leq N$. Here, $\mathcal{F}$ denotes a suitable feature space, e.g., a space of spectral, MFCC, or chroma vectors. On the basis of a suitable cost measure $c : \mathcal{F} \times \mathcal{F} \to \mathbb{R}_{\geq 0}$, one then computes an N-square *self-similarity matrix*[1] $\mathcal{S}$ defined by $\mathcal{S}(n, m) := c(v_n, v_m)$, effectively comparing all feature vectors v_n and v_m for $1 \leq n, m \leq N$ in a pairwise fashion. In the third step, the path structure is extracted from the resulting self-similarity matrix. Here, the underlying principle is that similar segments in the audio signal are revealed as paths along diagonals in the corresponding self-similarity matrix, where each such path corresponds to a pair of similar segments. Finally, in the fourth step, the global repetitive structure is derived from the information about pairs of similar segments using suitable clustering techniques.

To illustrate this approach, we consider two examples, which also serve as running examples throughout this chapter. The first example, for short referred to as *Brahms example*, consists of an Ormandy interpretation of the Hungarian Dance No. 5 by Johannes Brahms. This piece has the musical form $A_1A_2B_1B_2CA_3B_3B_4D$ consisting of three repeating A-parts A_1, A_2, and A_3, four repeating B-parts B_1, B_2, B_3, and B_4, as well as a C- and a D-part. Generally, we will denote musical parts of a piece of music by capital letters such as X, where all repetitions of X are enumerated as X_1, X_2, and so on. In the following, we distinguish between a *piece of music* (in an abstract sense) and a particular *audio recording* (a concrete interpretation) of the piece. Here, the term *part* is used in the context of the abstract music domain, whereas the term *segment* is used for the audio domain.

The self-similarity matrix of the Brahms recording (with respect to suitable audio features and a particular cost measure) is shown in Fig. 7.1. Here, the repetitions implied by the musical form are reflected by the path structure of the matrix. For example, the path starting at $(1, 22)$ and ending at $(22, 42)$ (measured in seconds) indicates that the audio segment represented by the time interval $[1 : 22]$ is similar to the segment $[22 : 42]$. Manual inspection reveals that the segment $[1 : 22]$ corresponds to part A_1, whereas $[22 : 42]$ corresponds to A_2. Furthermore, the curved path starting at (42,69) and ending at (69,89) indicates that the segment $[42 : 69]$ (corresponding to B_1) is similar to $[69 : 89]$ (corresponding to B_2). Note that in the Ormandy interpretation, the B_2-part is played much faster than the B_1-part. This fact is also revealed by the gradient of the path, which encodes the relative tempo difference between the two segments.

[1] Since c is assumed to be a local cost measure rather than a similarity measure the resulting matrix is strictly speaking a *cost matrix*. Nevertheless, we use the term *self-similarity matrix* according to the standard term used in previous work.

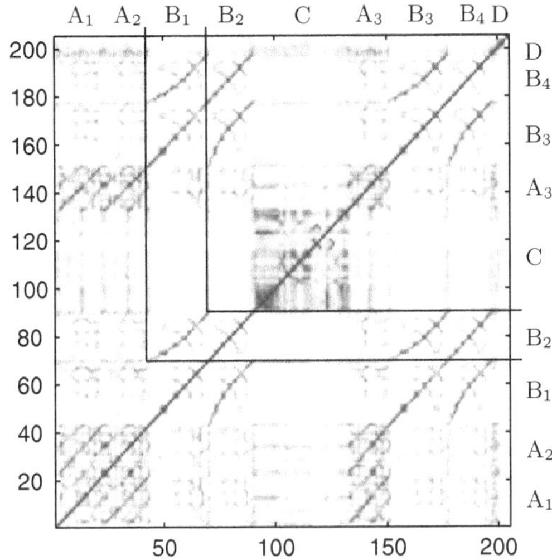

Fig. 7.1. Self-similarity matrix $\mathcal{S}[41, 10]$ of an Ormandy interpretation of Brahms' Hungarian Dance No. 5. Here, *dark colors* correspond to low values (high similarity) and *light colors* correspond to high values (low similarity). The musical form $A_1A_2B_1B_2CA_3B_3B_4D$ is reflected by the path structure. For example, the curved path marked by the *horizontal* and *vertical lines* indicates the similarity between the segments corresponding to B_1 and B_2

As a second example, in the following referred to as *Shostakovich example*, we consider a Chailly interpretation of the second Waltz of the Jazz Suite No. 2 by Dimitri Shostakovich. This piece has the musical form $A_1A_2BC_1C_2A_3A_4D$, where the theme, represented by the A-part, appears four times. However, there are significant variations in the four A-parts concerning instrumentation, articulation, as well as dynamics. For example, in A_1 the theme is played by a clarinet, in A_2 by strings, in A_3 by a trombone, and in A_4 by the full orchestra. As is illustrated by Fig. 7.2, these variations result in a fragmented path structure of low quality, making it hard to identify the musically similar segments $[4 : 40]$, $[43 : 78]$, $[145 : 179]$, and $[182 : 217]$ corresponding to A_1, A_2, A_3, and A_4, respectively.

In the subsequent sections, we introduce a new structure analysis algorithm following the four-stage strategy as described earlier. To account for musical variations, we incorporate invariance and robustness at all four stages simultaneously. The main ingredients of our procedure can be summarized as follows.

1. *Audio Features.* As audio features, we use the chroma-based CENS features as introduced in Sect. 3.3. Such features not only absorb variations in parameters such as dynamics, timbre, articulation, execution of note groups, and temporal microdeviations, but can also be efficiently processed in the subsequent steps due to their low resolution.

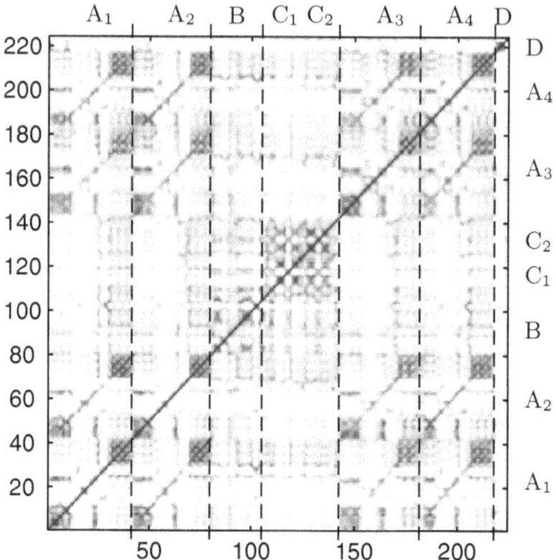

Fig. 7.2. Self-similarity matrix $\mathcal{S}[41,10]$ of an Chailly interpretation of Shostakovich's Waltz 2, Jazz Suite No. 2, having the musical form $A_1A_2BC_1C_2A_3A_4D$. Because of significant variations in the audio recording, the path structure is fragmented and of low quality. See also Fig. 7.4

2. *Local Cost Measure.* To enhance the path structure of a self-similarity matrix, we incorporate contextual information at various tempo levels into the local cost measure (Sect. 7.2). This accounts for local temporal variations and significantly smooths the path structures.
3. *Path Extraction.* On the basis of the enhanced matrix, we suggest a robust and efficient path extraction procedure using a greedy strategy (Sect. 7.3). This step takes care of relative differences in the tempo progression between musically similar segments.
4. *Global Structure.* Each path encodes a pair of musically similar segments. To determine the global repetitive structure, we describe a one-step transitivity clustering procedure, which balances out the inconsistencies introduced by inaccurate and incorrect path extractions (Sect. 7.4).

We have conducted experiments on a wide range of Western classical music including complex orchestral and vocal pieces (Sect. 7.5). The experimental results show that our method successfully identifies the repetitive structure – often corresponding to the musical form of the underlying piece – even in the presence of significant variations as in the case of the Brahms and Shostakovich examples. Our MATLAB implementation performs the structure analysis task within a couple of minutes even for long and versatile audio recordings such as Ravel's Bolero, which has a duration of more than 15 min and possesses a rich path structure.

7.2 Enhancing Similarity Matrices

In our structure analysis scenario, we consider audio segments as similar if they represent the same musical content regardless of the specific articulation and instrumentation. Therefore, as in the audio synchronization and audio matching scenario, we use the chroma-based $\mathrm{CENS}^{\mathbf{w}}_{\mathbf{d}}$-features (Sect. 6.3). These features are elements of the set $\mathcal{F} = \{v \in [0,1]^{12} \mid \|v\|_2 = 1\}$, see (3.4). Recall that CENS features strongly correlate to the short-time harmonic content of the underlying audio signal and constitute a good tradeoff between the two mutually conflicting goals of having robustness to admissible variations on the one hand and accuracy with respect to the relevant characteristics on the other hand. Figure 7.3 shows the CENS feature sequences for the audio segments corresponding to the parts B_1 and B_2 of our Brahms example.

We now introduce a strategy for enhancing the path structure of a self-similarity matrix by designing a suitable local cost measure. To this end, we proceed in three steps. As a starting point, let $c : \mathcal{F} \times \mathcal{F} \to [0,1]$ be the cost measure defined by

$$c(v,w) := 1 - \langle v,w \rangle \tag{7.1}$$

for $\mathrm{CENS}^{\mathbf{w}}_{\mathbf{d}}$-vectors $v, w \in \mathcal{F}$. Since v and w are normalized, the inner product $\langle v,w \rangle$ coincides with the cosine of the angle between v and w. For short, the resulting self-similarity matrix will also be denoted by $\mathcal{S}[\mathbf{w},\mathbf{d}]$ or simply by $\mathcal{S}$ if $\mathbf{w}$ and $\mathbf{d}$ are clear from the context.

To further enhance the path structure of $\mathcal{S}[\mathbf{w},\mathbf{d}]$, we incorporate contextual information into the local cost measure. A similar approach has been suggested in [14] or [127], where the self-similarity matrix is filtered along

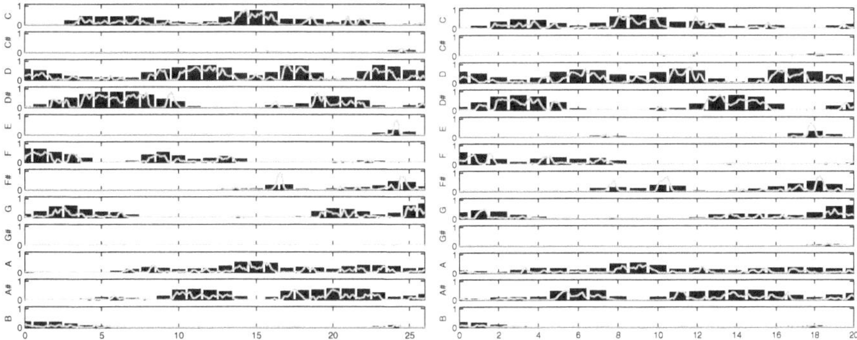

Fig. 7.3. Local chroma energy distributions (*light curves*, 10 feature vectors per second) and CENS^{41}_{10}-feature sequence (*dark bars*, 1 feature vector per second) of the segment [42 : 69] (*left*, corresponding to B_1) and segment [69 : 89] (*right*, corresponding to B_2) of the Brahms example (Fig. 7.1). Note that even though the relative tempo progression in the parts B_1 and B_2 is different, the harmonic progression at the low resolution level of the CENS features is very similar

diagonals assuming constant tempo. We will show later in this section how to remove this assumption by, intuitively speaking, filtering along various directions simultaneously, where each of the directions corresponds to a different local tempo. In [161], matrix enhancement is achieved by using HMM-based "dynamic" features, which model the temporal evolution of the spectral shape over a fixed time duration. For the moment, we also assume constant tempo and then, in a second step, describe how to get rid of this assumption. Let $L \in \mathbb{N}$ be a length parameter. We define the *contextual cost measure* c_L by

$$c_L(n,m) := \frac{1}{L} \sum_{\ell=0}^{L-1} c(v_{n+\ell}, v_{m+\ell}), \tag{7.2}$$

where $1 \leq n,m \leq N-L+1$. By suitably extending the CENS sequence $(v_1,\ldots,v_N)$, e.g., via zero-padding, one may extend the definition to $1 \leq n,m \leq N$. Then, the *contextual similarity matrix* $\mathcal{S}_L$ is defined by $\mathcal{S}_L(n,m) := c_L(n,m)$. In this matrix, a value $c_L(n,m) \in [0,1]$ close to zero implies that the entire sequence $(v_n,\ldots,v_{n+L-1})$ is similar to the sequence $(v_m,\ldots,v_{m+L-1})$, resulting in an enhancement of the diagonal path structure in the similarity matrix. This is also illustrated by our Shostakovich example (Fig. 7.4), where the diagonal path structure of $\mathcal{S}_{10}[41,10]$ is much smoother and clearer than the one of $\mathcal{S}_{10}[41,10]$. This matrix enhancement not only simplifies the extraction of structural information but also allows for further decreasing the

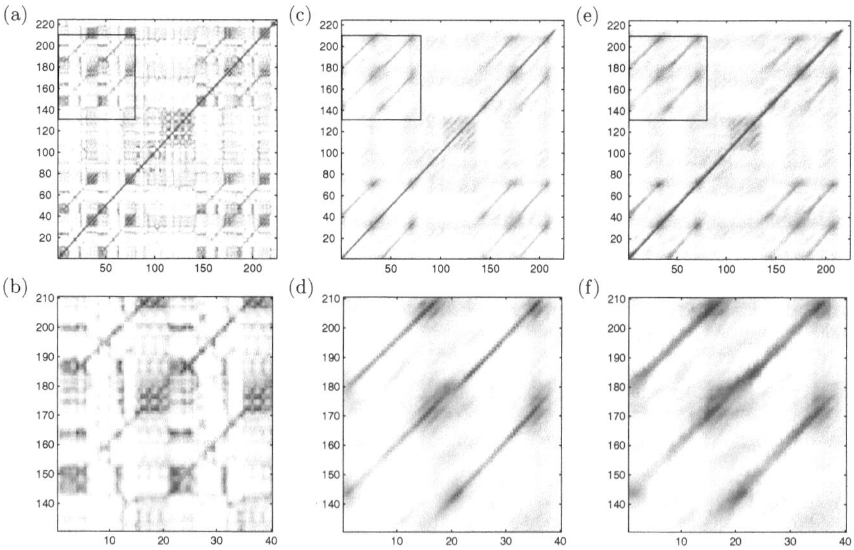

Fig. 7.4. Enhancement of the similarity matrix of the Shostakovich example (Fig. 7.2). **(a),(b)**: $\mathcal{S}[41,10]$ and enlargement. **(c),(d)**: $\mathcal{S}_{10}[41,10]$ and enlargement. **(e),(f)**: $\mathcal{S}_{10}^{\min}[41,10]$ and enlargement

feature sampling rate. Note that the contextual similarity matrix $\mathcal{S}_L$ can be efficiently computed from $\mathcal{S}$ by applying an averaging filter along the diagonals. More precisely, $\mathcal{S}_L(n,m) = \frac{1}{L}\sum_{\ell=0}^{L-1}\mathcal{S}(n+\ell, m+\ell)$ (with a suitable zero-padding of $\mathcal{S}$).

The enhancement procedure considered so far is problematic when similar segments do not have the same tempo. Such a situation frequently occurs in classical music – even within the same interpretation – as is shown by our Brahms example (Fig. 7.1). To account for tempo variations, we proceed as described in Sect. 6.1.3, we locally simulate tempo changes at the feature level by modifying the values of $\mathbf{w}$ and $\mathbf{d}$ of the $\mathrm{CENS}^{\mathbf{w}}_{\mathbf{d}}$-features and incorporate all information into a single, joint local cost measure. More precisely, let $V[\mathbf{w},\mathbf{d}] = (v[\mathbf{w},\mathbf{d}]_1, v[\mathbf{w},\mathbf{d}]_2, \ldots)$ denote the $\mathrm{CENS}^{\mathbf{w}}_{\mathbf{d}}$-feature sequence obtained from the audio data stream in question. We choose $\mathbf{w} = 41$ and $\mathbf{d} = 10$ as reference parameters, resulting in a feature sampling rate of 1 Hz. Furthermore, we use the eight pairs $(\mathbf{w}_j, \mathbf{d}_j)$ with $\mathbf{d}_j = j+6$ and $\mathbf{w}_j = 4\mathbf{d}_j+1$, $j \in [1:8]$, each corresponding to a different tempo. These pairs cover tempo variations of roughly -30 to $+40\%$ (Table 6.1). We then define a joint cost measure $c_L^{\min}$ by

$$c_L^{\min}(n,m) := \min_{j\in[1:8]} \frac{1}{L}\sum_{\ell=0}^{L-1} c\Big(v[41,10]_{n+\ell}, v[\mathbf{w}_j,\mathbf{d}_j]_{m_j+\ell}\Big), \tag{7.3}$$

where $m_j := \lceil m \cdot 10/\mathbf{d}_j \rceil$ and with a suitable zero-padding of the involved sequences. In other words, at position (n,m), the subsequence of $V[41,10]$ of length L starting at absolute time n (which also corresponds to feature position n having a feature sampling rate of 1 Hz) is compared with the subsequence of $V[\mathbf{w},\mathbf{d}]$ of length L starting at absolute time m (which corresponds to feature position $m_j = \lceil m \cdot 10/\mathbf{d}_j \rceil$ having a feature sampling rate of $10/\mathbf{d}$ Hz). From this we obtain the joint contextual similarity matrix $\mathcal{S}_L^{\min}$ defined by $\mathcal{S}_L^{\min}(n,m) := c_L^{\min}(n,m)$. Figure 7.5 shows that incorporating local tempo variations into contextual similarity matrices significantly improves the quality of the path structure, in particular for the case that similar audio segments exhibit different local relative tempi. For further examples we refer to [137] and the accompanying web site [138].

7.3 Path Extraction

On the basis of a smooth and structurally enhanced self-similarity matrix $\mathcal{S} = \mathcal{S}_L^{\min}[\mathbf{w},\mathbf{d}]$, we now describe a flexible and efficient path extraction procedure using a simple greedy strategy. Mathematically, we define a *path* to be a sequence $P = (p_1, p_2, \ldots, p_K)$ of pairs of indices $p_k = (n_k, m_k) \in [1:N]^2$, $1 \le k \le K$, satisfying the path constraints

$$p_{k+1} = p_k + \delta \quad \text{for some} \quad \delta \in \Delta, \tag{7.4}$$

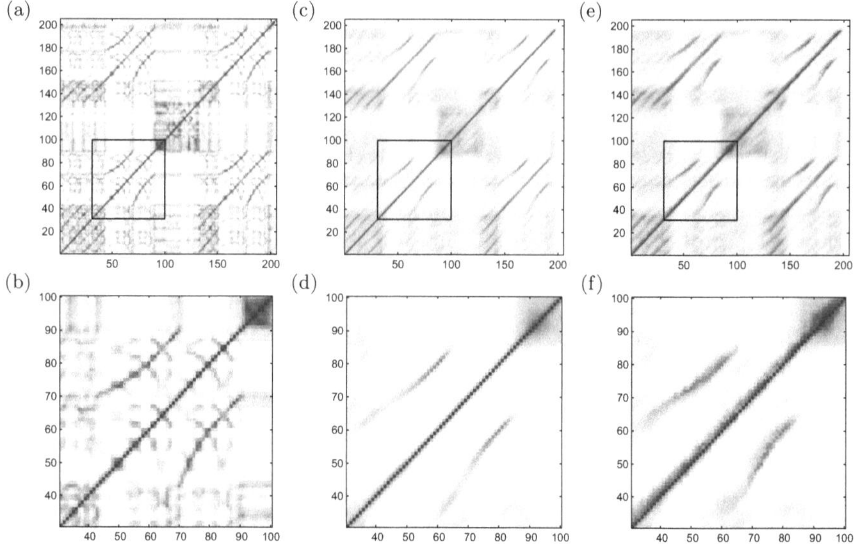

Fig. 7.5. Enhancement of the similarity matrix of the Brahms example, see Fig. 7.1. **(a)**,**(b)**: $\mathcal{S}[41,10]$ and enlargement. **(c)**,**(d)**: $\mathcal{S}_{10}[41,10]$ and enlargement. **(e)**,**(f)**: $\mathcal{S}_{10}^{\min}[41,10]$ and enlargement

where $\Delta := \{(1,1),(1,2),(2,1)\}$ and $1 \leq k \leq K-1$. The pairs p_k will also be called the *links* of P. Then the *cost* of link $p_k = (n_k, m_k)$ is defined as $\mathcal{S}(n_k, m_k)$. Now, it is the objective to extract long paths consisting of links having low costs. Our path extraction algorithm consists of three steps. In Step (1), we start with a link of minimal cost, referred to as *initial link*, and construct a path in a greedy fashion by iteratively adding links of low cost, referred to as *admissible links*. In Step (2), all links in a neighborhood of the constructed path are excluded from further considerations by suitably modifying $\mathcal{S}$. Then, Step (1) and Step (2) are repeated until there are no links of low cost left. Finally, the extracted paths are postprocessed in Step (3). The details are as follows:

(0) *Initialization.* Set $\mathcal{S} = \mathcal{S}_L^{\min}[\mathbf{w}, \mathbf{d}]$ and let $C_{\text{in}}, C_{\text{ad}} \in \mathbb{R}_{>0}$ be two suitable thresholds for the maximal cost of the initial links and the admissible links, respectively. (In our experiments, we typically chose $0.08 \leq C_{\text{in}} \leq 0.15$ and $0.12 \leq C_{\text{ad}} \leq 0.2$.) We modify $\mathcal{S}$ by setting $\mathcal{S}(n,m) = C_{\text{ad}}$ for $n \leq m$, i.e., the links below the diagonal will be excluded in the following steps. Similarly, we exclude the neighborhood of the diagonal path $P = ((1,1),(2,2),\ldots,(N,N))$ by modifying $\mathcal{S}$ using the path removal strategy as described in Step (2).

(1) *Path Construction.* Let $p_0 = (n_0, m_0) \in [1:N]^2$ be the indices minimizing $\mathcal{S}(n,m)$. If $\mathcal{S}(n_0, m_0) \geq C_{\text{in}}$, the algorithm terminates. Otherwise, we construct a new path P by extending p_0 iteratively, where all

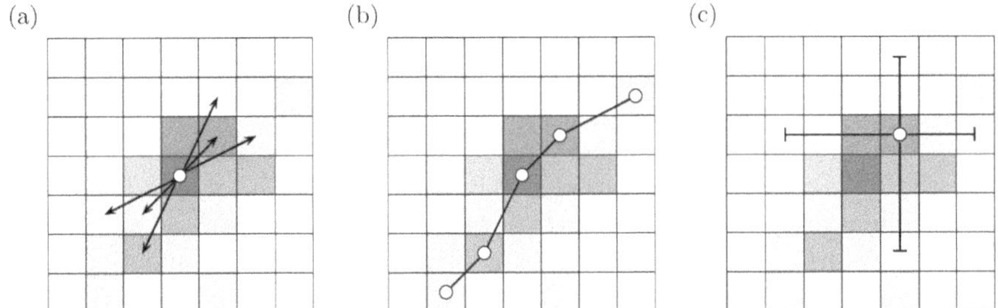

Fig. 7.6. **(a)** Initial link and possible path extensions. **(b)** Path resulting from Step (1). **(c)** Rays used for path removal in Step (2)

possible extensions are described by Fig. 7.6a. Suppose we have already constructed $P = (p_a, \ldots, p_0, \ldots, p_b)$ for $a \leq 0$ and $b \geq 0$. Then, if $\min_{\delta \in \Delta}(\mathcal{S}(p_b + \delta)) < C_{\mathrm{ad}}$, we extend P by setting

$$p_{b+1} := p_b + \mathrm{argmin}_{\delta \in \Delta}(\mathcal{S}(p_b + \delta)), \tag{7.5}$$

and, if $\min_{\delta \in \Delta}(\mathcal{S}(p_a - \delta)) < C_{\mathrm{ad}}$, extend P by setting

$$p_{a-1} := p_a - \mathrm{argmin}_{\delta \in \Delta}(\mathcal{S}(p_a - \delta)). \tag{7.6}$$

Figure 7.6b illustrates such a path. If there are no further extensions with admissible links, we proceed with Step (2). Shifting the indices by $a + 1$, we may assume that the resulting path is of the form $P = (p_1, \ldots, p_K)$ with $K = a + b + 1$.

(2) *Path Removal.* For a fixed link $p_k = (n_k, m_k)$ of P, we consider the maximal number $m_k \leq m^* \leq N$ with the property that $\mathcal{S}(n_k, m_k) \leq \mathcal{S}(n_k, m_k + 1) \leq \ldots \leq \mathcal{S}(n_k, m^*)$. In other words, the sequence (n_k, m_k), $(n_k, m_k + 1), \ldots, (n_k, m^*)$ defines a *ray* starting at position (n_k, m_k) and running horizontally to the right such that $\mathcal{S}$ is monotonically increasing. Analogously, we consider three other types of rays starting at position (n_k, m_k) running horizontally to the left, vertically upwards, and vertically downwards, see Fig. 7.6c for an illustration. We then consider all such rays for all links p_k of P. Let $\mathcal{N}(P) \subset [1 : N]^2$ be the set of all pairs (n, m) lying on one of these rays. Note that $\mathcal{N}(P)$ defines a neighborhood of the path P. To exclude the links of $\mathcal{N}(P)$ from further consideration, we set $\mathcal{S}(n, m) = C_{\mathrm{ad}}$ for all $(n, m) \in \mathcal{N}(P)$ and continue by repeating Step (1).

In our actual implementation, we made Step (2) more robust by softening the monotonicity condition on the rays. After the above algorithm terminates, we obtain a set of paths denoted by $\mathcal{P}$, which is postprocessed in a third step by means of some heuristics. For the following, let $P = (p_1, p_2, \ldots, p_K)$ denote a path in $\mathcal{P}$.

(3a) *Removing Short Paths.* All paths that have a length K shorter than a threshold $K_0 \in \mathbb{N}$ are removed. (In our experiments, we chose $5 \leq K_0 \leq 10$.) Such paths frequently occur as a result of residual links that have not been correctly removed by Step (2).

(3b) *Pruning Paths.* We prune each path $P \in \mathcal{P}$ at the beginning by removing the links $p_1, p_2, \ldots, p_{k_0}$ up to the index $0 \leq k_0 \leq K$, where k_0 denotes the maximal index such that the cost of each link $p_1, p_2, \ldots, p_{k_0}$ exceeds some suitably chosen threshold C_{pr} lying in between C_{in} and C_{ad}. Analogously, we prune the end of each path. This step is performed due to the following observation: introducing contextual information into the local cost measure results in a smoothing effect of the paths along the diagonal direction. This, in turn, results in a blurring effect at the beginning and end of such paths – as illustrated by Fig. 7.4f – unnaturally extending such paths at both ends in the construction of Step (1).

(3c) *Extending Paths.* We extend each path $P \in \mathcal{P}$ at its end by adding suitable links $p_{K+1}, \ldots, p_{K+L_0}$. This step is performed due to the following reason: since we have incorporated contextual information into the local cost measure, a low cost $\mathcal{S}(p_K) = c_L^{\min}(n_K, m_K)$ of $p_K = (n_K, m_K)$ implies that the entire sequence $(v_{n_K}[41,10], \ldots, v_{n_K+L-1}[41,10])$ is similar to $(v_{m_K}[\mathbf{w}_j, \mathbf{d}_j], \ldots, v_{m_K+L-1}[\mathbf{w}_j, \mathbf{d}_j])$ for the minimizing index $j \in [1:8]$, see (7.3). Here the length and direction of the extension $p_{K+1}, \ldots, p_{K+L_0}$ depends on the value $\mathbf{d}_j$. (In the case $(\mathbf{w}_j, \mathbf{d}_j) = (41, 10)$, we set $L_0 = L$ and $p_k = p_K + (k, k)$ for $k = 1, \ldots, L_0$.)

Figure 7.7 illustrates the steps of our path extraction algorithm for the Brahms example. Part (d) shows the resulting path set $\mathcal{P}$. Note that each path corresponds to a pair of similar segments and encodes the relative tempo progression between these two segments. Figure 7.8b shows the set $\mathcal{P}$ for the Shostakovich example. In spite of the matrix enhancement, the similarity between the segments corresponding to A_1 and A_3 has not been correctly identified, resulting in the aborted path P_1 (which should correctly start at link $(4, 145)$). Even though, as we will show in the next section, the extracted information is sufficient to correctly derive the global structure.

7.4 Global Structure Analysis

In this section, we propose an algorithm to determine the global repetitive structure of the underlying piece of music from the relations defined by the extracted paths. We first introduce some notation. A *segment* $\alpha = [s:t]$ is given by its starting point s and end point t, where s and t are given in terms of the corresponding indices in the feature sequence $V = (v_1, v_2, \ldots, v_N)$. A *similarity cluster* $\mathcal{A} := \{\alpha_1, \ldots, \alpha_M\}$ of size $M \in \mathbb{N}$ is defined to be a set of segments α_m, $1 \leq m \leq M$, which are considered to be mutually similar. Then, the *global structure* is described by a complete list of relevant similarity clusters of maximal size. In other words, the list should represent all

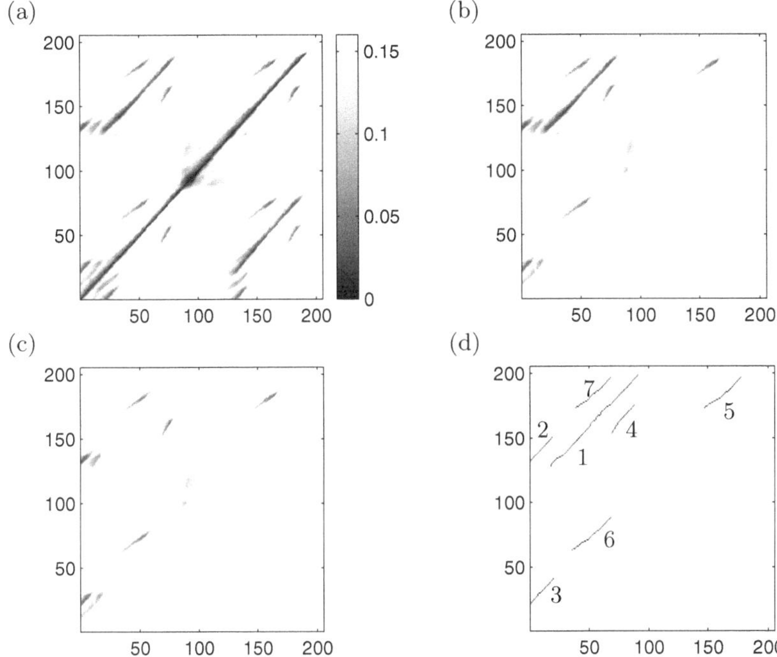

Fig. 7.7. Illustration of the path extraction algorithm for the Brahms example of Figure 7.1. **(a)** Self-similarity matrix $\mathcal{S} = \mathcal{S}_{16}^{\min}[41, 10]$. Here, all values exceeding the threshold $C_{\mathrm{ad}} = 0.16$ are plotted in white. **(b)** Matrix $\mathcal{S}$ after Step (0) (initialization). **(c)** Matrix $\mathcal{S}$ after performing Step (1) and Step (2) one time using the thresholds $C_{\mathrm{in}} = 0.08$ and $C_{\mathrm{ad}} = 0.16$. Note that a long path in the left upper corner was constructed, the neighborhood of which has then been removed. **(d)** Resulting path set $\mathcal{P} = \{P_1, \ldots, P_7\}$ after the postprocessing of Step (3) using $K_0 = 5$ and $C_{\mathrm{pr}} = 0.10$. The index m of P_m is indicated along each path, respectively

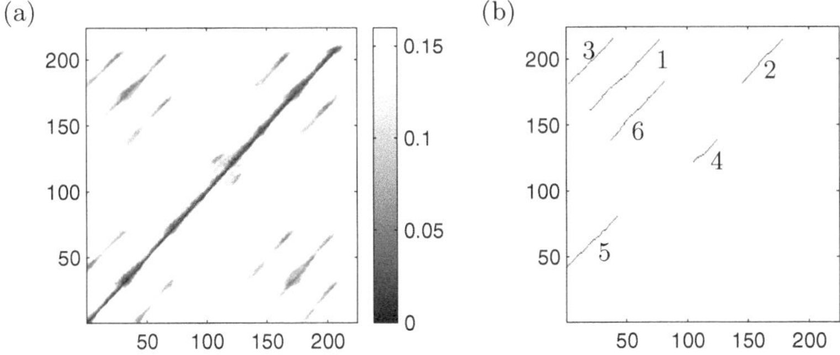

Fig. 7.8. Shostakovich example of Fig. 7.2. **(a)** $\mathcal{S}_{16}^{\min}[41, 10]$. **(b)** $\mathcal{P} = \{P_1, \ldots, P_6\}$ based on the same parameters as in the Brahms example of Fig. 7.7. The index m of P_m is indicated along each path, respectively

repetitions of musically relevant segments. Furthermore, if a cluster contains a segment α, then the cluster should also contain all other segments similar to α. For example, in our Shostakovich example of Fig. 7.2 the global structure is described by the clusters $\mathcal{A}_1 = \{\alpha_1, \alpha_2, \alpha_3, \alpha_4\}$ and $\mathcal{A}_2 = \{\gamma_1, \gamma_2\}$, where the segments α_k correspond to the parts A_k for $1 \leq k \leq 4$ and the segments γ_k to the parts C_k for $1 \leq k \leq 2$. Given a cluster $\mathcal{A} = \{\alpha_1, \ldots, \alpha_M\}$ with $\alpha_m = [s_m : t_m]$, $1 \leq m \leq M$, the *support* of $\mathcal{A}$ is defined to be the subset

$$\operatorname{supp}(\mathcal{A}) := \bigcup_{m=1}^{M} [s_m : t_m] \subset [1 : N]. \tag{7.7}$$

Recall that each path P indicates a pair of similar segments. More precisely, the path $P = (p_1, \ldots, p_K)$ with $p_k = (n_k, m_k)$ indicates that the segment $\pi_1(P) := [n_1 : n_K]$ is similar to the segment $\pi_2(P) := [m_1 : m_K]$. Such a pair of segments will also be referred to as a *path relation*. As an example, Fig. 7.9a shows the path relations of our Shostakovich example. In this section, we describe an algorithm that derives large and consistent similarity clusters from the path relations induced by the set $\mathcal{P}$ of extracted paths. From a theoretical point of view, one has to construct some kind of transitive closure of the path relations, see also [49]. For example, if segment α is similar to segment β, and segment β is similar to segment γ, then α should also be regarded as similar to γ resulting in the cluster $\{\alpha, \beta, \gamma\}$. The situation becomes more complicated when α overlaps with some segment β which, in turn, is similar to segment γ. This would imply that a subsegment of α is similar to some subsegment of γ. In practice, the construction of similarity clusters by iteratively continuing in the above fashion is problematic. Here, inconsistencies in the path relations due to semantic (vague concept of musical similarity) or to algorithmic (inaccurately extracted or missing paths) reasons may lead to meaningless clusters, e. g., containing a series of segments where each segment is a slightly shifted version of its predecessor. For example, let $\alpha = [1 : 10]$, $\beta = [11 : 20]$, $\gamma = [22 : 31]$, and $\delta = [3 : 11]$. Then similarity relations between α and β, β and γ, γ and δ would imply that $\alpha = [1 : 10]$ has to be regarded as similar to $\delta = [3 : 11]$, and so on. To balance out such inconsistencies, previous strategies such as [81] rely upon the constant tempo assumption. To achieve a robust and meaningful clustering even in the presence of significant local tempo variations, we suggest a new clustering algorithm, which proceeds in three steps. To this end, let $\mathcal{P} = \{P_1, P_2, \ldots, P_M\}$ be the set of extracted paths P_m, $1 \leq m \leq M$. In Step (1) (transitivity step) and Step (2) (merging step), we compute for each P_m a similarity cluster $\mathcal{A}_m$ consisting of all segments that are either similar to $\pi_1(P_m)$ or to $\pi_2(P_m)$. In Step (3), we then discard the redundant clusters. We exemplarily explain the procedure of Step (1) and Step (2) by considering the path P_1.

(1) *Transitivity Step.* Let T_{ts} be a suitable tolerance parameter measured in percent (in our experiments we used $T_{\mathrm{ts}} = 90$). First, we construct a cluster $\mathcal{A}_1^1$ for the path P_1 and the segment $\alpha := \pi_1(P_1)$. To this end, we check for all paths P_m whether the intersection $\alpha_0 := \alpha \cap \pi_1(P_m)$ contains more

than T_{ts} percent of α, i. e., whether $|\alpha_0|/|\alpha| \geq T_{ts}/100$. In the affirmative case, let β_0 be the subsegment of $\pi_2(P_m)$ that corresponds under P_m to the subsegment α_0 of $\pi_1(P_m)$. We add α_0 and β_0 to $\mathcal{A}_1^1$. Similarly, we check for all paths P_m whether the intersection $\alpha_0 := \alpha \cap \pi_2(P_m)$ contains more than T_{ts} percent of α and add in the affirmative case α_0 and β_0 to $\mathcal{A}_1^1$, where this time β_0 is the subsegment of $\pi_1(P_m)$ that corresponds under P_m to α_0. Note that β_0 generally does not have the same length as α_0. (Recall that the relative tempo variation is encoded by the gradient of P_m.) Analogously, we construct a cluster $\mathcal{A}_1^2$ for the path P_1 and the segment $\alpha := \pi_2(P_1)$. The clusters $\mathcal{A}_1^1$ and $\mathcal{A}_1^2$ can be regarded as the result of the first iterative step toward forming the transitive closure.

(2) *Merging Step.* The cluster $\mathcal{A}_1$ is constructed by basically merging the clusters $\mathcal{A}_1^1$ and $\mathcal{A}_1^2$. To this end, we compare each segment $\alpha \in \mathcal{A}_1^1$ with each segment $\beta \in \mathcal{A}_1^2$. In the case that the intersection $\gamma := \alpha \cap \beta$ contains more than T_{ts} percent of α and of β (i. e., α essentially coincides with β), we add the segment γ to $\mathcal{A}_1$. In the case that for a fixed $\alpha \in \mathcal{A}_1^1$ the intersection $\alpha \cap \mathrm{supp}(\mathcal{A}_1^2)$ contains less than $(100 - T_{ts})$ percent of α (i. e., α is essentially disjoint with all $\beta \in \mathcal{A}_1^2$), we add α to $\mathcal{A}_1$. Symmetrically, if for a fixed $\beta \in \mathcal{A}_1^2$ the intersection $\beta \cap \mathrm{supp}(\mathcal{A}_1^1)$ contains less than $(100 - T_{ts})$ percent of β, we add β to $\mathcal{A}_1$. Note that by this procedure, the first case balances out small inconsistencies, whereas the second case and the third case compensate for missing path relations. Furthermore, segments $\alpha \in \mathcal{A}_1^1$ and $\beta \in \mathcal{A}_1^2$ that do not fall into one of the above categories indicate significant inconsistencies and are left unconsidered in the construction of $\mathcal{A}_1$.

After Step (1) and Step (2), we obtain a cluster $\mathcal{A}_1$ for the path P_1. In an analogous fashion, we compute clusters $\mathcal{A}_m$ for all paths P_m, $1 \leq m \leq M$.

(3) *Discarding Clusters.* Let T_{dc} be a suitable tolerance parameter measured in percent (in our experiments we chose T_{dc} between 80 and 90%). We say that cluster $\mathcal{A}$ is a T_{dc}-*cover* of cluster $\mathcal{B}$ if the intersection $\mathrm{supp}(\mathcal{A}) \cap \mathrm{supp}(\mathcal{B})$ contains more than T_{dc} percent of $\mathrm{supp}(\mathcal{B})$. By pairwise comparison of all clusters $\mathcal{A}_m$ we successively discard all clusters that are T_{dc}-covered by some other cluster consisting of a larger number of segments. (Here the idea is that a cluster with a larger number of smaller segments contains more information than a cluster having the same support while consisting of a smaller number of larger segments.) In the case that two clusters are mutual T_{dc}-covers and consist of the same number of segments, we discard the cluster with the smaller support.

The steps of the clustering algorithm are also illustrated by Figs. 7.9 and 7.10. Recall from Sect. 7.3 that in the Shostakovich example, the significant variations in the instrumentation led to a defective path extraction. In particular, the similarity of the segments corresponding to parts A_1 and A_3

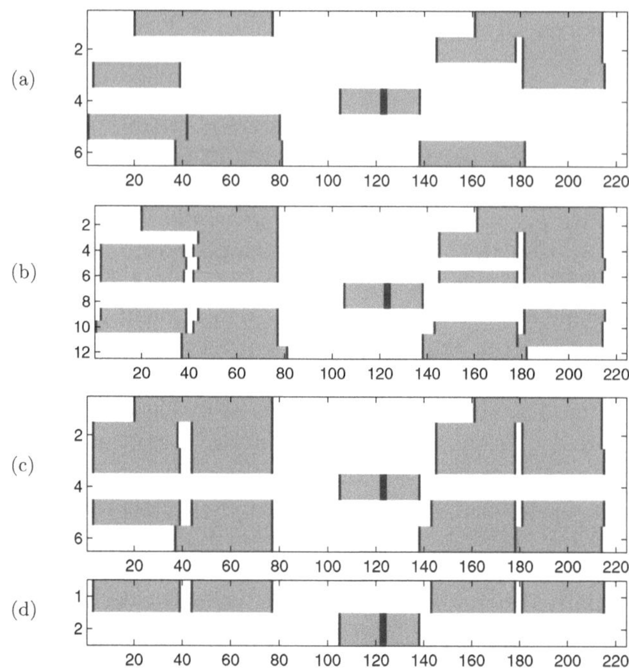

Fig. 7.9. Illustration of the clustering algorithm for the Shostakovich example. The path set $\mathcal{P} = \{P_1, \ldots, P_6\}$ is shown in Fig. 7.8b. Segments are indicated by *gray bars* and overlaps are indicated by *black regions*. **(a)** Illustration of the two segments $\pi_1(P_m)$ and $\pi_2(P_m)$ for each path $P_m \in \mathcal{P}$, $1 \leq m \leq 6$. Row m corresponds to P_m. **(b)** Clusters $\mathcal{A}_m^1$ and $\mathcal{A}_m^2$ (rows $2m-1$ and $2m$) computed in Step (1) with $T_{\mathrm{ts}} = 90$. **(c)** Clusters $\mathcal{A}_m$ (row m) computed in Step (2). **(d)** Final result of the clustering algorithm after performing Step (3) with $T_{\mathrm{dc}} = 90$. The derived global structure is given by two similarity clusters. The first cluster corresponds to the musical parts $\{A_1, A_2, A_3, A_4\}$ (first row) and the second cluster to $\{C_1, C_2\}$ (second row), cf. Fig. 7.2

could not be correctly identified as reflected by the truncated path P_1, see Figs. 7.8b and 7.9a. Nevertheless, the correct global structure was derived by the clustering algorithm, cf. Fig. 7.9d. Here, the missing relation was recovered by Step (1) (transitivity step) from the correctly identified similarity relation between segments corresponding to A_3 and A_4 (path P_2) and between segments corresponding to A_1 and A_4 (path P_3). The effect of Step (3) is illustrated by comparing (c) and (d) of Fig. 7.9. Since the cluster $\mathcal{A}_5$ is a 90% cover of the clusters $\mathcal{A}_1$, $\mathcal{A}_2$, $\mathcal{A}_3$, and $\mathcal{A}_6$, and has the largest support, the latter clusters are discarded.

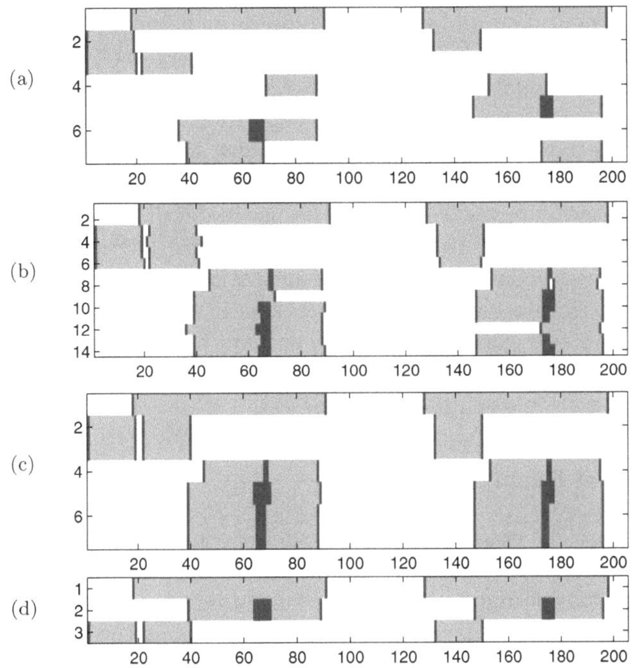

Fig. 7.10. Steps of the clustering algorithm for the Brahms example analogous to Fig. 7.9. The path set $\mathcal{P} = \{P_1, \ldots, P_7\}$ is shown in Fig. 7.7d. The final result indicated by (d) correctly represents the global structure. The cluster of the second row corresponds to $\{B_1, B_2, B_3, B_4\}$, while the cluster of the third row corresponds to $\{A_1, A_2, A_3\}$. Finally, the cluster of the first row expresses the similarity between $A_2B_1B_2$ and $A_3B_3B_4$ (cf. Fig. 7.1)

7.5 Experiments

We implemented our algorithm for audio structure analysis in MATLAB and tested it on about 100 audio recordings reflecting a wide range of mainly Western classical music, including pieces by Bach, Beethoven, Brahms, Chopin, Mozart, Ravel, Schubert, Schumann, Shostakovich, and Vivaldi. In particular, we used musically complex orchestral pieces exhibiting a large degree of variations in their repetitions with respect to instrumentation, articulation, and local tempo variations. From a musical point of view, the global repetitive structure is often ambiguous since it depends on the particular notion of similarity, on the degree of admissible variations, as well as on the musical significance and duration of the respective repetitions. Furthermore, the structural analysis can be performed at various levels: at a global level (e.g., segmenting a sonata into exposition, repetition of the exposition, development, and recapitulation), an intermediary level (e.g., further splitting up the exposition into first and second theme), or on a fine level (e.g., segmenting into repeating motifs). This makes the automatic structure extraction as well as an objective evaluation of the results a difficult and problematic task.

In our experiments, we looked for repetitions at a global to intermediary level corresponding to segments of at least 15–20 s of duration, which is reflected in our choice of parameters, see Sect. 7.5.1. In that section, we will also present some general results and discuss in detail two complex examples: Mendelssohn's Wedding March and Ravel's Bolero. In Sect. 7.5.2, we discuss the running time behavior of our implementation. It turns out that the algorithm is applicable to pieces even longer than 45 min, which covers essentially any piece of Western classical music. To account for transposed (pitch-shifted) repeating segments, we adopted the shifting technique suggested by Goto [81]. Some results will be discussed in Sect. 7.5.3.

7.5.1 General Results

To demonstrate the capability of our structure analysis algorithm, we discuss some representative results in detail. This will also illustrate the kind of difficulties generally found in music structure analysis. Our algorithm is fully automatic, in other words, no prior knowledge about the respective piece is exploited in the analysis. In all examples, we use the following fixed set of parameters. For the self-similarity matrix, we use $\mathcal{S}_{16}^{\min}[41, 10]$ with a corresponding feature resolution of 1 Hz, see Sect. 7.2. In the path extraction algorithm of Sect. 7.3, we set $C_{\mathrm{in}} = 0.08$, $C_{\mathrm{ad}} = 0.16$, $C_{\mathrm{pr}} = 0.10$, and $K_0 = 5$. Finally, in the clustering algorithm of Sect. 7.4 we set $T_{\mathrm{ts}} = 90$ and $T_{\mathrm{dc}} = 90$. The choice of the above parameters and thresholds constitutes a trade-off between being tolerant enough to allow relevant variations and being robust enough to deal with artifacts and inconsistencies.

As a first example, we consider a Varsi recording of Chopin's Etude Op. 10, No. 3 ("Tristesse"). The underlying piece has the musical form $A_1A_2B_1CA_3B_2D$. This structure has successfully been extracted by our algorithm, see Fig. 7.11a. Here, the first cluster $\mathcal{A}_1$ corresponds to the parts A_2B_1 and A_3B_2, whereas the second cluster $\mathcal{A}_2$ corresponds to the parts A_1, A_2, and A_3. For simplicity, we use the notation $\mathcal{A}_1 \sim \{A_2B_1, A_3B_2\}$ and $\mathcal{A}_2 \sim \{A_1, A_2, A_3\}$. The similarity relation between B_1 and B_2 is induced from cluster $\mathcal{A}_1$ by "subtracting" the respective A-part which is known from cluster $\mathcal{A}_2$. The small gaps between the segments in cluster $\mathcal{A}_2$ are due to the fact that the tail of A_1 (passage to A_2) is different from the tail of A_2 (passage to B_1).

The next example is a Barenboim interpretation of the second movement of Beethoven's Pathetique, which has the musical form $A_1A_2BA_3CA_4A_5D$. The interesting point of this piece is that the A-parts are variations of each other. For example, the melody in A_2 and A_4 is played one octave higher than the melody in A_1 and A_3. Furthermore, A_3 and A_4 are rhythmic variations of A_1 and A_2. Nevertheless, the correct global structure has

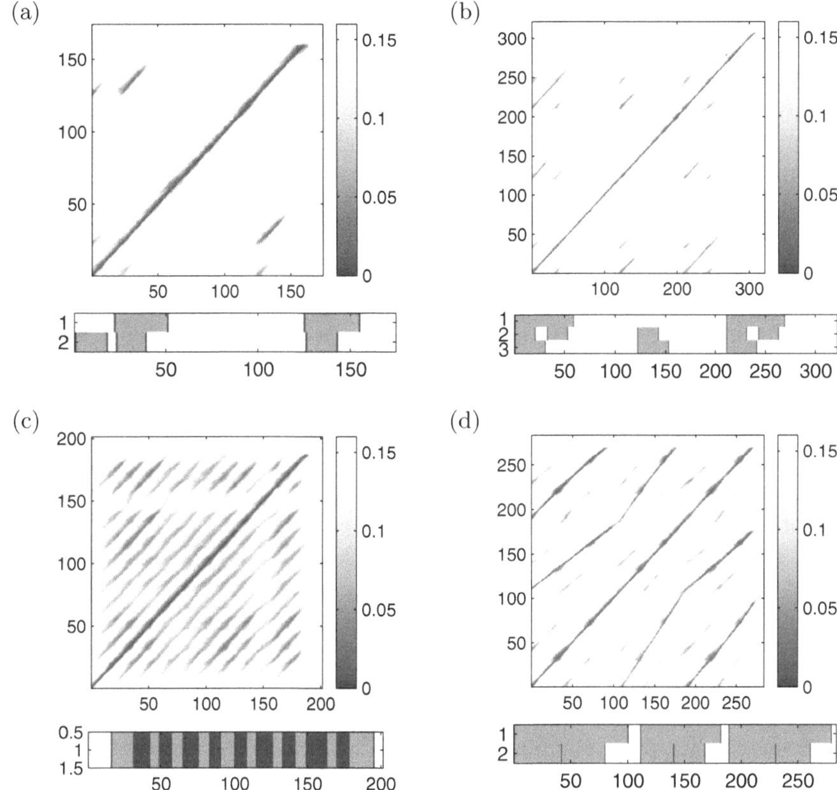

Fig. 7.11. **(a)** Chopin, "Tristesse," Etude Op. 10, No. 3, played by Varsi. **(b)** Beethoven, "Pathetique," second movement, Op. 13, played by Barenboim. **(c)** Gloria Gaynor, "I will survive." **(d)** Part A_1A_2B of the Shostakovich example of Fig. 7.2 repeated three times in modified tempi (normal tempo, 140% of normal tempo, accelerating tempo from 100% to 140%)

been extracted, see Fig. 7.11b. The three clusters are in correspondence with $\mathcal{A}_1 \sim \{A_1A_2, A_4A_5\}$, $\mathcal{A}_3 \sim \{A_1, A_3, A_5\}$, and $\mathcal{A}_2 \sim \{A'_1, A'_2, A'_3, A'_4, A'_5\}$, where A'_k denotes a truncated version of A_k. Hence, the segments A_1, A_3, and A_5 are identified as a whole, whereas the other A-parts are identified only up to their tail. This is due to the fact that the tails of the A-parts exhibit some deviations leading to higher costs in the self-similarity matrix, as illustrated by Fig. 7.11b.

The popular song "I will survive" by Gloria Gaynor consists of an introduction I followed by eleven repetitions A_k, $1 \leq k \leq 11$, of the chorus. This highly repetitive structure is reflected by the secondary diagonals in the self-similarity matrix, see Fig. 7.11c. The segments exhibit variations not only with respect to the lyrics but also with respect to instrumentation and tempo. For example, some segments include a secondary voice in the violin, others harp arpeggios or trumpet syncopes. The first chorus A_1 is played without percussion, whereas A_5 is a purely instrumental version. Also note that there is a significant ritardando in A_9 between seconds 150 and 160. In spite of these variations, the structure analysis algorithm works almost correctly. However, there are two artifacts that have not been ruled out by our strategy. Each chorus A_k can be split up into two subparts $A_k = A'_k A''_k$. The computed cluster $\mathcal{A}_1$ corresponds to the ten parts $A''_{k-1} A'_k A''_k$, $2 \leq k \leq 11$, revealing an overlap in the A''-parts. In particular, the extracted segments are "out of phase" since they start with subsegments corresponding to the A''-parts. This may be due to extreme variations in A'_1 making this part dissimilar to the other A'-parts. Since A'_1 constitutes the beginning of the extracted paths, it has been (mistakenly) truncated in Step (3b) (pruning paths) of Sect. 7.3.

To check the robustness of our algorithm with respect to global and local tempo variations, we conducted a series of experiments with synthetically time-stretched audio signals (i. e., we changed the tempo progression without changing the pitch). As it turns out, there are no problems in identifying similar segments that exhibit global tempo variations of up to 50% as well as local tempo variations such as ritardandi and accelerandi. As an example, we consider the audio file corresponding to the part $A_1 A_2 B$ of the Shostakovich example of Fig. 7.2. From this, we generated two additional time-stretched variations: a faster version at 140% of the normal tempo and an accelerating version speeding up from 100 to 140%. The musical form of the concatenation of these three versions is $A_1 A_2 B_1 A_3 A_4 B_2 A_5 A_6 B_3$. This structure has been correctly extracted by our algorithm, see Fig. 7.11d. The correspondences of the two resulting clusters are $\mathcal{A}_1 \sim \{A_1 A_2 B_1, A_3 A_4 B_2, A_5 A_6 B_3\}$ and $\mathcal{A}_2 \sim \{A_1, A_2, A_3, A_4, A_5, A_6\}$.

Next, we discuss an example with a musically more complicated structure. This will also illustrate some problems typically appearing in automatic structure analysis. The "Wedding March" by Mendelssohn has the musical form

$$A_1 B_1 A_2 B_2 C_1 B_3 C_2 B_4 \quad D_1 D_2 E_1 D_3 E_1 D_4 B_5 \quad F_1 G_1 G_2 H_1 \quad A_3 B_6 C_3 B_7 A_4 \quad I_1 I_2 J_1$$

Furthermore, each segment B_k for $1 \leq k \leq 7$ has a substructure $B_k = B'_k B''_k$ consisting of two musically similar subsegments B'_k and B''_k. However, the B''-parts reveal significant variations even at the note level. Our algorithm has computed seven clusters, which are arranged according to the lengths of their support, see Fig. 7.12a. Even though not visible at first glance, these clusters represent most of the musical structure accurately. Manual inspection reveals that the cluster segments correspond, up to some tolerance, to the musical parts as follows:

$$\begin{aligned}
\mathcal{A}_1 &\sim \{B_2 C_1 B'_3,\ B_3 C_2 B'_4, B_6 C_3 B'_7\} \\
\mathcal{A}_2 &\sim \{B_2 C_1 B_3{+},\ B_6 C_3 B_7{+}\} \\
\mathcal{A}_3 &\sim \{B_1,\ B_2,\ B_3,\ B_6,\ B_7\} \\
\mathcal{A}_4 &\sim \{B'_1,\ B'_2,\ B'_3,\ B'_4,\ B'_5,\ B'_6,\ B'_7\}
\end{aligned}$$

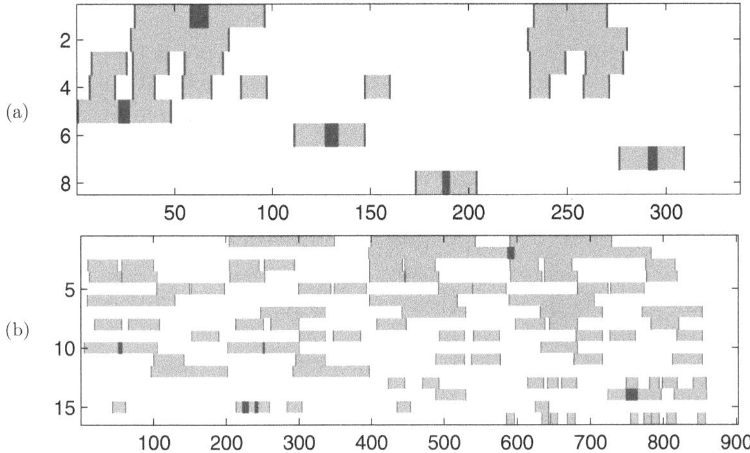

Fig. 7.12. (a) Mendelssohn, "Wedding March", Op. 21, No. 7, conducted by Tate. **(b)** Ravel, "Bolero", conducted by Ozawa

$$\mathcal{A}_5 \sim \{A_1 B_1 A_2,\ A_2 B_2+\}$$
$$\mathcal{A}_6 \sim \{D_2 E_1 D_3,\ D_3 E_2 D_4\}$$
$$\mathcal{A}_7 \sim \{G_1,\ G_2\}$$
$$\mathcal{A}_8 \sim \{I_1,\ I_2\}$$

In particular, all seven B'-parts (truncated B-parts) are represented by cluster $\mathcal{A}_4$, whereas $\mathcal{A}_3$ contains five of the seven B-parts. The missing and truncated B-parts can be explained as in the Beethoven example of Fig. 7.11b. Cluster $\mathcal{A}_1$ reveals the similarity of the three C-parts, which are enclosed between the B- and B'-parts known from $\mathcal{A}_3$ and $\mathcal{A}_4$. The A-parts, an opening motif, have a duration of less than 8 s – too short to be recognized by our algorithm as a separate cluster. Because of the close harmonic relationship of the A-parts with the tails of the B-parts and the heads of the C-parts, it is hard to exactly determine the boundaries of these parts. This leads to clusters such as $\mathcal{A}_2$ and $\mathcal{A}_5$, whose segments enclose several parts or only fragments of some parts (indicated by the + sign). Furthermore, the segments of cluster $\mathcal{A}_6$ enclose several musical parts. Because of the overlap in D_3, one can derive the similarity of D_2, D_3, and D_4 as well as the similarity of E_1 and E_2. The D- and E-parts are too short (less than 10 s) to be detected as separate clusters. This also explains the undetected part D_1. Finally, the clusters $\mathcal{A}_7$ and $\mathcal{A}_8$ correctly represent the repetitions of the G- and I-parts, respectively.

Another complex example, in particular with respect to the occurring variations, is Ravel's Bolero, which has the musical form $D_1 D_2 D_3 D_4 A_9 B_9 C$ with $D_k = A_{2k-1} A_{2k} B_{2k-1} B_{2k}$ for $1 \leq k \leq 4$. The piece repeats two tunes (corresponding to the A- and B-parts) over and over again, each time played in a different instrumentation including flute, clarinet, bassoon, saxophone, trumpet, strings, and culminating in the full orchestra. Furthermore, the volume gradually grows from quiet pianissimo to a vehement fortissimo. Note that playing an instrument in piano or in fortissimo not only makes a difference in volume but also in the relative energy distribution within the chroma bands, which is due to effects such as noise, vibration, and reverberation. Nevertheless, the CENS features absorb most of the resulting variations. The extracted clusters represent the global structure up to a few missing segments, see Fig. 7.12b. In particular, the cluster $\mathcal{A}_3 \sim \{A_k- \mid 1 \leq k \leq 9\}$ correctly identifies all nine A-parts in a slightly truncated form (indicated by the $-$ sign). Note that the truncation may result from Step (2) (merging step) of Sect. 7.4, where path inconsistencies are ironed out by segment intersections. The cluster $\mathcal{A}_4$ correctly identifies the full-size A-parts with only part A_4 missing. Here, an additional transitivity step might have helped to perfectly identify all nine A-parts in full length. The similarity of the B-parts is reflected by $\mathcal{A}_5$, where only part B_9 is missing. All other clusters reflect superordinate similarity relations (e. g., $\mathcal{A}_1 \sim \{A_3A_4B_3, A_5A_6B_5, A_7A_8B_7\}$ or $\mathcal{A}_2 = \{D_3+, D_4+\}$), or similarity relations of smaller fragments.

For other pieces of music – we manually analyzed the results for about 100 pieces – our structure analysis algorithm typically performs as indicated by the above examples and the global repetitive structure can be recovered to a high degree. We summarize some typical problems associated with the extracted similarity clusters. First, some clusters consist of segments that only correspond to fragments or truncated versions of musical parts. Note that this problem is not only due to algorithmic reasons such as the inconsistencies stemming from inaccurate path relations but also due to musical reasons such as extreme variations in tails of musical parts. Second, the set of extracted clusters is sometimes redundant as in the case of the Bolero – some clusters almost coincide while differing only by a missing part and by a slight shift and length difference of their respective segments. Here, a higher degree of transitivity and a more involved merging step in Sect. 7.4 could help to improve the overall result. (Because of the inconsistencies, however, a higher degree of transitivity may also degrade the result in other cases.) Third, the global structure is sometimes not given explicitly but is somehow hidden in the clusters. For example, the similarity of the B-parts in the Chopin example results from "subtracting" the segments corresponding to the A-parts given by $\mathcal{A}_2$ from the segments of $\mathcal{A}_1$. Or, in the Mendelssohn example, the similarity of the D- and E-parts can be derived from cluster $\mathcal{A}_6$ by exploiting the overlap of the segments in a subsegment corresponding to part D_3. It seems promising to exploit such overlap relations in combination with a subtraction strategy to further improve the cluster structure. Furthermore, we expect an additional improvement in expressing the global structure by means of some hierarchical approach as discussed in Sect. 7.6.

7.5.2 Running Time Behavior

In this section, we discuss the running time behavior of the MATLAB implementation of our structure analysis algorithm. Tests were run on an Intel Pentium IV, 3.6 GHz, with 2 GB RAM under Windows 2000. Table 7.1 shows the running times for several pieces sorted by duration.

The first step of our algorithm consists of the extraction of robust audio features, see Sect. 3.3. The running time to compute the CENS feature sequence is linear in the duration of the audio file under consideration – in our tests roughly one third of the duration of the piece, see the third column of Table 7.1. Here, the decomposition of the audio signal into the 88 frequency bands as described in Sect. 3.1 constitutes the bottleneck of the feature extraction, consuming far more than 99% of the entire running time. The subsequent computations to derive the CENS features from the filter subbands only take a fraction of a second even for long pieces such as Ravel's Bolero. In view of our experiments, we computed the chroma features of Sect. 3.3 at a resolution of 10 Hz for each piece in our music database and stored them on hard disk, making them available for the subsequent steps irrespective of the parameter choice made in Sects. 7.3 and 7.4.

The time and space complexity to compute a self-similarity matrix $\mathcal{S}$ is quadratic in the length N of the feature sequence. This makes the usage of such matrices infeasible for large N. Here, our strategy is to use coarse CENS features, which not only introduces a high degree of robustness toward admissible variations but also keeps the feature resolution low. In the above experiments, we used CENS^{41}_{10} features with a sampling rate of 1 Hz. Furthermore, incorporating the desired invariances into the features itself allows us to use a local distance measure based on the inner product that can be evaluated by a computationally inexpensive algorithm. This affords an efficient computation of $\mathcal{S}$ even for long pieces of up to 45 min of duration, see the fourth column of Table 7.1. For example, in case of the Bolero it took 4.36 s to compute $\mathcal{S}^{\mathrm{min}}_{16}[41,10]$ from a feature sequence of length $N = 901$, corresponding to 15 min of audio. Tripling the length N by using a threefold concatenation of the Bolero results in a running time of 37.9 s, showing an increase by a factor of nine.

The running time for the path extraction algorithm as described in Sect. 7.3 mainly depends on the structure of the self-similarity matrix below the threshold C_{ad} (rather than on the size of the matrix), see the fifth column of Table 7.1. Here, crucial parameters are the number as well as the lengths of the path candidates to be extracted, which influences the running time in a linear fashion. Even for long pieces with a very rich path structure – as is the case for the Bolero – the running time of the path extraction is only a couple of seconds.

Finally, the running time of the clustering algorithm of Sect. 7.4 is negligible, see the last column of Table 7.1. Only for a very large (and practically irrelevant) number of paths, the running time seems to increase significantly.

Table 7.1. Running-time behavior of the overall structure analysis algorithm

Piece	Length (s)	t(CENS) (s)	t($\mathcal{S}$) (s)	t(Pa.) (s)	#(Pa.)	t(Cl.) (s)
Chopin, "Tristesse" (Fig. 7.11a)	173.1	54.6	0.20	0.06	3	0.17
Gaynor, "I will survive" (Fig. 7.11c)	200.0	63.0	0.25	0.16	24	0.33
Brahms, "Hungarian Dance" (Fig. 7.1)	204.1	64.3	0.31	0.09	7	0.19
Shostakovich, "Waltz" (Fig. 7.2)	223.6	70.5	0.34	0.09	6	0.20
Beethoven, "Pathetique" (Fig. 7.11b)	320.0	100.8	0.66	0.15	9	0.21
Mendelssohn, "Wedding March" (Fig. 7.12a)	336.6	105.7	0.70	0.27	17	0.27
Schubert, "Unfinished" (Fig. 7.13a)	900.0	282.1	4.40	0.85	10	0.21
Ravel, "Bolero" (Fig. 7.12b)	901.0	282.7	4.36	5.53	71	1.05
2× "Bolero"	1802.0		17.06	84.05	279	9.81
3× "Bolero"	2703.0		37.91	422.69	643	97.94

All time durations are measured in seconds. The columns indicate the respective piece of music, the duration of the piece, the running time to compute the CENS features (Sect. 3.3), the running time to compute the self-similarity matrix $\mathcal{S} = \mathcal{S}_{16}^{\min}[41, 10]$ (Sect. 7.2), the running time for the path extraction (Sect. 7.3), the number of extracted paths, and the running time for the clustering algorithm (Sect. 7.4)

Basically, the overall performance of the structure analysis algorithm depends on the feature extraction step, which depends linearly on the input size.

7.5.3 Transpositions

It is often the case, in particular for classical music, that certain musical parts are repeated in another key. For example, the second theme in the exposition of a sonata is often repeated in the recapitulation transposed by a fifth (i. e., shifted by seven semitones upwards). To account for such modulations, we have adopted the idea of Goto [81], which is based on the observation that the 12 cyclic shifts of a 12-dimensional chroma vector naturally correspond to the 12 possible transpositions. In [81], similarity clusters (called line segment groups) are computed for all 12 transpositions separately, which are then suitably merged in a postprocessing step. In contrast to this, we incorporate all transpositions into a single self-similarity matrix, which then allows for performing a singly joint path extraction and clustering step only. The details of this procedure are as follows. Let $\sigma : \mathbb{R}^{12} \to \mathbb{R}^{12}$ denote the *cyclic shift* defined by

$$\sigma((v(1), v(2), \ldots, v(12))) := (v(2), \ldots, v(12), v(1)) \tag{7.8}$$

for $v := (v(1), \dots, v(12)) \in \mathbb{R}^{12}$. Then, for a given audio data stream with CENS feature sequence $V := (v_1, v_2, \dots, v_N)$, the *i-transposed self-similarity matrix* $\sigma^i(\mathcal{S})$ is defined by

$$\sigma^i(\mathcal{S})(n,m) := c(v_n, \sigma^i(v_m)), \tag{7.9}$$

$1 \leq n, m \leq N$. $\sigma^i(\mathcal{S})$ describes the similarity relations between the original audio data stream and the audio data stream transposed by i semitones, $i \in \mathbb{Z}$. Obviously, one has $\sigma^{12}(\mathcal{S}) = \mathcal{S}$. Taking the minimum over all 12 transpositions, we obtain the *transposition-invariant self-similarity matrix* $\sigma^{\min}(\mathcal{S})$ defined by

$$\sigma^{\min}(\mathcal{S})(n,m) := \min_{i \in [0:11]} \Big(\sigma^i(\mathcal{S})(n,m) \Big). \tag{7.10}$$

Furthermore, we store the minimizing shift indices in an additional N-square matrix $\mathcal{I}$:

$$\mathcal{I}(n,m) := \operatorname{argmin}_{i \in [0:11]} \Big(\sigma^i(\mathcal{S})(n,m) \Big). \tag{7.11}$$

Analogously, one defines $\sigma^{\min}(\mathcal{S}_L^{\min}[\mathbf{w}, \mathbf{d}])$. Now, replacing the self-similarity matrix by its transposition-invariant version one can proceed with the structure analysis as described in Sects.7.3 and 7.4. The only difference is that in Step (1) of the path extension (Sect. 7.3) one has to ensure that each path $P = (p_1, p_2, \dots, p_K)$ consists of links exhibiting the same transposition index: $\mathcal{I}(p_1) = \mathcal{I}(p_2) = \dots = \mathcal{I}(p_K)$.

We illustrate this procedure by means of two examples. The song "In the year 2525" by Zager and Evans is of the musical form

$$AB_1^0 B_2^0 B_3^0 B_4^0 C B_5^1 B_6^1 D B_7^2 E B_8^2 F,$$

where the chorus, the B-part, is repeated 8 times. Here, B_5^1 and B_6^1 are transpositions by one semitone and B_7^2 and B_8^2 are transpositions of the parts B_1^0 to B_4^0 by two semitones upwards. Figure 7.13a shows the similarity clusters derived from the structure analysis based on $\mathcal{S} = \mathcal{S}_{16}^{\min}[41, 10]$. Note that the transposed parts are separated into different clusters corresponding to $\mathcal{A}_1 \sim \{B_1^0, B_2^0, B_3^0, B_4^0\}$, $\mathcal{A}_2 \sim \{B_5^1, B_6^1\}$, and $\mathcal{A}_3 \sim \{B_7^2, B_8^2\}$. In contrast, the analysis based on $\sigma^{\min}(\mathcal{S})$ leads to a cluster $\mathcal{A}_1$ corresponding to all eight B-parts.

As a second example, we consider an Abbado recording of the first movement of Schubert's "Unfinished." This piece, which is composed in the sonata form, has the rough musical form $A_1^0 B_1^0 C_1^0 A_2^0 B_2^0 C_2^0 D \tilde{A}_3 B_3^7 C_3^4 E$, where $A_1^0 B_1^0 C_1^0$ corresponds to the exposition, $A_2^0 B_2^0 C_2^0$ to the repetition of the exposition, D to the development, $\tilde{A}_3 B_3^7 C_3^4$ to the recapitulation, and E to the coda. Note that the B_1^0-part of the exposition is repeated up a fifth as B_3^7 (shifted by 7 semitones upwards) and the C_1^0-part is repeated up a third as C_3^4 (shifted by 4 semitones upwards). Furthermore, the A_1^0-part is repeated as $\tilde{A}_3$, however, in form of a multilevel transition from the tonic to the dominant.

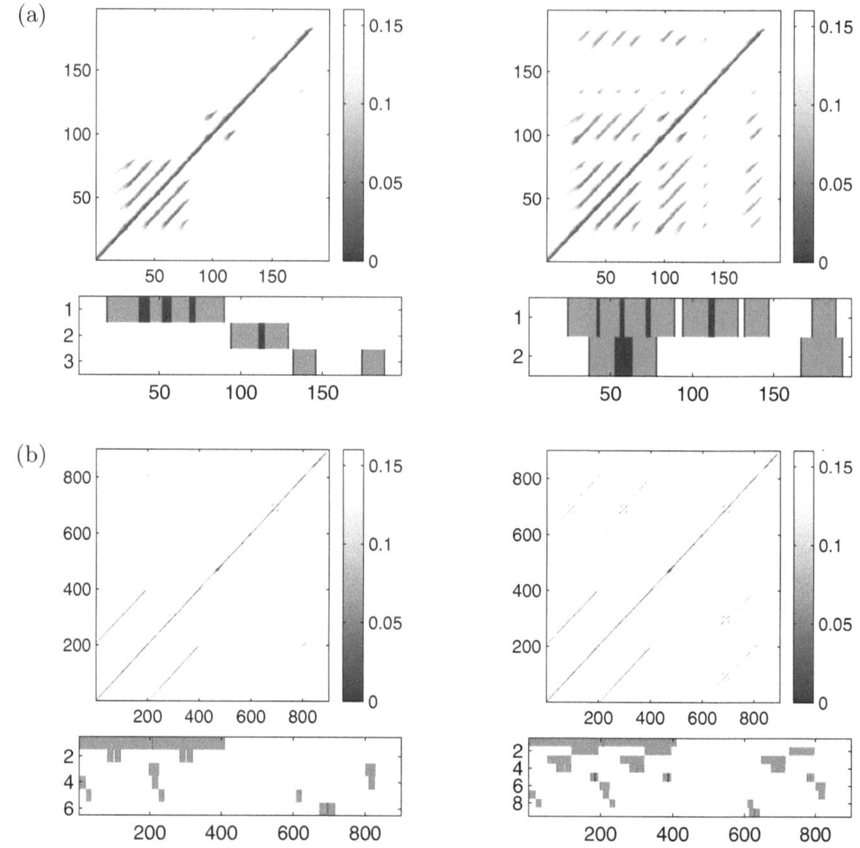

Fig. 7.13. (a) Zager and Evans, "In the year 2525." *Left*: $\mathcal{S}$ with the resulting similarity clusters. *Right*: $\sigma^{\min}(\mathcal{S})$ with the resulting similarity clusters. The parameters are fixed as described in Sect. 7.5.1. **(b)** Schubert, "Unfinished", first movement, D759, conduced by Abbado. Left and right part are analogous to (a)

Again the structure is revealed by the analysis based on $\sigma^{\min}(\mathcal{S})$, where one has, among others, the correspondences $\mathcal{A}_1 \sim \{A_1^0 B_1^0 C_1^0, A_2^0 B_2^0 C_2^0\}$, $\mathcal{A}_2 \sim \{B_1^0, B_2^0, B_3^7\}$, and $\mathcal{A}_3 \sim \{C_1^0, C_2^0, C_3^4\}$. The other clusters correspond to further structures on a finer level.

Finally, since the transposition-invariant similarity matrix $\sigma^{\min}(\mathcal{S})$ is derived from the 12 i-transposed matrices $\sigma^i(\mathcal{S})$, $i \in [0:11]$, the resulting running time to compute $\sigma^{\min}(\mathcal{S})$ is roughly 12 times longer than the time to compute $\mathcal{S}$. For example, it took 51.4 s to compute $\sigma^{\min}(\mathcal{S})$ for the Schubert's "Unfinished" as opposed to 4.4 s needed to compute $\sigma(\mathcal{S})$ (cf. Table 7.1).

7.6 Further Notes

In this chapter, we have described a robust and efficient algorithm for extracting the repetitive structure of an audio recording [139]. As opposed to previous methods, our approach is robust to significant variations in the repetitions concerning instrumentation, execution of note groups, dynamics, articulation, transposition, and tempo. For the first time, detailed experiments have been conducted for a wide range of Western classical music. The results show that the extracted audio structure often closely corresponds to the musical form of the underlying piece, even though no a priori knowledge on the musical structure has been used. In our approach, we converted the audio signal into a sequence of coarse, harmony-related CENS features. Such features are well-suited to characterize pieces of Western classical music, which often exhibit prominent harmonic progressions. Furthermore, instead of relying on complicated and delicate path extraction algorithms, we suggested a different approach by dealing with local variations at the feature and cost measure levels. This way we improved the path structure of the self-similarity matrix [137], which then allowed for an efficient robust path extraction.

Most of the previous approaches to structural audio analysis focus on the detection of repeating patterns in popular music based on the strategy as described in Sect. 7.1. The concept of similarity matrices has been introduced to the music context by Foote to visualize the temporal structure of audio and music [69]. On the basis of these matrices, Foote and Cooper [46] report on first experiments on automatic audio summarization using mel-frequency cepstral coefficients (MFCCs). To allow for small variations in performance, orchestration and lyrics, Bartsch and Wakefield [13,14] introduced chroma-based audio features to structural audio analysis. Chroma features, representing the spectral energy of each of the 12 traditional pitch classes of the equal-tempered scale, were also used in subsequent work such as [49,81]. Goto [81] describes a method that detects the chorus sections within audio recordings of popular music. Important contributions of this work are, among others, the automatic identification of both ends of a chorus section (without prior knowledge of the chorus length) and the introduction of some shifting technique to deal with transpositions. Furthermore, Goto introduces a technique to cope with missing or inaccurately extracted candidates of repeating segments. In their work on repeating pattern discovery, Lu et al. [127] suggest a local distance measures that is invariant with respect to harmonic intervals, introducing some robustness to variations in instrumentation. Furthermore, they describe a postprocessing technique to optimize boundaries of the candidate segments. At this point, we note that the above-mentioned approaches, while exploiting that repeating segments are of the same duration, are based on the constant tempo assumption. Dannenberg and Hu [49] describe several general strategies for path extraction, which indicate how to achieve robustness to small local tempo variations. There are also several approaches to structural analysis based on learning techniques such as hidden Markov models (HMMs), which are used to cluster similar segments into groups, see [126, 161] and the references therein. In the context of *music summarization*, where the aim is to generate a list of the most representative musical segments without considering musical structure, Xu et al. [221] use support vector machines (SVMs) for classifying audio recordings into segments of pure and vocal music. Maddage et al. [129] exploit some heuristics on the typical structure of popular music for both determining candidate segments and deriving the musical structure of a particular recording based on those segments. Their approach to structure analysis relies on the assumption that the analyzed recording follows a typical *verse-chorus pattern repetition*. As opposed to the general strategy introduced in Sect. 7.1, their approach only requires to implicitly calculate parts of a self-similarity matrix by restricting themselves to only considering certain candidate segments.

In summary, there have been several recent approaches to audio structure analysis that work well for types of music where the repetitions largely agree with respect to instrumentation, articulation, and tempo progression – as is often the case for popular music. In particular, most of the proposed strategies assume constant tempo throughout the piece (i. e., the path candidates have gradient $(1,1)$ in the self-similarity matrix), which is then exploited in the path extraction and clustering procedure. For example, this assumption is used by Goto [81] in his strategy for segment recovery, by Lu et al. [127] in their boundary refinement, and by Chai et al. [36, 38], in the step of segment merging. The reported experimental result almost entirely refers to popular music. For this genre, the proposed structure analysis algorithms report on good results even in presence of variations with respect to instrumentation and lyrics. In this chapter, we have introduced several fundamental techniques, which allow for efficiently performing structural audio analysis even in presence of significant local tempo variations.

As another problem, the high time and space complexity of $O(N^2)$ to compute and store a self-similarity matrix makes it problematic or infeasible to use this concept for large N [49]. Here, our idea was to work with relatively low feature sampling rates leading to feasible algorithms for pieces of up to 45 min of duration. Furthermore, enhancing the path structure of the similarity-matrix allowed for a very efficient path extraction procedure while keeping the number of path candidates small. To further reduce the running time as well as the memory requirements, it seems promising to employ index-based methods as introduced in Sect. 6.4.

First audio interfaces have been developed to facilitate intuitive audio browsing based on the extracted audio structure. The SmartMusicKIOSK system [82] integrates functionalities for jumping to the chorus section and other key parts of a popular song as well as for visualizing song structure. The system constitutes the first interface that allows the user to easily skip sections of low interest even within a song. The SyncPlayer system [114] allows a multimodal presentation of audio and associated music-related data. Here, a recently developed audio structure plug-in not only allows for an efficient audio browsing but also for a direct comparison of musically related segments, which constitutes a valuable tool in music research. For further details on the SyncPlayer system we refer to Chap. 8.

To obtain a more comprehensive representation of audio structure, obvious extensions of this work consist of combining harmony-based features with other types of features describing the rhythm, dynamics, or timbre of music. Another extension concerns the hierarchical nature of music. So far, our analysis aimed at recognizing repetitions at a global to intermediary level corresponding to segments of at least 15–20 s of duration. As has also been noted by other researches, musical structure can often be expressed in a hierarchical manner, starting with the coarse musical form and proceeding to finer substructures such as repeating themes and motifs. Here, one typically tolerates larger variations in the analysis of coarser structures than in the analysis of finer structures. For future work, we suggest a hierarchical approach to structure analysis by simultaneously computing and combining structural information at various temporal resolutions. To this end, we conducted first experiments based on the self-similarity matrices $\mathcal{S}_{16}^{\min}[41,10]$, $\mathcal{S}_{16}^{\min}[21,5]$, and $\mathcal{S}_{16}[9,2]$ with corresponding feature resolutions of 1, 2, and 5 Hz, respectively. The resulting similarity clusters are shown in Fig. 7.14 for the Shostakovich

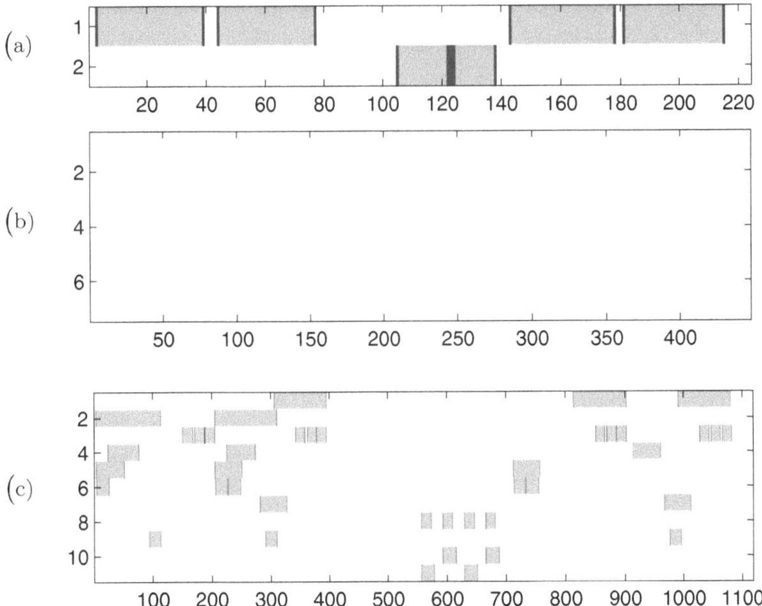

Fig. 7.14. Similarity clusters for the Shostakovich example of Fig. 7.2 resulting from a structure analysis using **(a)** $\mathcal{S}_{16}^{\min}[41,10]$, **(b)** $\mathcal{S}_{16}^{\min}[21,5]$, and **(c)** $\mathcal{S}_{16}[9,2]$

example. Note that the musical form $A_1A_2BC_1C_2A_3A_4D$ has been correctly identified at the low resolution level, see (a). Increasing the feature resolution has two effects. On the one hand, finer repetitive substructures are revealed as illustrated by (c). On the other hand, the algorithm becomes more sensitive toward local variations, resulting in fragmentation and incompleteness of the coarser structures. A difficult problem that has to be solved is to integrate the extracted similarity relations at all resolutions into a single hierarchical model that best describes the musical structure.

8

SyncPlayer: An Advanced Audio Player

In the previous chapters, we have discussed various MIR techniques and algorithms for automatically generating annotations and linking structures of interrelated music data. The generated data can be used to support inter- and intradocument browsing and retrieval in complex and inhomogeneous music collections, thus allowing users to discover and explore music in an intuitive and multimodal way. To demonstrate the potentials of our MIR techniques, we have developed the SyncPlayer system [114], which is a client-server based advanced audio player. The SyncPlayer integrates novel functionalities for multimodal presentation of audio as well as symbolic data and comprises a search engine for lyrics and other metadata. In Sect. 8.1, we give an overview of the SyncPlayer system, which consists of a server as well as a client component. The server component, as will be described in Sect. 8.2, includes functionalities such as audio identification, data retrieval, and data delivery. In contrast, the client component constitutes the user front end of the system and provides the user interfaces for the services offered by the server (Sect. 8.3). A discussion of related work and possible extensions of our system can be found in Sect. 8.4. A demo version of the SyncPlayer is available at [199].

8.1 Overview

Classical MIR techniques and applications typically work with a single type of music representation. Examples include most query-by-example scenarios like audio identification (where a query that consists of an audio fragment is to be identified within a large audio database), melodic queries (where a note sequence is to be matched to a melody database), polyphonic search (where an excerpt of a score is searched in a score database), or text-based search (where a sequence of query terms is searched in a database of lyrics or textual metadata). In such a scenario, all steps including the query formulation, the actual retrieval process, as well as the display of the query results are often based on the same data type. However, in general, there is no unique data representation that is equally suitable for all involved steps. As an example, consider a user who queries a melody database. Although it may be the most natural option to just hum a melody resulting in a *waveform-based* query, most successful algorithms for melody-based retrieval work in the domain of *symbolic* music. On the other hand, retrieval results are most naturally presented by playing back an actual *audio recording* of the piece of music containing the melody, while a *musical score* or a *piano-roll* representation may be the most appropriate form for visually displaying the query results.

It is the objective of the SyncPlayer system to overcome the above-mentioned limitations by systematically integrating, exploiting, and synchronizing various music representations. One typical application scenario realized by the client-server based SyncPlayer system can be sketched as follows: a user plays back an audio recoding (CD, MP3) selected from his personal music collection using the audio player offered by our system (the SyncPlayer client). During playback, the client connects to the SyncPlayer server, which in turn identifies the particular piece of audio as well as the current playback position. Subsequently, the server on-the-fly delivers available metadata (e. g., song title, composer, interpreter) as well as position-specific information on the audio piece (e. g., current score parameters, lyrics, tablature) to the client. The client then synchronously displays the received information during acoustic playback using a multimodal visualization plug-in. As a second application scenario, assume that a user only remembers parts of some lyrics such as "our yellow submarine." Using a Google-like interface provided by the SyncPlayer client (LyricsSearch plug-in), the user can submit a text-based query. The server delivers all available songs that contain the query phrase as well as the exact time positions within each of these songs. The time information can then be used to play back the audio recordings at the positions containing the query phrase, see Fig. 8.3.

The SyncPlayer system basically consists of three software components: a server component, a client component, and some data administration tools. Most parts of these components are implemented in Java – only some of the more time-critical tasks are realized as C++-modules. The role of the three components can be summarized as follows (see also Fig. 8.1):

1. The user operates the *client component*, which provides functionalities for acoustic playback, query interfaces, as well as various visualization tools.
2. A remote computer system runs the *server component*, which performs the audio identification and delivers metadata and annotations. Furthermore, the server comprises a search engine for symbolic metadata and annotations (which is currently extended to facilitate audio matching, see Chap. 6).
3. Several server-side *administration tools* are available for creating and updating the databases and indexes that underly the SyncPlayer system.

The SyncPlayer scenario is based on the assumption that, on the server-side, there exists a large collection of music documents. Here, for each given

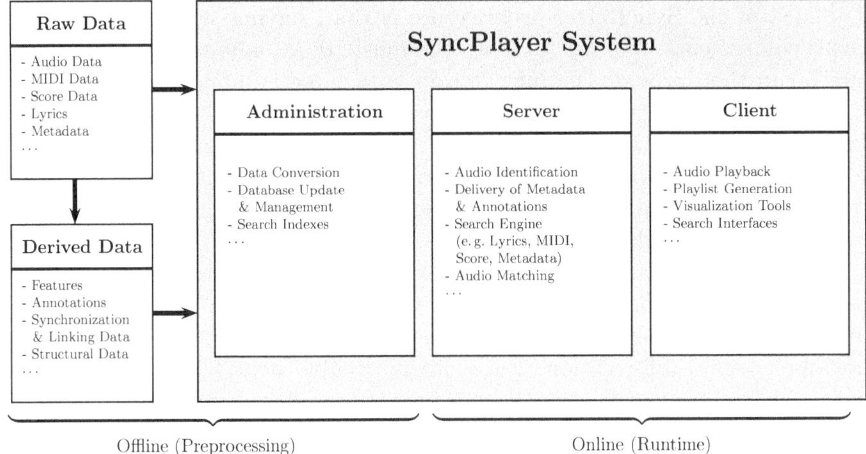

Fig. 8.1. Overview of the SyncPlayer system

piece of music the server should possess various digital representations (e. g., audio, MIDI, MusicXML, scanned images of sheet music) as well as associated metadata (e. g. lyrics, tablature). In the following, these kind of data will be referred to as *raw data*. The raw data are further processed to generate what we refer to as *derived data*. The derived data comprise high-level audio features (Chap. 3), various kinds of synchronization and linking data (Chap. 5), or structural data (Chap. 7). As we have seen in the previous chapters, such data may be generated efficiently in a purely automatic fashion by means of MIR techniques. Other types of derived data may include textual annotations of audio recordings such as synchronized lyrics of some recorded song or opera. The synchronization of such textual information with an audio recoding is currently done semimanually[1] [114]. Using the SyncPlayer administration tools, the raw data as well as the derived data are indexed and stored in databases, which can then be efficiently accessed by the SyncPlayer server. The generation of the derived data as well as the data organization and indexing can be done offline in some preprocessing step (Fig. 8.1). For further technical details concerning the data administration and the SyncPlayer implementation, we refer to [72,114]. In the following two sections, we describe the main functionalities of the server and client.

8.2 SyncPlayer Server

In the preprocessing step, music data of various types, including the raw data as well as the derived data, have to be collected, organized, and indexed.

[1] The automatic lyrics-audio synchronization is yet a more or less unsolved research problem. A first approach under strong structural assumptions has been described in [213].

The idea of the SyncPlayer architecture is that, having separate server and client components, one has to store the music data only once on the server, which can then provide the information for a large number of users (clients) simultaneously. When a client connects to the SyncPlayer server through the Internet or local networks, the main task of the SyncPlayer server is to identify, retrieve, and deliver the information requested by the client. In the following, we summarize the main services provided by the SyncPlayer server and refer to [72, 114] for details.

When a user starts playback of a specific audio recording and requests particular annotations, the server has to automatically identify the recording as well as the current playback position prior to retrieving and delivering position-specific information. To avoid noticeable delays in displaying this data, the audio identification has to be performed very efficiently. Furthermore, as audio recordings are available in different qualities or may have been edited prior to playback, the identification should be robust against basic signal processing operations such as resampling, lossy compression (e. g., MP3), cropping, or equalization. In the last few years, several powerful audio identification methods have been proposed [3, 28, 92, 112, 114, 177, 194, 211]. In the SyncPlayer system, we use the identification method described in [114, 177]. In this approach, an audio signal is transformed into a set of features as follows: first, a short time Fourier transform (STFT) is computed yielding a 2D time-frequency representation of the signal. Next, one extracts time-frequency pairs that correspond to local maxima in the STFT according to absolute values. Then, a feature is basically such a pair consisting of a time as well as a (quantized) positive real-valued frequency parameter (Fig. 8.2). In the preprocessing step, each audio recording of the data collection is transformed into such a set of features, also referred to as *fingerprints*, which are then indexed using an inverted file index [42].

Now, during playback, the SyncPlayer client locally extracts fingerprints from the selected audio recording (using a small audio excerpt of a few seconds of duration) and sends them to the server. On the basis of the fingerprint index, the server can efficiently identify the audio recording (assuming that the recording is contained in the collection) as well as the current position of the excerpt within the recording. To make audio identification robust to the above-mentioned signal distortions, the system employs various mechanisms of fault tolerance including mismatches and fuzzy search. We refer to [114, 177] for further details.

Once the audio recording as well as the current playback position has been identified, the SyncPlayer server retrieves the data that is available for the current playback position (e. g., lyrics, score parameters, structural data, linking data, or general metadata) from annotation database. Using a suitable communication protocol, the information is then transferred to the client. During subsequent playback, the client may constantly request further position-specific annotations on the identified piece of audio. In such a case, using a suitable synchronization protocol between client and server, a repeated

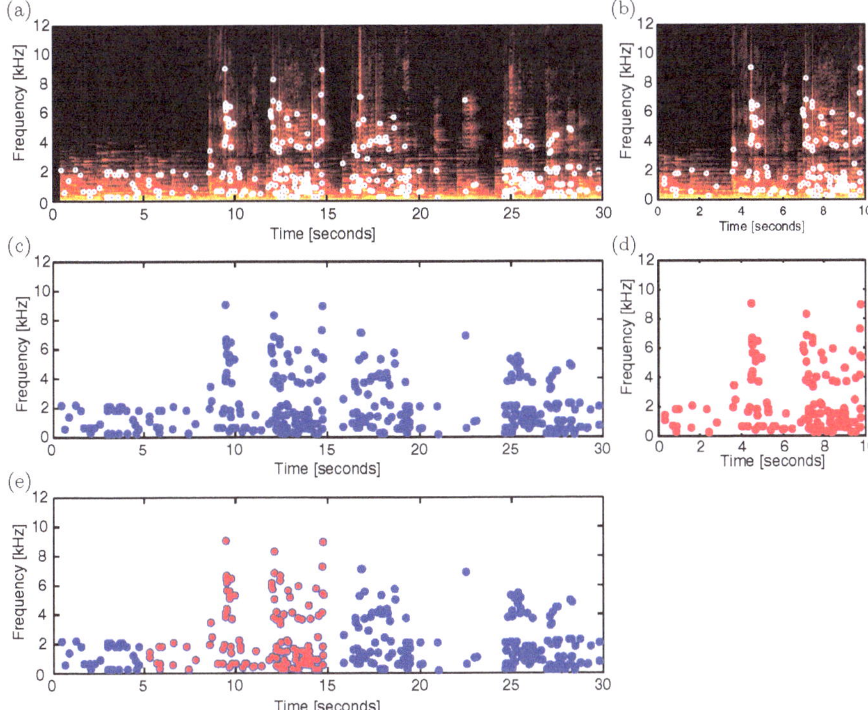

Fig. 8.2. Illustration of the audio identification process using fingerprints (from [72]). STFT of **(a)** an example database document and **(b)** a query excerpt. The extracted fingerprints are indicated by *blue* and *red dots*, respectively. **(c)** Database fingerprints, which are included in the audio index. **(d)** Query fingerprints, which are sent by the client to the server. **(e)** Superposition of the database fingerprints and time-shifted query fingerprints indicating that the query excerpt has been identified as a segment of the database document starting at time position 5 (s)

fingerprint-based identification of the current playback position may not be necessary.

As another service, the SyncPlayer server provides a search engine that allows a user to search for fragments of lyrics that occur in a song or an opera. Our technique for lyrics retrieval is term-based and, as our audio identification technique described earlier, uses an inverted file index, which is computed in the preprocessing step. The retrieval algorithm is capable of finding exact matches as well as fuzzy matches with a given number of mismatches [42]. The special feature of the SyncPlayer is that it not only retrieves the lyrics containing the query terms, but also the time positions at which the query occurs in the retrieved audio recordings. This allows the user to directly jump to the desired audio excerpts. As an example, the right part of Fig. 8.3 shows

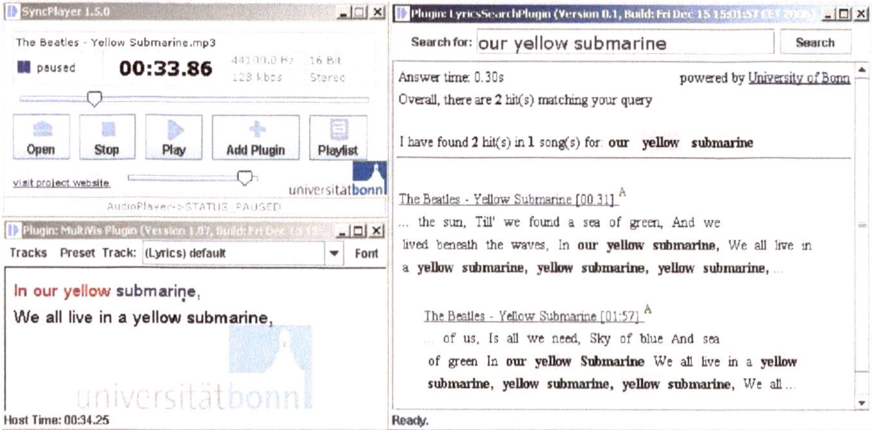

Fig. 8.3. *Right*: LyricsSearch plug-in of the SyncPlayer client showing a text-based query (*top*) and two retrieval results including the time position in the audio (*bottom*). *Left*: Main window of the SyncPlayer client used for playback (*top*) and MultiVis plug-in for lyrics visualization (*bottom*)

the *LyricsSearch* plug-in of the client component (see Sect. 8.3) for textual queries. After the user has submitted a query phrase, the server delivers a ranked list of query results, where the lyrics positions of matched query terms are highlighted. Upon selecting one of the query results contained in this list, the visualization plug-in (bottom left of Fig. 8.3) for lyrics is launched and playback starts at the position where the corresponding lyrics occur within the audio recording. A more detailed discussion of the lyrics search engine is given in [114].

8.3 SyncPlayer Client

The client component is the user front-end of the SyncPlayer system and in its basic mode conceptually behaves like a standard audio player. Furthermore, the SyncPlayer client provides several user interfaces, which are available as plug-ins connecting to the various services offered by the SyncPlayer server. In the current version of the SyncPlayer, plug-ins are available for the visualization of audio data, for the synchronous display of annotations, for structure-based audio navigation, for lyrics retrieval, and for audio switching. In this section, we give a short overview of these functionalities and refer to [72] for a more detailed account.

The main graphical user interface of the SyncPlayer client offers standard functionalities of an audio player including file handling and playback controls (left part of Fig. 8.4). Currently, the SyncPlayer supports playback of `*.wav` and `*.mp3` files. The main window displays the filename of the currently

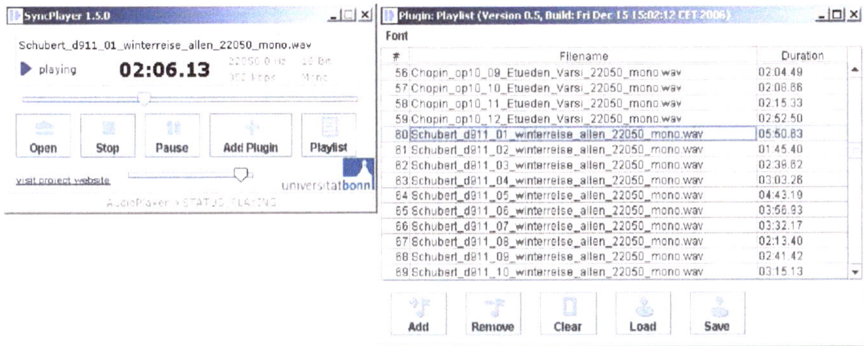

Fig. 8.4. Basic graphical user interface of the SyncPlayer client (*left*) and the Playlist plug-in (*right*)

selected audio file as well as the current playback time. Furthermore, the button "Add Plugin" allows the user to open an arbitrary number of plug-ins for the various functionalities of the SyncPlayer. Finally, through the button "Playlist," the user may activate the *Playlist* plug-in for quick and convenient access to audio collections (right part of Fig. 8.4). This plug-in allows the user to load, create, rearrange, and save playlists according to his needs.

The waveform or a time-frequency representation (spectrogram) of the audio recording can be visualized by means of the *AudioPlotter* plug-in (Fig. 8.5). The plug-in supports zooming, which enables the user to select arbitrary sections of the audio being displayed at the desired resolution. The AudioPlotter may be easily extended for visualizing other types of audio representations such as pitch, chroma, or MFCC representations (Chap. 3).

Most of the current audio players only allow for handling the audio data itself or for visualizing derived representations and audio features. The SyncPlayer client additionally facilitates visualization of symbolic annotations alongside the audio playback by means of the *MultiVis* plug-in (Multimodal Visualization plug-in). The annotations are organized on a per-piece basis and can be loaded from locally available XML annotation files or retrieved from the SyncPlayer server during playback. For each recording there may be multiple annotation tracks of various types – the current system supports lyrics annotations, MIDI annotations, and annotations for the audio structure. For each type, in turn, there may be several different useful display strategies, depending upon the desired application. Currently, the MultiVis plug-in offers only one display strategy for each type, which are now discussed in more detail.

As a first example, Fig. 8.6 shows an instance of the MultiVis plug-in for displaying lyrics annotations. The text is displayed similar as in karaoke applications. During audio playback, several text lines are displayed in advance. When the time position of a specific word is reached, the word is

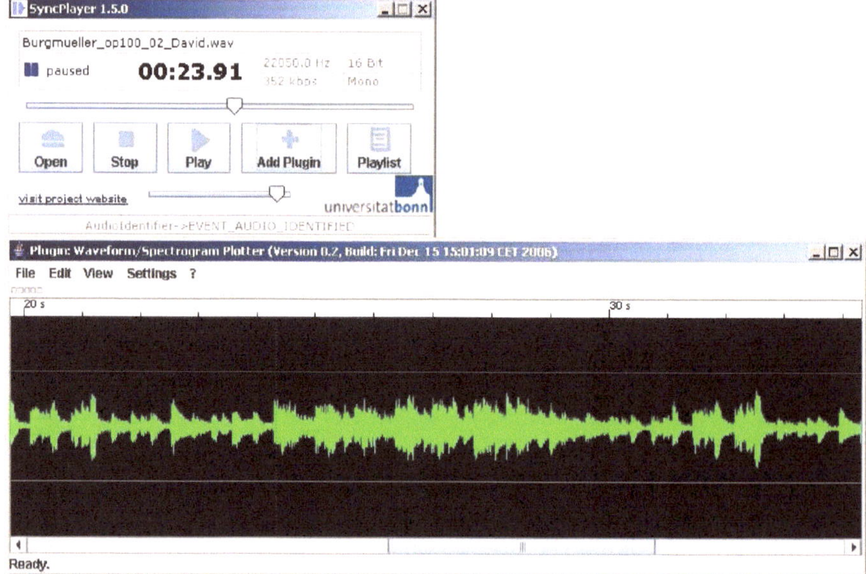

Fig. 8.5. AudioPlotter plug-in for visualizing audio representations. The current version supports the visualization of the waveform (as shown in this figure) and of a spectrogram

Fig. 8.6. MultiVis plug-in shown in the lyrics mode (*right*). The text is displayed as in typical karaoke applications

highlighted by changing color: future words are displayed in black, the active word is displayed in blue, and past words are displayed in red. A small blue box that is moving underneath the text indicates the estimated current playback position within the active annotation region. This position is calculated by linear interpolation between the start and end points of the word event within the audio. Furthermore, the user may choose from a list of fonts and sizes used to display the lyrics.

The second mode of the MultiVis plug-in is used for displaying synchronized MIDI information (Fig. 8.7). In the current version, the MIDI

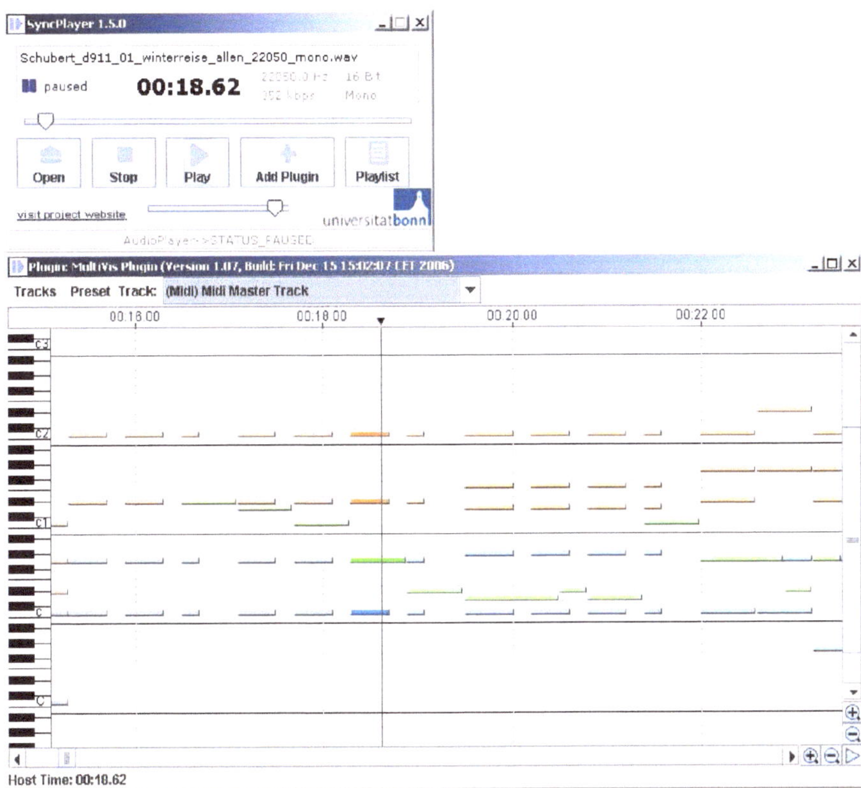

Fig. 8.7. MultiVis plug-in in the piano-roll mode for a recording of Schubert's Winterreise D911 No. 1 ("Gute Nacht") for piano and voice. The three available MIDI tracks are shown in different colors: vocals (*green*), piano left hand (*blue*), and piano right hand (*orange*)

information is displayed in the piano-roll representation as introduced in Sect. 2.1.3. Each MIDI note is represented by a horizontal bar within a two-dimensional diagram. The vertical location of the bar indicates the MIDI pitch, whereas the start and end points in horizontal direction reflect the note onset and offset, respectively. The vertical black line indicates the current playback position. Notes that are hit by the playback indicator are currently audible in the audio playback. The piano-roll display allows the observer to visually understand the onsets and durations of past, active, and future musical notes. Furthermore, the MIDI annotations used in the MultiVis plug-in are adjusted to the particular audio recording, thus reflecting the local tempo variations due to musical interpretation. The MultiVis plug-in in the piano-roll mode supports several functionalities for zooming and audio navigation. For example, by clicking on the timeline indicator on top of the piano-roll

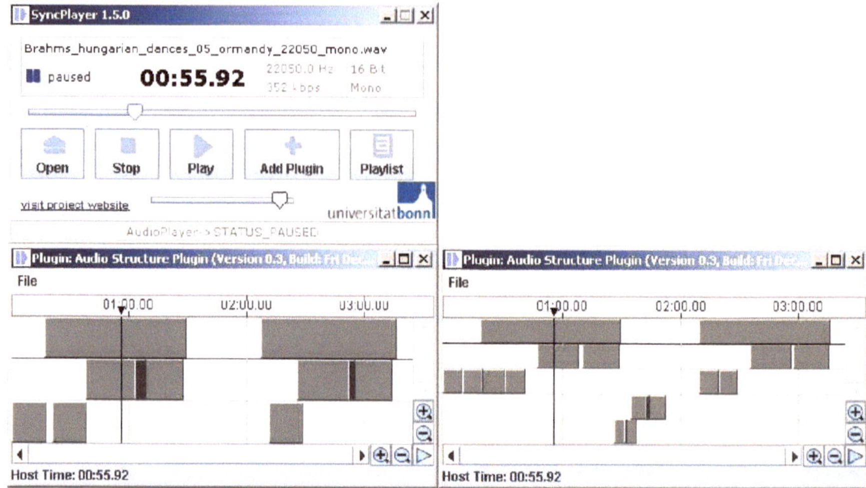

Fig. 8.8. MultiVis plug-in in the audio structure mode for the Ormandy recording of Brahms' Hungarian Dance No. 5. The *bottom left* shows an instance of the MultiVis plug-in for the audio structure as described in Fig. 7.10. The *bottom right* shows another instance for the audio structure computed at a finer resolution level (cf. Fig. 7.14)

representation during playback, one can directly jump to the corresponding time position in the audio recording.

A third mode of the MultVis plug-in allows for displaying the audio structure, which has been determined in a preprocessing step as described in Chap. 7. Figure 8.8 shows two instances of the MultiVis plug-in in the audio structure mode for an Ormandy recording of Brahms' Hungarian Dance No. 5 having the musical form $A_1A_2B_1B_2CA_3B_3B_4D$ (Sect. 7.1). In the graphical representation, each audio segment is represented by a grey horizontal bar. Each row consists of a list of bars encoding a cluster of similar audio segments. Time segments that are contained in the same cluster may overlap, which is encoded by the dark-gray. The display controls are essentially the same as for the piano-roll display. Additionally, the user may set the playback position to the beginning of any audio segment simply by clicking on the corresponding bar, see also [82] for a similar functionality. As illustrated by Fig. 8.8, the user may simultaneously open several instances of the MultiVis plug-in showing the audio structure at different resolution levels.

Next, we discuss the *AudioSwitcher* plug-in for simultaneous playback of synchronized audio recordings (Fig. 8.9). The AudioSwitcher plug-in allows the user to select several recordings of the same piece of music, which have been previously synchronized as described in Sect. 5.2. Each of the selected recordings is represented by a slider bar indicating the current playback position with respect to the recording's particular time scale. The audio recording

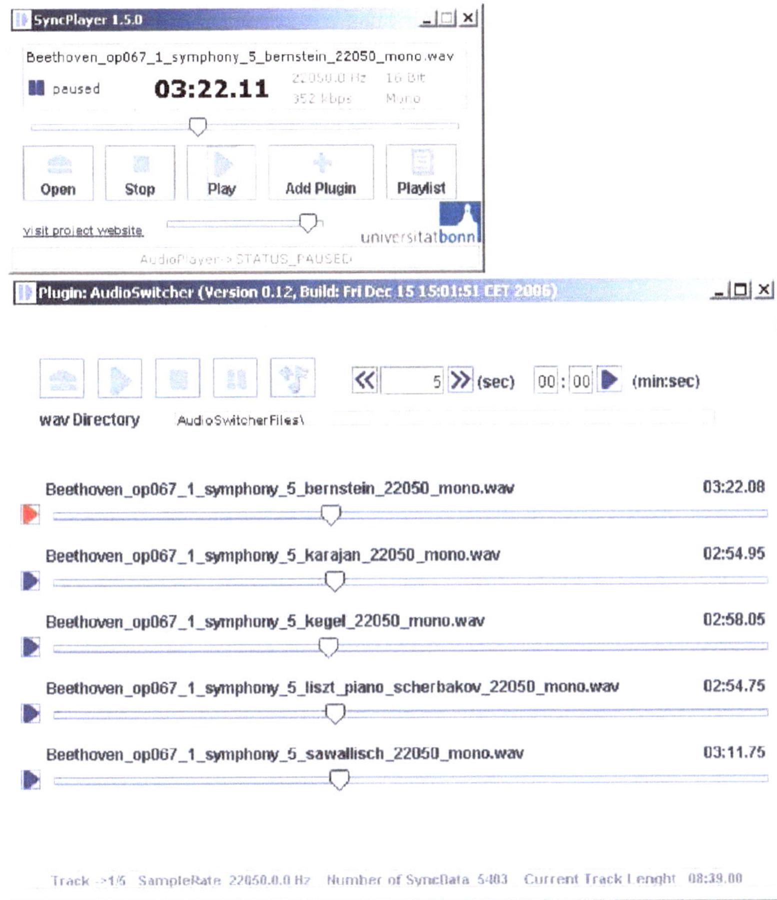

Fig. 8.9. AudioSwitcher plug-in for synchronous playback of different audio recordings of the same piece of music. In this example, five different interpretations of Beethoven's Fifth Symphony are loaded

that is currently used for audio playback, in the following referred to as reference recording, is represented by a red marker. The slider of the reference recording moves at constant speed while the sliders of the other recordings move according to the relative tempo variations with respect to the reference. The reference recording may be changed at any time simply by clicking on the respective marker located to the left of each slider. The playback of the new reference recording then starts at the time position that musically corresponds to the last playback position of the former reference. One can also jump to any position within any of the recordings by directly selecting a position of the respective slider. This will automatically initiate a switch of reference to the respective recording.

Finally, the SyncPlayer client supplies a *LyricsSearch* plug-in (Fig. 8.3), as has already been mentioned in Sect. 8.2, implementing a user interface for text-based queries in lyrics. The user enters a search phrase consisting of a few words, which is then transmitted to the SyncPlayer server. The server efficiently performs the search on the lyrics collection and delivers a list of retrieval results, also referred to as hits. This list is presented to the user by the LyricsSearch plug-in. Each hit is indicated by the title of the recording along with the exact time position where the query phrase occurs in the audio. To support the accessibility of the retrieval results, the following mechanism has been added: making use of the audio identification service offered by the SyncPlayer Server, the LyricsSearch plug-in is capable of creating a list of identified pieces that are available in the user's music collection. Simply by clicking on a hit, the user may then directly access the corresponding audio excerpt as long as it is contained in his personal music collection. Here, the SyncPlayer starts playback at a position located a few seconds before the actual match position containing the query phrase. Using this mechanism, the user can quickly browse through the audio files containing the retrieved hits.

Finally, we note that the SyncPlayer plug-in concept allows any number of plug-in instances to be opened at the same time (Fig. 8.10) affording multimodal visualization of music data. Further functionalities and technical details can be found in [72,114]. A demo version of the SyncPlayer is available at [199].

8.4 Further Notes

In this chapter, we have discussed the SyncPlayer system, which constitutes an integrated MIR system for multimodal visualization, browsing, and retrieval of music data. The client–server architecture makes this system suitable for distributed environments. Our prototypical SyncPlayer implementation illustrates the impact and potentials of MIR research including audio analysis tasks such as music synchronization, audio identification, or audio structure analysis. The visualization and browsing functionalities of the SyncPlayer have turned out to be beneficial for manually evaluating (e. g., by listening to the audio while simultaneously checking the annotations) the performance of our MIR algorithms. We hope that our advanced audio player opens new and unprecedented ways of music listening and experience, provides novel browsing and retrieval strategies, and constitutes a valuable tool for music education and music research. For the future, large-scale software evaluations and systematic user studies have to be conducted to identify user needs and to develop the SyncPlayer system into a marketable commodity.

Until now, the integration and usage of different types of music representation within a single MIR application has been mainly considered in the retrieval scenario. For example, in the *query-by-humming* scenario, (see,

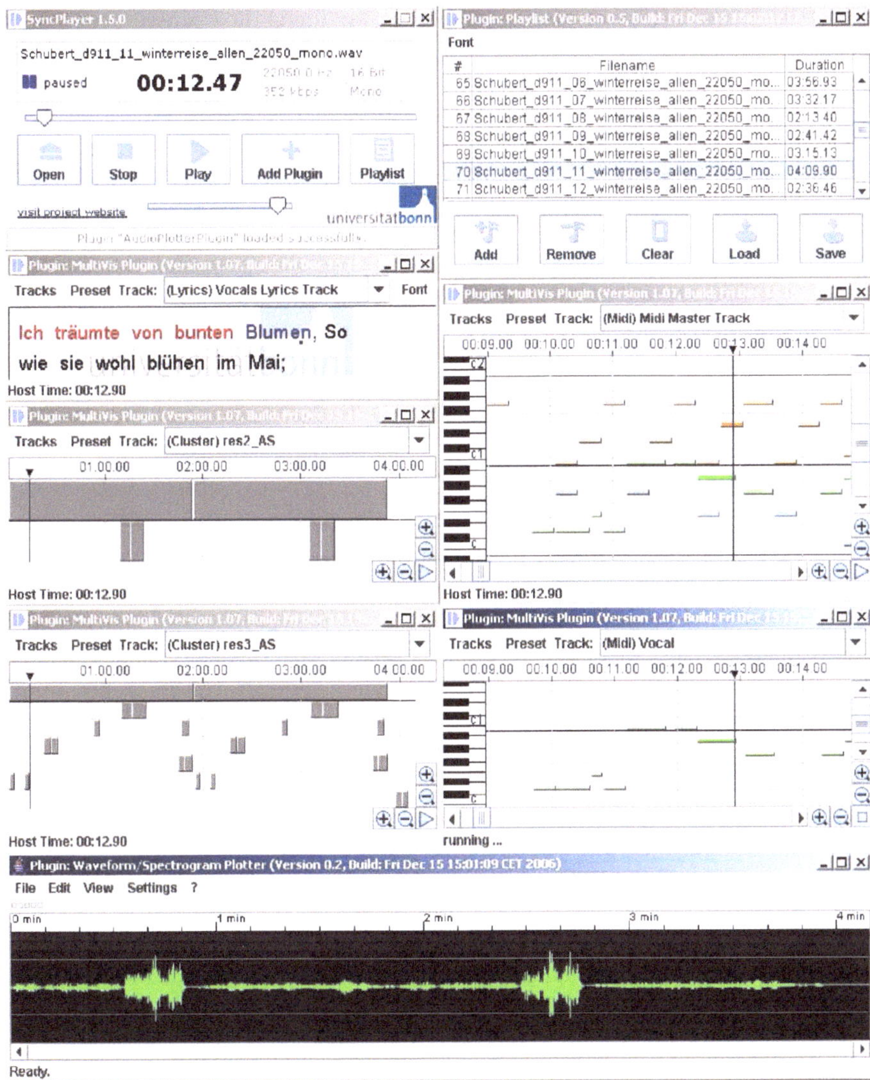

Fig. 8.10. The SyncPlayer client with multiple instances of available plug-ins

e. g., [160]) the user hums a melody fragment into a microphone. The resulting audio excerpt is then converted into the symbolic domain and used as a query to a melody database. Other methods have been proposed to query music in a cross-domain fashion. For example, Pickens et al. [163] present a system for querying polyphonic audio recordings in a database of polyphonic music given in the symbolic domain.

Most of the MIR systems developed so far focus on a single MIR application – in contrast to the SyncPlayer system, which integrates several MIR techniques. For example, Gracenote [85] and `freedb.org` [71] offer services based upon a database of metadata for music CDs. The database contains global information on the CD such as artist, year, and publisher, as well as the names and durations of the audio tracks. By checking the track durations of the requested CD, the service identifies the CD and delivers the metadata to the client. Both projects also offer web-based search interfaces that allow the user to search for individual songs or CDs via keywords. In recent years, services such as *song identification* have become commercially available for pop music and are offered by several mobile providers. Here, the user transfers an excerpt of about 10 s of an unknown song to the service provider, for example, through a voice connection established by a cell phone. The service, in turn, identifies the song using audio identification techniques and sends back metadata such as the song's title or the artist's name. Other services offer identification through an Internet connection such as Tunatic [201]. Minilyrics [135] is a service for the automatic delivery of lyrics annotations to a given song. The service is realized through plug-ins, which are available for several common audio players (e. g., WinAmp, iTunes, Windows Media Player). Filename or tags containing artist and title are transferred to a server, which in turn sends back available lyrics annotations for the given song. In contrast to the SyncPlayer, the Minilyrics services requires the identified audio file to be properly tagged by using ID3-tags[2]. Most of the available lyrics are roughly synchronized to the audio recording on a per-line-of-text basis. The Sonic Visualizer [27] is an application for visualizing and analyzing the contents of music audio files. It allows the user to load and create various derived audio representations and supports the use of external feature extractors to generate audio annotation. The display of MIDI files is supported as well. In contrast to the SyncPlayer, however, the MIDI files are assumed to correspond to the audio files – the MIDI data cannot be automatically synchronized to the audio data. The SmartMusicKIOSK system [82, 84] is an advanced music playback interface for trial listening as can be used in music stores. Similar to the SyncPlayer system, the start and end points of repeating audio segments of popular songs, especially of the chorus sections, are extracted in a preprocessing step [81]. As in Fig. 8.8, the audio structure is then displayed on a screen enabling the user to navigate through the structural representation of the song.

The current version of the SyncPlayer system only represents a snapshot of how various MIR techniques may be combined to build up a powerful, multimodal MIR system. Obviously, there are many meaningful ways for extending the SyncPlayer system. Exemplarily, we sketch two extensions that are currently under development. One of our current research projects is concerned

[2] ID3 is a format to store metadata such as the title, the artist, the track number or other information within an MP3 file.

with the alignment of the pixels of scanned sheet music with corresponding time positions of an audio file (Sect. 5.4). This synchronization data can then be used to highlight the current position in the scanned score or to automatically turn pages during playback of a CD recording. The need of such a functionality has also been expressed in [61]. A second extension concerns the current audio identification technique, which is too restrictive when considering different interpretations for one and the same piece of music. In particular in the case of classical music, the assumption that the server's audio database contains all possible recordings of a given piece of music seems unrealistic. Here, the identification of a piece of music regardless of the specific interpretation would overcome this limitation. To this end, it is the goal of an ongoing research project to integrate the audio matching techniques (Chap. 6) into the SyncPlayer system to achieve independence of specific performances of a piece of music.

Part II

Analysis and Retrieval Techniques for Motion Data

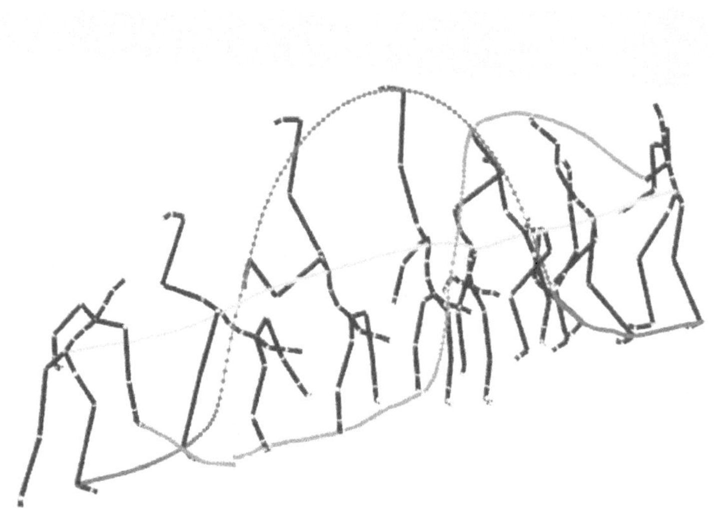

9

Fundamentals on Motion Capture Data

The second part of this monograph deals with content-based analysis and retrieval of 3D motion capture data as used in computer graphics for animating virtual human characters. In this chapter, we provide the reader with some fundamental facts on motion representations. We start with a short introduction on motion capturing and introduce a mathematical model for the motion data as used throughout the subsequent chapters (Sect. 9.1). We continue with a detailed discussion of general similarity aspects that are crucial in view of motion comparison and retrieval (Sect. 9.2). Then, in Sect. 9.3, we formally introduce the concept of kinematic chains, which are generally used to model flexibly linked rigid bodies such as robot arms or human skeletons. Kinematic chains are parameterized by joint angles, which in turn can be represented in various ways. In Sect. 9.4, we describe and compare three important angle representations based on rotation matrices, Euler angles, and quaternions. Each of these representations has its strengths and weaknesses depending on the respective analysis or synthesis application.

9.1 Motion Capture Data

There are many ways to generate motion capture data using, e.g., mechanical, magnetic, or optical systems, each technology having its own strengths and weaknesses. For an overview and a discussion of the pros and cons of such systems we refer to Wikipedia [215]. We exemplarily discuss an optical marker-based technology, which yields very clean and detailed motion capture data. Here, the actor is equipped with a set of 40–50 retro-reflective markers attached to a suit. These markers are tracked by an array of 6–12 calibrated high-resolution cameras at a frame rate of up to 240 Hz, see Fig. 9.1. From the recorded 2D images of the marker positions, the system can then reconstruct the 3D marker positions with high precision (present systems have a resolution of less than a millimeter). Then, the data are cleaned with the aid of semi-automatic gap filling algorithms exploiting kinematic constraints. Cleaning is

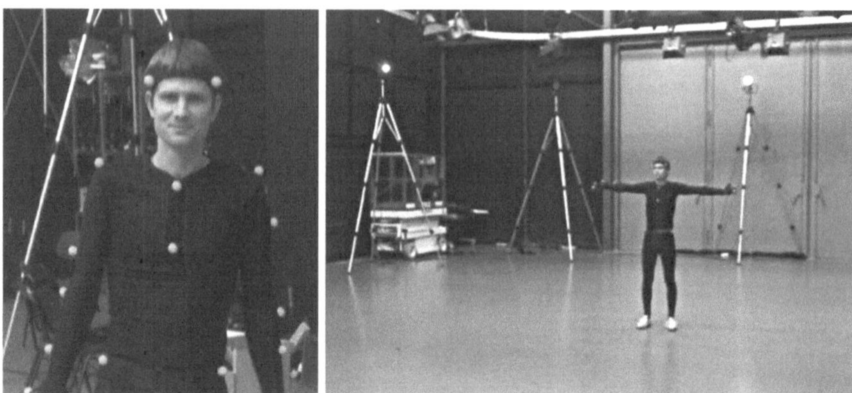

Fig. 9.1. Optical motion capture system based on retro-reflective markers attached to the actor's body. The markers are tracked by an array of 6–12 calibrated high-resolution cameras, typically arranged in a circle

necessary to account for missing and defective data, where the defects are due to marker occlusions and tracking errors. In many applications, the 3D marker positions can be directly used for further processing. For example, in computer animation one often directly maps the 3D marker positions to corresponding positions of the animated characters. For other applications, the 3D marker positions have to be converted to a skeletal kinematic chain representation using appropriate fitting algorithms [54, 155]. Such an abstract model has the advantage that it does not depend on the specific number and the positions of the markers used for the recording. However, the mapping process from the marker data onto the abstract model can introduce significant artifacts that are not due to the marker data itself. Here, one major problem is that skeletal models are only approximations of the human body that often do not account for biomechanical issues, see Zatsiorsky [223].

In our scenario, we assume that the mocap data are modeled using a *kinematic chain*, which may be thought of as a simplified copy of the human skeleton. A kinematic chain consists of *body segments* (the *bones*) that are connected by *joints* of various types, see Fig. 9.2a. Let J denote the set of joints, where each joint is referenced by an intuitive term such as "root," "lankle" (for "left ankle"), "rankle" (for "right ankle"), "lknee" (for "left knee"), and so on. For simplicity, end effectors such as toes or fingers are also regarded as joints. In the following, a *motion capture data stream* is thought of as a sequence of *frames*, each frame specifying the 3D coordinates of the joints at a certain point in time. Moving from the technical background to an abstract geometric context, we also speak of a *pose* instead of a frame. Mathematically, a pose can be regarded as a matrix $P \in \mathbb{R}^{3\times|J|}$, where $|J|$ denotes the number of joints. The jth column of P, denoted by P^j, corresponds to the 3D coordinates of joint $j \in J$. A motion capture data stream (in information

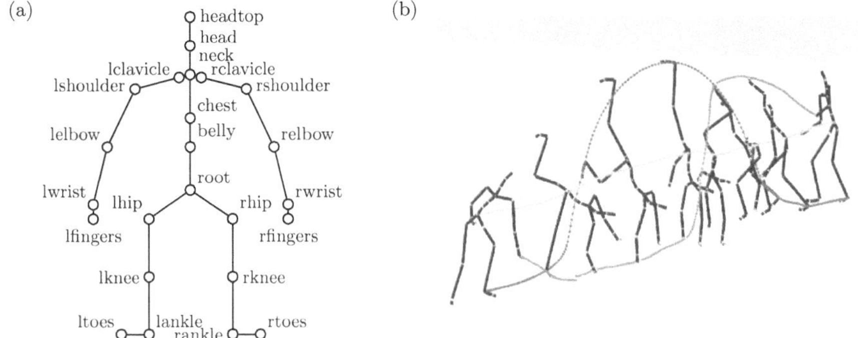

Fig. 9.2. (a) Skeletal kinematic chain model consisting of rigid *bones* that are flexibly connected by *joints*, which are highlighted by circular markers and labeled with joint names. **(b)** Motion capture data stream of a cartwheel represented as a sequence of poses. The figure shows the 3D trajectories of the joints "root" (*green*), "rfingers" (*red*), and "lankle" (*blue*)

retrieval terminology also referred to as a *document*) can be modeled as a function

$$D : [1 : T] \to \mathcal{P} \subset \mathbb{R}^{3\times|J|}, \tag{9.1}$$

where $T \in \mathbb{N}$ denotes the number of poses, $[1 : T] := \{1, 2, \ldots, T\}$ corresponds to the time axis (for a fixed sampling rate), and $\mathcal{P}$ denotes the set of poses. A subsequence of consecutive frames is also referred to as a *motion clip*. Finally, the curve described by the 3D coordinates of a single body joint is termed *3D trajectory*. This definition is illustrated by Fig. 9.2b.

9.2 Similarity Aspects

One central task in motion analysis is the design of suitable similarity measures to compare two given motion sequences in a semantically meaningful way. The notion of similarity, however, is an ill-defined term that depends on the respective application or on a person's perception. For example, a user may be interested only in the rough course of the motion, disregarding motion style or other motion details such as the facial expression. In other situations, a user may be particularly interested in certain nuances of motion patterns, which allows him to distinguish, e.g., between a front kick and a side kick, see Fig. 1.1. In the following, we discuss some similarity aspects that play an important role in the design of suitable similarity measures or distance functions.

Typically, two motions are regarded as similar if they only differ by certain *global transformations* as illustrated by Fig. 9.3a. For example, one may leave the absolute position in time and space out of consideration by using a similarity measure that is invariant under temporal and spatial translations.

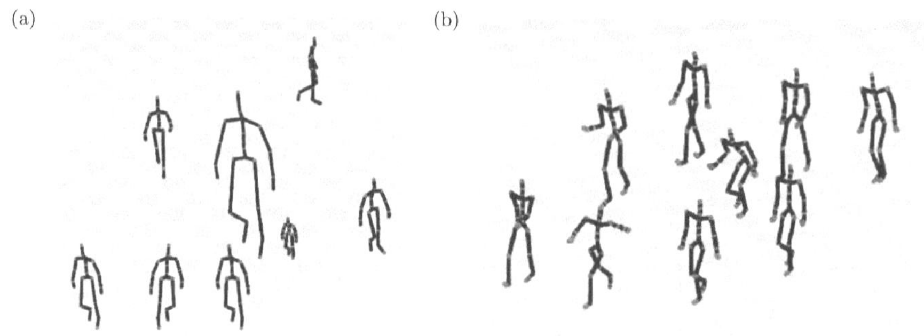

Fig. 9.3. **(a)** Different global transformations applied to a walking motion. **(b)** Different styles of walking motions

Often, two motions are identified when they differ with respect to a global rotation about the vertical axis or with respect to a global reflection. Furthermore, the size of the skeleton or the overall speed of the motions may not be of interest – in such a case, the similarity measure should be invariant to spatial or temporal scalings.

More complex are variations that are due to different motion styles, see Fig. 9.3b. For example, walking motions may differ by performance (e.g., limping, tiptoeing, or marching), by emotional expression or mood (e.g., "cheerful walking," "furious walking," "shy walking"), and by the complex individual characteristics determined by the motion's performer. The abstract concept of *motion style* appears in the literature in various forms and is usually contrasted by some notion of *motion content*, which is related to the semantics of the motion. In the following, we give an overview of how motion style and motion content are treated in the literature.

In the context of gait recognition, Lee and Elgammal [118] define motion style as the time-invariant, personalized aspects of gait, whereas they view motion content as a time-dependent aspect representing different body poses during the gait cycle. Similarly, Davis and Gao [53] view motions as depending on style, pose, and time. In their experiments, they use PCA on expert-labeled training data to derive those factors (essentially linear combinations of joint trajectories) that best explain differences in style. Rose et al. [180] group several example motions that differ only by style into *verb* classes, each of which corresponds to a certain motion content. They synthesize new motions from these verb classes by suitable interpolation techniques, where the user can control interpolation parameters for each verb. These parameters are referred to as *adverbs* controlling the style of the verbs. To synthesize motions in different styles, Brand and Hertzmann [19] use example motions to train so-called *style machines* that are based on hidden Markov models (HMMs). Here, motion style is captured in certain parameters of the style machine such as average state dwell times and emission probability distributions for each state. On the other hand, motion content is encoded as the most likely state sequence of the style machine. Hsu et al. [93] propose a system for *style translation* that is capable of changing motions performed in a specific input style into new motions with the same content but a different output style. The characteristics of the input and output styles are learned from example data and are abstractly encoded in a linear dynamic system. A physically-based approach to grasping the stylistic characteristics of a motion performance is proposed by Liu et al. [122]. They use a complex physical model of the human body including bones, muscles, and tendons, the biomechanical properties of which (elasticity, stiffness, muscle activation preferences) can be learned from training data to achieve different motion styles in a synthesis step. Troje [200] trains linear PCA classifiers to recognize the gender of a person from recorded gait sequences, where the "gender" attribute seems to be located in the first three principal components of a suitable motion representation. Using a Fourier expansion of 3D locomotion data, Unuma et al. [205] identify certain *emotional* or *mood* aspects of locomotion style (for instance, "tired," "brisk," "normal") as gain factors for certain frequency bands.

Pullen and Bregler [169] also use a frequency decomposition of motion data, but their aim is not to pinpoint certain parameters that describe specific styles. Instead, they try to extract those details of the data that account for the natural look of captured motion by means of multiresolution analysis (MRA) on mocap data [22]. These details are found in certain high-frequency bands of the MRA hierarchy and are referred to as *motion texture* in analogy to the texture concept in computer graphics, where photorealistic surfaces are rendered with texture mapping. The term "motion texture" is also used by Li et al. [121] in the context of motion synthesis, but their concept is in no way related to the signal processing approach of Pullen and Bregler [169]. In their parlance, motion textures are generative statistical models describing an entire class of motion clips. Similar to style machines [19], these models consist of a set of *motion textons* together with transition probabilities encoding typical orders in which the motion textons can be traversed. Each motion texton is a linear dynamic system (see also Hsu et al. [93]) that specializes in generating certain subclips of the modeled motion. Parameter tuning at the texton level then allows for manipulating stylistic details.

Inspired by the performing arts literature, Neff and Fiume [153, 154] explore the aspect of *expressiveness* in synthesized motions. Their system enables the user to describe motion content in a high-level scripting language. The content can be modified globally and locally by applying procedural *character sketches* and *properties*, which implement expressive aspects such as "energetic," "dejected," or "old man,"

Returning to the walking example of Fig. 9.3b, we are faced with the question of how a walking motion can be characterized and recognized irrespective of motion style or motion texture. Video-based motion recognition systems such as [20, 87] tackle this problem by using hierarchical HMMs to

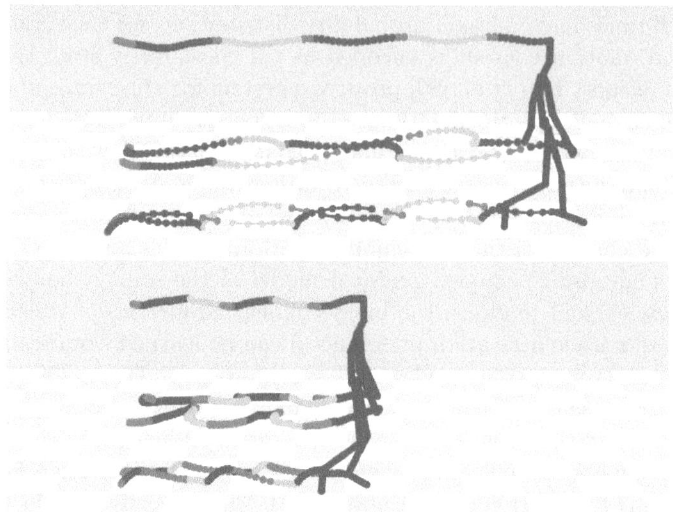

Fig. 9.4. Two walking motions performed in different speeds and styles. The figure shows the 3D trajectories for "headtop," "rfingers," "lfingers," "rankle," and "lankle." Semantically corresponding segments between the two walking motions are indicated by the same colors

model the motion content. The lower levels of the hierarchy comprise certain HMM building blocks representing fundamental components of full-body human motion such as "turning" or "raising an arm." In analogy to *phonemes* in speech recognition, these basic units are called *dynemes* by Green and Guan [87] or *movemes* by Bregler [20]. Dynemes/movemes and higher-level aggregations of these building blocks are capable of absorbing some of the motion variations that distinguish different executions of a motion.

The focus of this chapter is the automatic analysis of motion content. How can one grasp the gist of a motion? How can semantically similar motions be identified even in the presence of significant spatial and temporal variations? How can one determine and encode characteristic aspects that are common to all motions contained in some given motion class? As was mentioned earlier, the main problem in motion comparison is that semantically related motions need not be numerically similar as was illustrated by the two kicking motions of Fig. 1.1. As another example, the two walking motions shown in Fig. 9.4 can be regarded as similar from a semantic point of view even though they differ considerably in speed and style. Here, using techniques such as dynamic time warping, one may compensate for spatio-temporal deformations between related motions by suitably warping the time axis to establish frame correspondences, see Kovar and Gleicher [107]. Most features and local similarity measures used in this context, however, are based on numerical comparison of spatial or angular coordinates and cannot

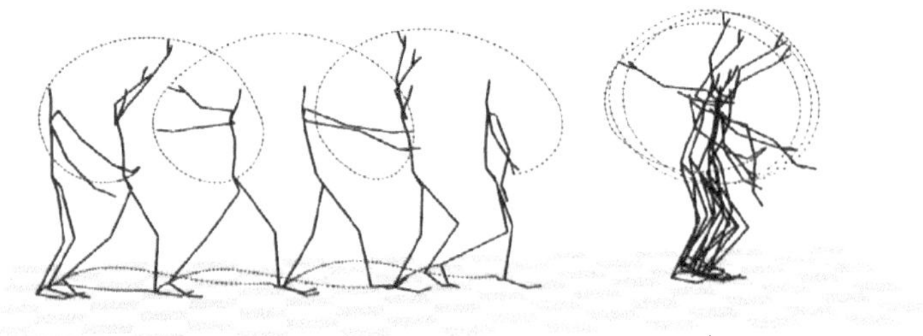

Fig. 9.5. Three repetitions of "rotating both arms forwards." The character on the left is walking while rotating the arms (2.7 s), whereas the character on the right is standing on one spot while rotating the arms (2.3 s). The trajectories of the joints "rankle," "lankle," and "lfingers" are shown

deal with qualitative variations. Besides spatio-temporal deformations, differences between semantic and numerical similarity can also be due to *partial similarity*. For example, the two instances of "rotating both arms forwards" as shown in Fig. 9.5 are almost identical as far as the arm movement is concerned, but differ with respect to the movement of the legs. Numerically, the resulting trajectories are very different – compare, for example, the cycloidal and the circular trajectories of the hands. Semantically, the two motions could be considered as similar.

Even worse, numerical similarity does not necessarily imply semantic similarity. For example, the two actions of picking up an object and placing an object on a shelf are very hard to distinguish numerically, even for a human being [107]. Here, the context of the motion or information about interaction with objects would be required, see also Kry and Pai [111]. Often, only minor nuances or partial aspects of a motion account for semantic differences. Think of the motions "standing on a spot" compared to "standing accompanied by weak waving with one hand": such inconspicuous, but decisive details are difficult for a full-body similarity measure to pick up unless the focus of the similarity measure is primarily on the motion of the hands. As a further example, consider the difference between walking and running. These motions may of course be distinguished by their absolute speed. Yet, the overall shape of most joints' trajectories is very similar in both motions. A better indicator would be the occurrence of simultaneous air phases for both feet, which is a discriminative feature of running motions.

Last but not least, noise is a further factor that may interfere with a similarity measure for motion clips. Mocap data may contain significant high-frequency noise components as well as undesirable artifacts such as sudden "flips" of a joint or systematic distortions due to wobbling mass or skin shift [116]. For example, consider the toe trajectory shown in the ballet motion of Fig. 9.6, where the noise shows as extremely irregular sample spacing. Such noise is usually due to adverse recording conditions, occlusions, improper setup

Fig. 9.6. A 500-frame ballet motion sampled at 120 Hz, adopted from the CMU mocap database [44]. The motion comprises two 180° right turns, the second of which is jumped. The trajectory of the joint "ltoes" is shown

or calibration, or data conversion faults. On the left hand side of the figure, there is a discontinuity in the trajectory, which results from a three-frame flip of the hip joint. Such flips are either due to confusions of trajectories in the underlying marker data or due to the fitting process. Ren et al. [176] have developed automatic methods for detecting "unnatural" movements in order to find noisy clips or clips containing artifacts within a mocap database. Noise and artifacts are also a problem in markerless, video-based mocap systems, see Rosenhahn [181]. In view of such scenarios, it is important to design noise-tolerant similarity measures for the comparison of mocap data.

9.3 Kinematic Chains

In fields such as robotics or biomechanics, kinematic chains are employed to model complex two- or three-dimensional movable objects. In particular, such chains are playing an important role in computer animation for representing the human skeleton in a compact and standardized form, see Fig. 9.2. In Sect. 9.3.1, we provide a formal definition of a kinematic chain, which is tailored to our needs. We then summarize some basic facts on forward kinematics (Sect. 9.3.2) and motion parameterization (Sect. 9.3.3). Further links to the literature can be found in Sect. 9.3.4.

9.3.1 Formal Definition

Basically, a kinematic chain is a hierarchical system of *rigid bodies* or *segments*, connected by joints of various degrees of freedom. One joint is marked as the *root*. A kinematic chain is called *open* if there are no cycles in the connection hierarchy. In this case, the hierarchy can be represented by a rooted tree, where the joints correspond to the vertices and the segments to the edges. Since the segments are rigid, the possible motions of the entire kinematic chain are determined by the motions of the joints. In general, one considers different types of joints including revolute, prismatic, or helical joints. In the following, we only consider *revolute* or *ball joints*, which reasonably well approximate the degrees of freedom of a human skeleton. Such joints allow for rotating a segment by any angle about any axis in $\mathbb{R}^3$ thus revealing three rotational degrees of freedom. The following definition provides a formal framework for a kinematic chain as used in our motion analysis and retrieval context.

Definition 9.1 (Kinematic chain). *A (skeletal)* kinematic chain *is a rooted, directed tree with a set of parameters represented by a tuple* $C = (J, r, B, (t_b)_{b\in B}; t_r, R_r, (R_b)_{b\in B})$, *where*

- J *is a set of vertices (the* joints*)*
- $r \in J$ *is the* root
- $B \subset J \times J$ *is a set of edges that are directed away from the root (the* bones*)*
- $t_b \in \mathbb{R}^3$ *describes the length as well as the standard direction of bone* $b \in B$ *(relative joint translation of a bone)*
- $t_r \in \mathbb{R}^3$ *is the* root translation
- $R_r \in \mathrm{SO}(3)$ *is the* root rotation
- $R_b \in \mathrm{SO}(3)$ *is the relative joint rotation of bone* $b \in B$

For a bone $(j_1, j_2) \in B$ *we say that* j_1 *is the* proximal joint *(closer to the root) and* j_2 *is the* distal joint *(further away from the root). The parent of a joint* $j \in J \setminus \{r\}$ *is denoted by* $p(j)$*. The leaves of a kinematic chain are also referred to as* end effectors.

A kinematic chain $C = (J, r, B, (t_b)_{b\in B}; t_r, R_r, (R_b)_{b\in B})$ provides hierarchical instructions about assembling and positioning the bones and joints in 3D, a process known as *forward kinematics*. Here, the parameters J, r, B, $(t_b)_{b\in B}$ are referred to as *skeletal parameters* and t_r, R_r, $(R_b)_{b\in B}$ as *free parameters* or *degrees of freedom* (DOF). The skeletal parameters J and B describe the topology (tree structure) of the kinematic chain, whereas the vectors $(t_b)_{b\in B}$ describe the lengths and standard directions of the bones. The root r serves as the reference point of the kinematic chain. The free parameters t_r and R_r determine the global 3D position and global orientation of the entire kinematic chain, whereas the $(R_b)_{b\in B}$ describe the direction of the bone b relative to the standard direction and its predecessor bone. The state of the skeletal kinematic chain, where all rotations are the identity, is also referred to as *standard pose*. In practice, this standard pose is often chosen as in Fig. 9.2 or as a so-called T-pose where both arms are stretched out to the side, see right side of Fig. 9.1.

9.3.2 Forward Kinematics

Starting from a *root* object, child objects are attached, which may in turn have further child objects, and so on. If a parent object moves by means of its degrees of freedom, the entire subtree below the parent object, treated as

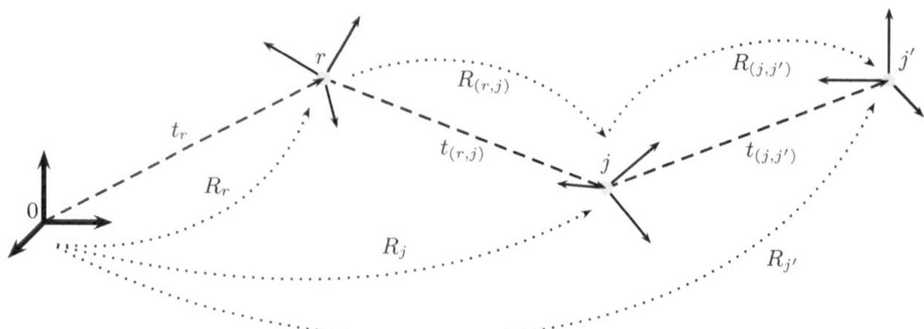

Fig. 9.7. The principle of forward kinematics, illustrating (9.2) and (9.3). The origin of the global coordinate system (left, bold coordinate axes) is marked as 0. The position of the joints r, j, and j' are shown as gray dots along with their respective orientation. *Dotted arrows* stand for rotations, while *dashed arrows* represent the segments (bones)

a single rigid object, moves along. In other words, transformations of a child object are meant to take place *relative* to a coordinate system that is fixed at the parent object. This is the idea of forward kinematics, which will now be discussed in detail.

Given a kinematic chain $C = (J, r, B, (t_b)_{b\in B}; t_r, R_r, (R_b)_{b\in B})$, the starting point of forward kinematics is provided by the root rotation R_r and the root translation t_r, which establish global 3D position and global orientation (often referred to as *root coordinate system*), see Fig. 9.7. Next, the bones that are incident to the root joint, i. e., all bones in the set $B_r := \{(r, j) \mid j \in J\} \cap B$, can be placed relative to the root coordinate system. For a bone $(r, j) \in B_r$, the 3D coordinates of the distal joint j are computed as

$$t_j = \underbrace{R_r R_{(r,j)}}_{=:R_j} t_{(r,j)} + t_r, \tag{9.2}$$

where the rotation $R_{(r,j)}$ defines the orientation of the distal joint relative to the root coordinate system and the composite rotation R_j defines the absolute rotation. This scheme is then continued in an analogous fashion for deeper levels of the hierarchy. For example, the 3D coordinates of the distal joint of a second-level bone $(j, j') \in B_j$ are computed as

$$t_{j'} = R_j R_{(j,j')} t_{(j,j')} + t_j = \underbrace{R_r R_{(r,j)} R_{(j,j')}}_{=:R_{j'}} t_{(j,j')} + R_r R_{(r,j)} t_{(r,j)} + t_r. \tag{9.3}$$

This bone placement scheme generalizes to the following algorithm.

Algorithm: FORWARDKINEMATICS

Input: Kinematic chain $C = (J, r, B, (t_b)_{b\in B}; t_r, R_r, (R_b)_{b\in B})$.
Output: $(t_j)_{j\in J} \in \mathbb{R}^{3\times|J|}$, where t_j is the 3D position of joint $j \in J$

Procedure: Starting at the root, use depth-first search or breadth-first search to traverse the kinematic chain while evaluating equations (9.4) and (9.5) for each $j \in J$ that is visited.

$$R_j := \begin{cases} R_r, & \text{if } j = r, \\ R_{p(j)} R_{(p(j),j)}, & \text{otherwise.} \end{cases} \tag{9.4}$$

$$t_j := \begin{cases} t_r, & \text{if } j = r, \\ R_j t_{(p(j),j)} + t_{p(j)}, & \text{otherwise.} \end{cases} \tag{9.5}$$

For each $j \in J$, the tree structure and the traversal method ensure that $R_{p(j)}$ and $t_{p(j)}$ have been computed in a previous step. The entire procedure requires a total of $|B|$ matrix-matrix multiplications, $|B|$ matrix-vector multiplications, and $|B|$ vector additions.

In the following, the skeletal parameters of a kinematic chain are assumed to be fixed. Then, the tuple $(t_r, R_r, (R_b)_{b\in B})$ will be denoted as *joint configuration*. The set of all possible joint configurations, the *joint space*, is given by $\mathcal{J} := \mathbb{R}^3 \times (\mathrm{SO}(3))^{|B|+1}$. From an abstract point of view, forward kinematics then provides a mapping f from the joint space $\mathcal{J}$ to the pose space $\mathcal{P}$:

$$f : \mathcal{J} \to \mathcal{P}, \quad (t_r, R_r, (R_b)_{b\in B}) \mapsto (t_j)_{j\in J}. \tag{9.6}$$

9.3.3 Animated Kinematic Chains

So far, we have considered a fixed kinematic chain represented by the tuple $C = (J, r, B, (t_b)_{b\in B}; t_r, R_r, (R_b)_{b\in B})$ yielding 3D joint coordinates. If we let any of the free parameters t_r, R_r, $(R_b)_{b\in B}$ vary over time, we obtain a time-dependent sequence of kinematic chains, which in turn induces a time-dependent sequence of 3D joint coordinates. Observe from (9.2) and (9.3) that varying the parameters (t_r, R_r) of the root coordinate system changes the global position and orientation of the entire kinematic chain since t_r and R_r always appear last in the transformation hierarchy. Variations of $(R_b)_{b\in B}$ induce rotations of the respective joints.

As in Sect. 9.1, we assume a discretized time interval $[1 : T]$ for our data stream. Then, an *animated kinematic chain* consists of a fixed set of skeletal parameters $(J, r, B, (t_b)_{b\in B})$ and a function

$$D_{\mathcal{J}} : [1 : T] \to \mathcal{J}. \tag{9.7}$$

In other words, for a fixed skeleton the free parameters vary over time. Applying the forward kinematics algorithm for all $t \in [1 : T]$, we can compute a motion capture data stream from an animated skeleton as

$$D := f \circ D_{\mathcal{J}}, \quad t \mapsto f(D_{\mathcal{J}}(t)) \tag{9.8}$$

for $t \in [1 : T]$. Finally, fixing a joint of an animated kinematic chain, we obtain a curve $[1 : T] \to \mathrm{SO}(3)$, which will be referred to as *angle trajectory*. Some examples are given in Fig. 9.8.

9.3.4 Further Notes

We have seen that motion capture data streams can be represented by animated kinematic chains in some very efficient way. Such a representation facilitates an intuitive and direct manipulation of motion data by modifying the free parameters of the kinematic chain and have become an indispensable tool in editing, blending, and morphing of motions. Using kinematic chains, one can easily impose constraints on the skeletal poses by simply limiting joint angles. Furthermore, constraints such as segment lengths are automatically fulfilled even when manipulating the free parameters ([133,150,223] and the references therein).

However, it is important to note that the kinematic chain as used in the subsequent chapters is a very coarse and simplified model of the real human body. For example, the actual number of joints of the human skeleton is much larger then the 24 joints shown in Fig. 9.2. Furthermore, the joints are generally not purely revolute and may also contain some translational components. This implies that the joint centers are generally not fixed for real motions [223]. As an example, consider the motion of the human knee, which we model as a 1 DOF hinge joint. From the point of view of a biomechanist, the knee joint has 6 DOF, three of which are translational and three of which are rotational. Therefore, improved knee joint models are used for enhanced realism in computer animation [10]. Further biomechanical issues such as skin deformations, wobbling mass, or muscle forces are not considered in our model. One major problem in the generation of skeleton-based mocap data arises from the fact that the markers are typically attached to the surface of the human body. In a postprocessing step one then employs optimization algorithms to fit in the skeletal model into the 3D coordinates of the recorded marker positions. This fitting process is problematic due to unrealistic simplifications regarding the skeletal model and due to skin deformations, which may cause significant shifts of the attached markers [116, 175].

In this section, we have introduced the Algorithm ForwardKinematics, which allows us to compute the 3D coordinates of the joints from the free (root and angle) parameters of a given kinematic chain [214]. The inverse problem, also known as *inverse kinematics*, deals with the process of adjusting the free parameters of a kinematic chain when fixing the 3D coordinates of certain joints [214]. The question of inverse kinematics is of great importance in fields such as robotics, where one manipulates the robot arm in terms of joint angles for, e.g., a predetermined 3D trajectory of the end effector. Similarly, inverse kinematics is used in motion synthesis to enforce certain kinematic constraints such as footplants, which require the feet to remain stationary on the ground at certain points in time [109].

Generally spoken, *kinematics* deals with the motion of the kinematic chain without regarding the external and internal forces and torques that cause the motion, whereas it is the object of *dynamics* to analyze these forces and torques. Similar to kinematics, one distinguishes between *forward dynamics*, which yields the 3D position or angle trajectories of the kinematic chain from the given forces and torques, and *inverse dynamics*, which is used to compute the associated forces and torques that cause the given movement. The study of forces is at the heart of biomechanical analysis, where one is interested in the muscle activity, including the timing of muscle contractions or the amount of generated force. For further details, we refer to the literature ([63, 65, 105] and the references therein).

9.4 Rotations

In Sect. 9.3, we have seen that rotations of the three-dimensional Euclidean space play an important role in representing human motions that are parameterized by kinematic chains. Rotations do not only describe the global orientation of a pose but also encode the joint angles in the kinematic chain. In this section, we discuss important properties and various representations of rotations, particularly matrix representations (Sect. 9.4.1), Euler angle representations (Sect. 9.4.2), and representations based on quaternions (Sect. 9.4.3). In computer animation, a large number of techniques for the analysis and synthesis of motions work with some kind of rotational data. As it turns out, each rotation representation has its assets and drawbacks and the choice of some specific parametrization depends on the respective applications. In Sect. 9.4.4, we give some references to the literature, where one can find a discussion of further rotation representations and its applications.

9.4.1 Basic Definitions and Properties

Mathematically, a proper *3D rotation* is a linear, bijective mapping of $\mathbb{R}^3$ onto itself that preserves angles, lengths, and orientation, thus mapping right-handed orthonormal bases into right-handed orthonormal bases. The set of all orientation preserving 3D rotations forms a subgroup of the automorphism group $\mathrm{Aut}(\mathbb{R}^3)$, the so-called *special orthogonal group* of $\mathbb{R}^3$,

$$\mathrm{SO}(3) := \{F \in \mathrm{Aut}(\mathbb{R}^3) \mid F^* = F^{-1},\ \det(F) = 1\}, \tag{9.9}$$

where F^* is the adjoint of F. Unlike $\mathrm{SO}(2)$, the group $\mathrm{SO}(3)$ is non-abelian.

Fixing a right-handed orthonormal basis of $\mathbb{R}^3$, any $F \in \mathrm{SO}(3)$ can be uniquely represented as an element of the matrix group

$$\mathrm{SO}_3 := \{R \in \mathbb{R}^{3\times 3} \mid R^\top = R^{-1},\ \det(R) = 1\}. \tag{9.10}$$

The condition $R^\top = R^{-1}$ implies that the row and the column vectors of a matrix $R \in \mathrm{SO}_3$ form an orthonormal basis, and the condition $\det(R) = 1$ implies that the orientation of this basis is right-handed. Since the columns of any matrix A are the image of the chosen basis under the action of the endomorphism represented by A, these conditions are a restatement of the property that rotations map right-handed orthonormal bases into right-handed orthonormal bases.

The action of a rotation matrix becomes clear by the following considerations. It can be shown that every $R \in \mathrm{SO}_3$ has the eigenvalue 1 with a corresponding one-dimensional R-invariant subspace, the *axis of rotation*. The orthogonal complement of the axis of rotation forms a two-dimensional R-invariant subspace, the *plane of rotation*. Within the plane of rotation, R describes a two-dimensional rotation by an angle of α. This implies that an axis of rotation and an angle describing the amount of rotation about this axis can be found for every $R \in \mathrm{SO}_3$.

Performing a change of basis by a change matrix B, the columns of which form an orthonormal eigenbasis composed of the axis of rotation and an arbitrary orthonormal basis of the plane of rotation, the matrix representation of R becomes

$$R' = BRB^{-1} = \begin{pmatrix} 1 & 0 & 0 \\ 0 & \cos\alpha & -\sin\alpha \\ 0 & \sin\alpha & \cos\alpha \end{pmatrix}, \tag{9.11}$$

describing a rotation about the x axis by an angle of α according to the right-hand rule.[1] Thus, every 3×3 rotation matrix is similar to a matrix of the form (9.11). Such *basic rotations* about the coordinate axes will play an important role in the Euler angle parametrization, see Sect. 9.4.2.

Even though a rotation is represented by a 3×3-matrix, only three of the nine entries are actually free.[2] Informally, it is straightforward to imagine why the dimension of SO_3 must be three: the rotation axis is described by a normalized 3D vector accounting for two free real parameters and the angle of rotation accounts for a third free real parameter. In spite of their redundancy, rotation matrices are widely used in different applications, including computer graphics and computer animation.

Further drawbacks attached to the matrix representation appear when one has to manipulate functions that depend on angle parameters. Doing numerical operations (e.g., numerical differentiation, integration, optimization) on such functions typically results in or involves matrices that are not

[1] Pointing the thumb of the right hand in the direction of the axis, the curled fingers indicate the direction of rotation.

[2] $\mathrm{SO}(3)$ and SO_3 are so-called three-dimensional Lie groups.

elements of SO_3. This, in turn, necessitates additional nonlinear constraints or frequent reorthonormalization, which introduces numerical errors. For the same reason, interpolation of rotations are difficult with rotation matrices. In view of applications such as editing, blending, or morphing of motions based on kinematic chains, there are more suitable ways of representing rotations. This is discussed in the following sections.

9.4.2 Euler Angles

Leonhard Euler (1707–1783) proved that any 3D rotation can be expressed as a sequence of three basic rotations about the coordinate axes. The three angles of rotation are referred to as *Euler angles*. To use Euler angles as a parametrization for SO_3, one has to make some choices:

(a) Which axes to rotate about
(b) In which order to rotate
(c) Whether to express rotations relative to a fixed coordinate system or relative to a coordinate system that moves along with the basic rotations

We emphasize that many different conventions are used in the literature – even within the same field such as computer graphics and robotics – leading to different formulas. In the following, we consider only one such convention.

(a) As the axes we use the coordinate axes of our right-handed coordinate system denoted as x, y, and z axis.
(b) We first rotate by an angle of α_1 about the x axis, then by an angle of α_2 about the y axis, and finally by an angle of α_3 about the z axis.
(c) All rotations are expressed relative to the fixed coordinate system.

Using the fixed coordinate system as basis, we obtain the following matrix representations of the three basic rotations denoted by R_x, R_y, and R_z, respectively:

$$R_x(\alpha_1) = \begin{pmatrix} 1 & 0 & 0 \\ 0 & \cos\alpha_1 & -\sin\alpha_1 \\ 0 & \sin\alpha_1 & \cos\alpha_1 \end{pmatrix}, \tag{9.12}$$

$$R_y(\alpha_2) = \begin{pmatrix} \cos\alpha_2 & 0 & \sin\alpha_2 \\ 0 & 1 & 0 \\ -\sin\alpha_2 & 0 & \cos\alpha_2 \end{pmatrix}, \tag{9.13}$$

$$R_z(\alpha_3) = \begin{pmatrix} \cos\alpha_3 & -\sin\alpha_3 & 0 \\ \sin\alpha_3 & \cos\alpha_3 & 0 \\ 0 & 0 & 1 \end{pmatrix}, \tag{9.14}$$

with $\alpha_i \in \mathbb{R}$ for $i \in \{1,2,3\}$. By convention (b), the final rotation, denoted by $R(\alpha_1,\alpha_2,\alpha_3)$, is obtained as composition of $R_x(\alpha_1)$, $R_y(\alpha_2)$, and $R_z(\alpha_3)$. Introducing the abbreviations $c_i := \cos\alpha_i$ and $s_i := \sin\alpha_i$, $R(\alpha_1,\alpha_2,\alpha_3)$ can be worked out to be

$$R(\alpha_1,\alpha_2,\alpha_3) := R_z(\alpha_3)R_y(\alpha_2)R_x(\alpha_1) = \begin{pmatrix} c_3c_2 & -s_3c_1 + c_3s_2s_1 & s_3s_1 + c_3s_2c_1 \\ s_3c_2 & c_3c_1 + s_3s_2s_1 & -c_3s_1 + s_3s_2c_1 \\ -s_2 & c_2s_1 & c_2c_1 \end{pmatrix} \quad (9.15)$$

As an example, let us consider the case $\alpha_i = \frac{\pi}{2}$ for $i = 1, 2, 3$. To demonstrate the effect of the resulting rotations, place a book on the table in front of you so that the front cover faces upwards and the spine of the book faces you. We assume that the x axis points to the right, the y axis points upwards, and the z axis points straight at you.[3] The first rotation is $R_x(\frac{\pi}{2})$, and so rotate the book by 90° about the x axis. The book now sits on its spine, the front cover facing towards you. The second rotation is $R_y(\frac{\pi}{2})$, which rotates the book counter-clockwise by 90° on its spine so the front cover now faces to the right. The third rotation is $R_z(\frac{\pi}{2})$, which once more brings the front cover to the top, with the spine facing to the right. In effect, we have rotated the book by 90° about the y axis. More generally, using $\alpha_2 = \frac{\pi}{2}$ and arbitrary angles α_1 and α_3, one obtains the following identity:

$$\begin{aligned} R(\alpha_1,\tfrac{\pi}{2},\alpha_3) &= \begin{pmatrix} \cos\alpha_3 & -\sin\alpha_3 & 0 \\ \sin\alpha_3 & \cos\alpha_3 & 0 \\ 0 & 0 & 1 \end{pmatrix} \begin{pmatrix} 0 & 0 & 1 \\ 0 & 1 & 0 \\ -1 & 0 & 0 \end{pmatrix} \begin{pmatrix} 1 & 0 & 0 \\ 0 & \cos\alpha_1 & -\sin\alpha_1 \\ 0 & \sin\alpha_1 & \cos\alpha_1 \end{pmatrix} \\ &= \begin{pmatrix} 0 & \cos\alpha_3\sin\alpha_1 - \sin\alpha_3\cos\alpha_1 & -\cos\alpha_3\cos\alpha_1 + \sin\alpha_3\sin\alpha_1 \\ 0 & \sin\alpha_3\sin\alpha_1 + \cos\alpha_3\cos\alpha_1 & \sin\alpha_3\cos\alpha_1 - \cos\alpha_3\sin\alpha_1 \\ -1 & 0 & 0 \end{pmatrix} \\ &= \begin{pmatrix} 0 & \sin(\alpha_1-\alpha_3) & \cos(\alpha_1-\alpha_3) \\ 0 & \cos(\alpha_1-\alpha_3) & -\sin(\alpha_1-\alpha_3) \\ -1 & 0 & 0 \end{pmatrix}, \end{aligned} \quad (9.16)$$

where we used the usual trigonometric identities in the computation. From (9.16), one obtains the identity

$$R(\alpha_1,\tfrac{\pi}{2},\alpha_3) = R_y(\tfrac{\pi}{2})R_x(\alpha_3-\alpha_1) = R_z(\alpha_1-\alpha_3)R_y(\tfrac{\pi}{2}). \quad (9.17)$$

In other words, if $\alpha_2 = \frac{\pi}{2}$, then a rotation about the x axis by α has the same effect as a rotation about the z axis by $-\alpha$, which amounts to losing one degree of rotational freedom. Equivalently, the one-dimensional set $\{(\alpha, \frac{\pi}{2}, \alpha) \mid \alpha \in \mathbb{R}\}$ of angles all describe the same rotation. This effect is generally referred to as *gimbal lock*.

The gimbal lock phenomenon is the result of a more fundamental problem attached to any parametrization of SO_3 based on Euler angles. The function $R : \mathbb{R}^3 \to SO_3$ as defined in (9.15) is a continuous and surjective function. To obtain a unique Euler representation for a rotation, one restricts the set of angles to a subset $D \subset \mathbb{R}^3$. Let $R_D : D \to SO_3$ denote the restriction of

[3] It is helpful to use the thumb, the index finger, and the middle finger of your right hand to imagine the x, y, and z axis, respectively.

R to D. Then it can be shown that there is no choice for D such that R_D is bijective with the inverse function $(R_D)^{-1} : \mathrm{SO}_3 \to D$ being a continuous function.[4] The discontinuity of the parametrization's inverse can then lead to a situation where similar rotations are encoded by very different Euler angles, thus precluding a meaningful comparison of rotations by their Euler parametrization, see Fig. 9.8. Such discontinuities typically appear at gimbal lock angle configurations as discussed earlier.

In summary, Euler angles are a very compact and intuitive representation of rotations, which have a direct geometric interpretation and facilitate a semantically meaningful way to constrain rotations by simply constraining the domain of the Euler angles. On the other hand, because of the discontinuities in the inverse of the parametrization, computing with Euler angles can lead to unforeseen artifacts and comparing rotations via their Euler angles may be problematic – in particular near gimbal lock angle configurations. Also, interpolations of rotations via Euler angles may lead to strong geometric distortions. In view of such tasks, there are more suitable representations as discussed in the next section.

9.4.3 Quaternions

The concept of quaternions allows for a compact parametrization of rotations, which not only avoids problems such as the gimbal lock but also constitutes a well-suited representation in particular for interpolations. Therefore, quaternions have become an indispensable tool for the analysis and synthesis of human motions. In this section, we first summarize some basic mathematical properties about quaternions and then describe their relation to rotations. For further details, we refer to the literature such as [133, 166, 192].

Quaternions have been introduced by Sir William Rowan Hamilton in 1843 as a noncommutative extension of the complex numbers. Recall that based on the basis $\{1, i\}$ of $\mathbb{R}^2$ one defines $i^2 := -1$, which extends to a multiplication on $\mathbb{R}^2$ and results in the field $\mathbb{C}$ of complex numbers. Similarly, one can introduce a multiplication on $\mathbb{R}^4$. To this end, we denote the standard basis of $\mathbb{R}^4$ by the symbols $1, i, j, k$. Then an element $q \in \mathbb{R}^4$, also denoted as *quaternion*, can be written uniquely as $q = w + xi + yj + zk$ for suitable $w, x, y, z \in \mathbb{R}$. The part $\mathrm{Re}(q) := w \in \mathbb{R}$ – a scalar – is called *real part* of q, whereas the part $\mathrm{Im}(q) := (x, y, z)^\mathrm{T} \in \mathbb{R}^3$ – a vector – is called *imaginary part* of q. A quaternion q is called *pure* if $w = 0$. For two quaternions $q_1 = w_1 + x_1 i + y_1 j + z_1 k$ and $q_2 = w_2 + x_2 i + y_2 j + z_2 k$ one defines the product quaternion $q_1 \cdot q_2$ by

$$\begin{aligned} q_1 \cdot q_2 := {} & w_1 w_2 - x_1 x_2 - y_1 y_2 - z_1 z_2 \\ & + (w_1 x_2 + w_2 x_1 + y_1 z_2 - z_1 y_2) i \end{aligned} \tag{9.18}$$

[4] This is a well-known topological fact stating that SO_3 is not homeomorphic to any subset of $\mathbb{R}^3$. Actually, SO_3 is homeomorphic to the three-dimensional real projective space $\mathbb{R}P(3)$, which is not even embeddable in $\mathbb{R}^4$.

$$+(w_1y_2 + w_2y_1 + z_1x_2 - x_1z_2)j$$
$$+(w_1z_2 + w_2z_1 + x_1y_2 - y_1x_2)k.$$

Now, it is an easy but tedious exercise to show that this indeed defines an associative and distributive multiplication on $\mathbb{R}^4$ with neutral element $e = 1$. Usually, one simply writes q_1q_2 for $q_1 \cdot q_2$. For a quaternion $q = w + xi + yj + zk$, the *conjugate quaternion* is defined as $\bar{q} := w - xi - yj - zk$ and the *norm* of q is defined as $\|q\| := \sqrt{w^2 + x^2 + y^2 + z^2}$. A straightforward computation shows that $q^{-1} := \bar{q}/\|q\|^2$ defines a right and left inverse in the case $q \neq 0$, i. e., $qq^{-1} = q^{-1}q = 1$. Note that the multiplication is not commutative. For example, one has $ij = -ji$. Altogether, we have seen that $\mathbb{R}^4$ equipped with vector addition and the multiplication defined by (9.18) satisfies all axioms of a field except for commutativity. Mathematically, such an object is called *skew field*. In honor of Hamilton, this skew field is also denoted by the symbol $\mathbb{H}$.

For the imaginary numbers $i, j, k \in \mathbb{H}$, the multiplication induces the following famous relations:

$$i^2 = j^2 = k^2 = -1, \quad ij = k, \quad jk = i, \quad ki = j. \tag{9.19}$$

Actually, the multiplication in (9.18) is uniquely determined by these relations if one requires the multiplication to be associative and distributive. Note that one also has $ij = -ji$, $jk = -kj$, and $ki = -ik$. To obtain more concise formulas, one often writes a quaternion $q \in \mathbb{H}$ as a tuple (s, v) with $s = \mathrm{Re}(q)$ (a scalar) and $v = \mathrm{Im}(q)$ (a vector). In the following, we identify a vector $v = (x, y, z)^\mathrm{T} \in \mathbb{R}^3$ with the pure quaternion $xi + yj + zk \in \mathbb{H}$ and simply write $v \in \mathbb{H}$. For two quaternions $q_1 = (s_1, v_1)$ and $q_2 = (s_2, v_2)$ with $v_1 = x_1i + y_1j + z_1k$ and $v_2 = x_2i + y_2j + z_2k$, let $\langle v_1|v_2\rangle := x_1x_2 + y_1y_2 + z_1z_2$ denote the inner product and $v_1 \times v_2 := (y_1z_2 - z_1y_2, z_1x_2 - x_1z_2, x_1y_2 - y_1x_2)^\mathrm{T} \in \mathbb{R}^3$ the cross product of v_1 and v_2. The following formulas can be verified by some straightforward computation:

$$q_1q_2 = (s_1s_2 - \langle v_1|v_2\rangle,\ s_1v_2 + s_2v_1 + v_1 \times v_2) \tag{9.20}$$
$$\|q\|^2 = q\bar{q} = s^2 + \langle v|v\rangle \tag{9.21}$$
$$q^{-1} = \frac{(s, -v)}{\|q\|^2} \tag{9.22}$$

A quaternion of norm one is also referred to as *unit quaternion*. The set of unit quaternions form a hypersphere $S^3 \subset \mathbb{H}$. It is not difficult to see that any unit quaternion q can be written as $q = (\cos\theta, v\sin\theta)$ for some suitable angle $\theta \in [-\pi, \pi]$ and some unit vector $v \in \mathbb{R}^3$. As we see later, such a quaternion can be used to describe a rotation in $\mathbb{R}^3$ about the axis v and angle 2θ. Before we formulate the main theorem of this section, we prove some basic properties of quaternions. The *center* $\mathrm{Center}(\mathbb{H})$ is defined to be the set of all those quaternions that commute with any other quaternion.

Lemma 9.2. *Let $q = (s, v)$, $q_1 = (s_1, v_1)$, and $q_2 = (s_2, v_2)$ be quaternions.*

(a) q is pure if and only if $q^2 \in \mathbb{R}$ and $q^2 \leq 0$.
(b) If v_1 and v_2 are orthogonal, then $v_1 v_2 = -v_2 v_1$.
(c) Any quaternion is expressible as the product of two pure quaternions.
(d) $\mathrm{Center}(\mathbb{H}) = \mathbb{R}$.

Proof. From (9.20), one obtains $q^2 = (s^2 - \langle v|v\rangle, 2sv)$ implying (a). If $\langle v_1|v_2\rangle = 0$, then $v_1 v_2 = (0, v_1 \times v_2) = -(0, v_2 \times v_1) = -v_2 v_1$, which proves (b). For arbitrary q, let b be a pure quaternion orthogonal to q (such a b obviously exists). Then qb is pure since $\mathrm{Re}(qb) = s \cdot 0 - \langle v|b\rangle = 0$. Furthermore, b^{-1} is pure by (9.22). Assertion (c) follows from $q = (qb)b^{-1}$. Assertion (d) is left as an exercise. □

Recall that we identify $\mathbb{R}^3$ with the subset of pure quaternions in $\mathbb{H}$. To prove that the rotations can be parameterized by quaternions, we proceed in three steps. First, we show that a pure quaternion defines a reflection in $\mathbb{R}^3$. Then, we prove that any rotation of $\mathbb{R}^3$ can be expressed as the composition of two reflections. From this we obtain our main result, see Theorem 9.5.

Lemma 9.3. *Let $q \neq 0$ be a pure quaternion. Then $-\rho_q : \mathbb{R}^3 \to \mathbb{R}^3$, $-\rho_q(v) := -qvq^{-1}$, v being a pure quaternion, defines a reflection in the plane orthogonal to q.*

Proof. By Lemma 9.2, one has $v^2 \in \mathbb{R}$ and $v^2 \leq 0$ for a pure quaternion v and hence $v^2 \in \mathrm{Center}(\mathbb{H})$. From this follows that $(-\rho_q(v))^2 = qvq^{-1}qvq^{-1} = qv^2q^{-1} = v^2qq^{-1} = v^2$. Therefore, again by Lemma 9.2, $-\rho_q(v)$ is pure and $-\rho_q$ indeed defines a map $\mathbb{R}^3 \to \mathbb{R}^3$. Next, from the distributivity of quaternion multiplication it follows that $-\rho_q$ defines a linear map. Obviously, $-\rho_q(q) = -q$. Furthermore, it follows from Lemma 9.2 (b) that $-\rho_q(v) = -qvq^{-1} = vqq^{-1} = v$ for all pure quaternions v with $\langle v|q\rangle = 0$. This proves the assertion of the Lemma. □

Lemma 9.4. *Any rotation of $\mathbb{R}^3$ can be represented as the composition of two reflections.*

Proof. Let $v_1, v_2 \in \mathbb{R}^3$ be two vectors that span a two-dimensional plane E through the origin. Then an arbitrary vector $x \in \mathbb{R}^3$ has a unique decomposition of the form $x = x_{\|} + x_{\perp}$ such that $x_{\|}$ lies in E and $x_{\perp}$ is orthogonal to E. Let E_1 and E_2 be the planes through the origin orthogonal to v_1 and v_2, respectively. Then E_1 and E_2 are both orthogonal to E. Note that a reflection in any plane orthogonal to E has no effect on $x_{\perp}$. Therefore, we only have to consider $x_{\|}$ in the following discussion. Reflecting $x_{\|}$ in the plane E_1 leads to some vector $x'_{\|}$ in E. The angle between $x_{\|}$ and $x'_{\|}$ equals $2\omega_1$, where ω_1 is the difference between 90° (E_1 is orthogonal to q_1) and the angle between v_1 and $x_{\|}$. Next, reflecting $x_{\|}$ on the plane E_2 leads to some vector $x''_{\|}$ in E. The angle between $x'_{\|}$ and $x''_{\|}$ equals $2\omega_2$, where ω_2 is the difference between 90° and the angle between v_2 and $x'_{\|}$. In other words, the composition of the two reflections transform the vector $x = x_{\|} + x_{\perp}$ into the vector $x = x''_{\|} + x_{\perp}$. Since this composition is a length-preserving linear map with positive determinant, it must be a rotation about the axis orthogonal to E. One easily checks that the rotation angle is $2\omega_1 + 2\omega_2 = 2\theta$, where θ denotes the angle between v_1 and v_2.

Conversely, suppose an arbitrary rotation is given with respect to some specified axis and angle. Then, from the above discussion it is obvious how to choose two suitable vectors $v_1, v_2 \in \mathbb{R}^3$ such that the resulting composition of reflections equals the given rotation. □

Theorem 9.5. *For every nonzero quaternion $q \in \mathbb{H}^* := \mathbb{H} \setminus \{0\}$, the map ρ_q defined by $\rho_q(v) := qvq^{-1}$, $v \in \mathbb{R}^3$, is a rotation in $\mathbb{R}^3$. Furthermore, the map*

$$\rho : \mathbb{H}^* \to \mathrm{SO}(3), \quad q \mapsto \rho_q \tag{9.23}$$

is a surjective group homomorphism from the multiplicative group $\mathbb{H}^*$ into the group $\mathrm{SO}(3)$ with kernel $\mathbb{R}^* := \mathbb{R} \setminus \{0\}$. The restriction of ρ to $S^3 \subset \mathbb{H}^*$ is also a surjective group homomorphism with kernel $\{\pm 1\}$.

Proof. By Lemma 9.2 (c), any $q \in \mathbb{H}^*$ can be written as product $q = v_1 v_2$ of two nonzero pure quaternions $v_1, v_2 \in \mathbb{R}^3$. Since $\rho_q = (-\rho_{v_1})(-\rho_{v_2})$, it follows from the proof of Lemma 9.4 that $\rho_q \in \mathrm{SO}(3)$. In particular, the map ρ is well-defined. An easy calculation shows $\rho_{q_1 q_2} = \rho_{q_1}\rho_{q_2}$ for $q_1, q_2 \in \mathbb{H}^*$, i. e., ρ is a group homomorphism. The surjectivity of ρ is a direct consequence of Lemma 9.4. Now, suppose that ρ_q is the identity in $\mathrm{SO}(3)$ for some $q \in \mathbb{H}^*$. Then $qvq^{-1} = v$ and hence $qv = vq$ for all $v \in \mathbb{R}^3$. In other words, q is in the kernel of ρ if and only if $q \in \mathrm{Center}(\mathbb{H})$. Then Lemma 9.2 (d) implies that the kernel of ρ is given by $\mathbb{R}^*$. Finally, it is obvious that $\rho_q = \rho_{\lambda q}$ for any $\lambda \in \mathbb{R}^*$ and $q \in S^3$. This implies the last assertion of the theorem. □

Corollary 9.6. *Let $q \in S^3$ be a unit quaternion, which can be written in the form $q = (\cos\theta, v \sin\theta)$ for some suitable angle $\theta \in [-\pi, \pi]$ and unit vector $v \in \mathbb{R}^3$. Then ρ_q describes a rotation about the axis v by an angle of 2θ.*

Proof. Let $v_1, v_2 \in \mathbb{R}^3$ be two pure unit quaternions that span the plane orthogonal to v and exhibit an angle θ. Then, $\cos(\theta) = \langle v_1 | v_2 \rangle$, $v = (v_1 \times v_2)/\|v_1 \times v_2\|$, and $\|v_1 \times v_2\| = \sin\theta$. This implies $q = (\cos\theta, v \sin\theta) = (\langle v_1 | v_2 \rangle, v_1 \times v_2) = -(0, v_1)(0, v_2)$. Since q and $-q$ induce the same rotation, the assertion follows from the proof of Lemma 9.4. □

Example 9.7. As an illustration, we consider the rotation of the vector $v = (1, 0, 1)^{\mathrm{T}} \in \mathbb{R}^3$ about the y axis by 180°. Regarded as quaternion, v is given by $(0, 1, 0, 1)$. By Cor. 9.6, the rotation is given by the quaternion $q = (\cos(\pi/2), (\sin(\pi/2)(0, 1, 0))) = (0, 0, 1, 0)$. The inverse is given by $q^{-1} = (0, 0, -1, 0)$. Then $\rho_q(v) = qvq^{-1} = (0, 0, 1, 0)(0, 1, 0, 1)(0, 0, -1, 0) = (0, 1, 0, -1)(0, 0, -1, 0) = (0, -1, 0, -1)$.

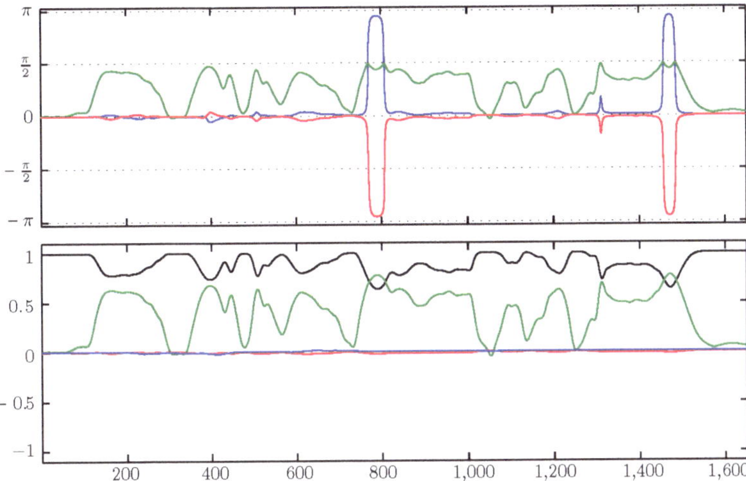

Fig. 9.8. Angle trajectory of some motion data stream (right knee angle) represented by Euler angles (*top*) and quaternions (*bottom*). The horizontal axis represents time in frames. The parameters are color-coded. Euler angles: blue (x), green (y), and red (z). Quaternions: black (w), blue (x), green (y), red (z). The peaks around frame 790 and around frame 1,470 in the Euler representation are due to the gimbal lock phenomenon, where the angle α_2 of the y-coordinate is close to $\frac{\pi}{2}$

The map ρ of Theorem 9.5 describes a parametrization of rotations using S^3 as parameter space (opposed to a subset of $\mathbb{R}^3$ as in the case of Euler angles). Actually, identifying antipodal points on S^3, one obtains the space $\mathbb{R}P(3) := S^3/\{\pm 1\}$ known as real three-dimensional *projective space*. Then Theorem 9.5 implies that ρ induces a bijective map $\mathbb{R}P(3) \to \mathrm{SO}(3)$. Even more, one can show that this map is continuous with a continuous inverse. From a differential geometric point of view, ρ exhibits the spherical geometry of SO(3). The practical consequence of this parametrization is that a small deformation of a rotation in SO(3) also results in a small deformation of the corresponding quaternion – the parametrization of SO(3) based on quaternions is free from gimbal lock. This fact is also illustrated by Fig. 9.8. Furthermore, any smooth path in SO(3) corresponds to a smooth path in $\mathbb{R}P(3)$ and vice versa.

Quaternions are a compact representation of rotations using four real parameters with one constraint (normalization of unit quaternion). The composition and inversion of rotations correspond to multiplication and inversion in $\mathbb{H}$, yielding easy and direct formulas. For example, multiplying two quaternions requires 16 scalar multiplications and 10 scalar additions, cf. (9.18), opposed to 27 scalar multiplications and 18 scalar additions needed to multiply two rotation matrices. Inversion of a unit quaternion boils down to simple conjugation, cf. (9.22). The equivalent to reorthonormalization as required for rotation matrices during numerical computations is a simple renormalization for unit quaternions. This operation is much simpler to compute and does not change the rotation that is described by the quaternion.

Constraining rotations to a specified range using quaternions is possible by the technique due to Liu and Prakash [124]. One of the main applications of quaternions is the *quaternion interpolation*, also known as spherical linear interpolation (Slerp), for the purpose of animating 3D rotation. This results in very smooth interpolation between two rotations having constant speed along a unit radius great circle arc. For details, we refer to Shoemake [192]. As a drawback, quaternions are not as intuitive as other parametrizations and are therefore typically not found in the front end of applications involving 3D rotations.

We close this section with a direct conversion formula, which assigns to a unit quaternion $q = (w + xi + yj + zk)$ the corresponding 3×3 rotation matrix $R_q \in \mathrm{SO}_3$ (shown by Cayley in 1845):

$$R_q = \begin{pmatrix} 1 - 2y^2 - 2z^2 & 2xy - 2wz & 2xz + 2wy \\ 2xy + 2wz & 1 - 2x^2 - 2z^2 & 2yz - 2wx \\ 2xz - 2wy & 2yz + 2wx & 1 - 2x^2 - 2y^2 \end{pmatrix}. \tag{9.24}$$

9.4.4 Further Notes

We have studied three different representations of the rotations in $\mathbb{R}^3$, which play a crucial role in describing orientations and angle parameters of kinematic chains. Matrix representations of elements in SO(3) are intuitive but very redundant, using nine real parameters with six nonlinear constraints. In the Euler angle representation, one describes a rotation by a composition of rotations in fixed order about three "basis axes." Euler angles constitute a very compact representation and allow the user to place limits on the legal range of angles in a straightforward manner. As main disadvantages, Euler angles have poor interpolation properties and suffer from gimbal lock, which leads to discontinuities in the inverse of the parametrization. Finally, we have seen that the group of unit quaternions $S^3 \subset \mathbb{H}$ form a twofold covering of SO(3). Since the local geometry of S^3 and SO(3) coincide, the quaternion parametrization is good for interpolation. However, quaternions may not be as intuitive as other parametrizations. In conclusion, one can say that there is no "best" parametrization of SO(3) – the choice for a specific parametrization depends on the respective applications. A detailed comparison amongst various representations that are popular in computer graphics can be found in Grassia [86].

Quaternions, even though having a long history in mathematics, have been introduced to the graphics community by Shoemake [192, 193], where one also finds an introduction to quaternion interpolation as well as conversion formulas between different rotation representations. A mathematical treatment of quaternions and their relation to Clifford Algebras can be found in

Porteuous [166]. Lee and Shin [119] introduce filter techniques for quaternion-based orientation data, which respect the spherical geometry of the rotation group. Results about spherical averages and applications to interpolation can be found in Buss and Fillmore [23].

There are plenty of useful representations of SO(3) that have not been described in this section. In conclusion, we mention two further prominent representations. Recall that any 3D rotation can be characterized by a directed *axis of rotation*, a unit vector in $S^2 \subset \mathbb{R}^3$, and an *angle of rotation* $\alpha \in (-\pi, \pi]$. The resulting *axis/angle parametrization* of SO_3 maps pairs of unit-length vectors and angles to SO_3. Benjamin Rodrigues (1794–1851) has found an explicit conversion formula from axis/angle parameters to rotation matrices [179]. A similar parametrization has been introduced by Grassia [86] to the graphics community, which is based on the exponential map. Here, $\mathbb{R}^3$ is mapped into SO(3) by summing an infinite series of exponentiated pure quaternions. In this parametrization, a single vector $v \in \mathbb{R}^3$ describes the axis (direction of v) as well as the angle (magnitude of v) of the rotation. Even though there are – as for the Euler angle parametrization – singularities in the inverse of the exponential map, these singularities can be avoided through a simple technique of dynamic reparametrization.

In computer animation, a large number of motion editing, warping, and blending techniques work with different rotation representations, for example [22, 165, 169]. This also holds true for numerical comparison of mocap data [93, 183, 220]. Finally, we want to mention that most of the common mocap file formats use a motion representation involving rotational data. For example, the *ASF/AMC format* is a skeleton-based mocap file format that was developed by the computer game producer Acclaim. The mocap data in this format are described by two separate ASCII-coded files: an ASF file contains the fixed skeleton information, while an AMC file encodes the free rotational parameters [8, 149]. Even though the usage of this format seems to have been discontinued, there is a large corpus of ASF/AMC data available to the public, for example, the CMU mocap database [44]. The *BVH format* is also a skeleton-based mocap format that was developed by the mocap services company Biovision. A short documentation can be found at the web page [24].

10

DTW-Based Motion Comparison and Retrieval

As we have seen in Chap. 4, dynamic time warping is a flexible tool for comparing time series in the presence of nonlinear time deformations. In this context, the choice of suitable local cost or distance measures is of crucial importance, since they determine the kind of (spatial) similarity between the elements (frames) of the two sequences to be aligned. For the mocap domain, we introduce two conceptually different local distance measures – one based on joint angle parameters and the other based on 3D coordinates – and discuss their respective strengths and weaknesses (Sect. 10.1). The importance of DTW is then illustrated by some synthesis and analysis applications (Sect. 10.2). By comparing a motion data stream to itself, one obtains a cost or distance matrix that exhibits self-similarities within the motion. In Sect. 10.3, we describe how this idea can be exploited for motion retrieval. Finally, in Sect. 10.4, we discuss some work related to DTW-based motion retrieval.

10.1 Local Distance Measures

10.1.1 Quaternion-Based Pose Distance

Recall from Sect. 9.3.3 that a mocap data stream can be modeled as an animated kinematic chain consisting of fixed set of skeletal parameters $(J, r, B, (t_b)_{b\in B})$ and a function $D_{\mathcal{J}} : [1 : T] \to \mathcal{J}$ that encodes the joint configurations over time. Each configuration $(t_r, R_r, (R_b)_{b\in B}) \in \mathcal{J}$ includes a translational parameter t_r and rotational parameter R_r that determine the root coordinate system. The remaining rotational parameters $(R_b)_{b\in B}$ describe the joint angles within the skeletal model. One possible way of encoding the joint rotations is to use unit quaternions as described in Sect. 9.4.3. It turns out that a quaternion-based motion representation is well-suited for the comparison of motions for the following reasons:

1. Using only the joint rotations and leaving the root parameters unconsidered, one can design local distance measures that are invariant under global transformations such as translations, rotations, and scalings of the skeleton.
2. The geodesic (spherical) distance between unit quaternions implied by the spherical distance on the hypersphere $S^3 \subset \mathbb{H}$ provides a natural cost measure for rotations.
3. The distance between unit quaternions can be efficiently computed by using the inner product of $\mathbb{R}^4$. This, in turn, gives rise to an efficient procedure for computing cost matrices by a simple multiplication of suitably defined matrices.

Note that some of the joint rotations have a much greater overall effect on the character's pose than others. For example, small rotations in the shoulder joint can lead to large changes in the position of the entire arm. Conversely, even large rotations in the wrist joint do not have much influence on the overall pose. To account for this, one can introduce suitable weights $(w_b)_{b\in B}$ with $\sum_{b\in B} w_b = 1$, where the positive number w_b expresses the relative importance of rotation R_b.

We now define a local distance measure

$$c^{\mathrm{Quat}} : \mathcal{J} \times \mathcal{J} \to [0,1], \tag{10.1}$$

which will be referred to as *quaternion-based pose distance*. To this end, we fix two joint configurations $j := (t_r, R_r, (R_b)_{b\in B}) \in \mathcal{J}$ and $j' := (t'_r, R'_r, (R'_b)_{b\in B}) \in \mathcal{J}$. Let q_b and q'_b, $b \in B$, denote the unit quaternions describing the relative joint rotations R_b and R'_b, respectively. Then we set

$$c^{\mathrm{Quat}}(j,j') := \sum_{b\in B} w_b \cdot \frac{2}{\pi} \cdot \arccos\left|\langle q_b|q'_b\rangle\right|, \tag{10.2}$$

where $\langle\cdot|\cdot\rangle$ denotes the standard inner product of $\mathbb{R}^4$. Here, note that the inner product $\langle q_b|q'_b\rangle$ gives the cosine of the angle $\varphi \in [0,\pi]$ enclosed between the unit vectors q_b and q'_b. Then the value $\varphi = \arccos\langle q_b|q'_b\rangle$ is the geodesic or spherical distance on S^3, i.e., the length of the shortest connecting path between the points q_b and q'_b on the four-dimensional unit sphere S^3. But this is not yet what we want, since one would obtain a large value φ in the case that q'_b equals $-q_b$. However, recall that antipodal unit quaternions correspond to the same rotation, see Theorem 9.5. We therefore take the absolute value $|\langle q_b|q'_b\rangle|$, which is invariant under antipodal identification, i.e., $|\langle q_b|q'_b\rangle| = |\langle q_b|-q'_b\rangle| = |\langle -q_b|q'_b\rangle|$. This allows us to define a distance on the space $\mathbb{R}P(3) = S^3/\{\pm 1\}$ via $\arccos|\langle q_b|q'_b\rangle|$, which is known as the geodesic distance of the real three-dimensional projective space. Note that $\arccos|\langle q_b|q'_b\rangle|$ assumes its maximal value $\frac{\pi}{2}$ in the case that q_b and q'_b are orthogonal. The factor $\frac{2}{\pi}$ in (10.2) is introduced for normalization. Finally, the geodesic distances of all quaternions q_b and q'_b are weighted by w_b and added up to yield the pose distance $c^{\mathrm{Quat}}(j,j') \in [0,1]$.

Even though the proposed quaternion-based pose distance offers the above mentioned advantages, there are also some drawbacks as is also noted in Kover et al. [108].

1. The assignment of the weights $(w_b)_{b\in B}$ is problematic since the effect and hence the importance of certain rotational joint parameters may change with the course of the motion. For example, the influence of the shoulder rotation on the overall pose is larger with a stretched elbow than with a bent elbow.
2. The distance measure c^{Quat} is not only invariant under any global 3D Euclidean transformation, but is also leaves the progression of the 3D coordinates in space more or less unconsidered – this may be an overkill in abstraction concerning motion identification. For example, a motion where an actor in horizontal position is doing a press-up could not be distinguished well from a motion where an actor in vertical position is pushing something in front of his body with both hands kept in parallel.
3. To obtain smooth transitions in blending applications, one requires more information than can be obtained at individual frames. To this end, one has to design local distance measures that not only account for differences in body postures, but also in joint velocities, accelerations, and possibly higher-order derivatives [108].

In the next section, we describe a local distance measure that overcomes some of these drawbacks.

10.1.2 3D Point Cloud Distance

In their work on motion synthesis [106–108], Kovar and Gleicher propose, as one of their contributions, a technique to identify motion clips within a given database that can be concatenated without visible artifacts at the transition points. To this end, they introduce a local distance measure that is based on a comparison of certain pose-driven 3D point clouds. The principle of their distance measure is explained in Fig. 10.1. Here, the goal is to compare the two poses from a running sequence that are shown in Fig. 10.1a. First, a temporal context around both frames is incorporated in the comparison, as depicted in Fig. 10.1b. This is required because the notion of smoothness implicitly depends on temporal derivatives, which in turn depend on a local temporal context. Next, the two short pose sequences are converted to 3D point clouds based on the joints of the underlying kinematic chain, see Fig. 10.1c. Finally, to achieve invariance under certain global translations and rotations, an optimal alignment of the two point clouds is computed with respect to a suitable transformation group. Then, the resulting total distance between corresponding points constitutes the *3D point cloud distance* between the two original poses.

We will now formally define the 3D point cloud distance, which is denoted by $c^{3\mathrm{D}}$. Since we want to incorporate temporal information into the distance

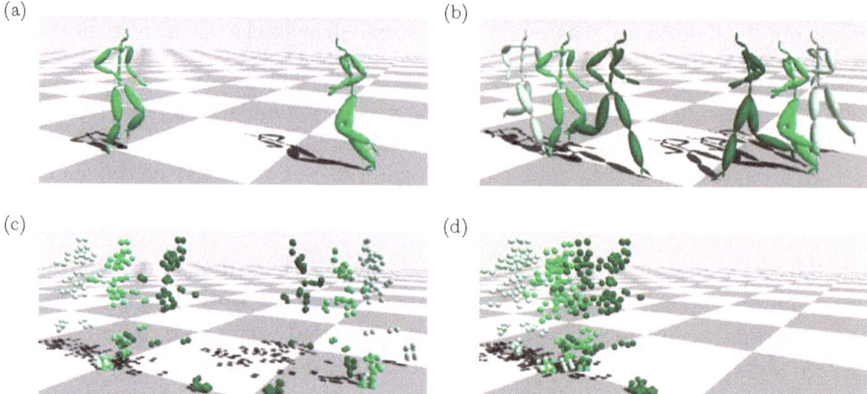

Fig. 10.1. Principle of the 3D point cloud distance, adopted from Kover and Gleicher [106]. (**a**) The two poses that are to be compared. (**b**) Inclusion of a temporal context for each of the poses. (**c**) Conversion to point clouds. (**d**) Optimal alignment of point clouds to determine pose distance as sum of distances of corresponding points

measure, we need to consider the two poses to be compared within the context of some motions. Without loss of generality, we may assume that the two poses are given as $D(n) \in \mathcal{P}$ and $D(m) \in \mathcal{P}$ contained within the same mocap data stream $D : [1 : T] \to \mathcal{P}$, $n, m \in [1 : T]$. Here, recall that $\mathcal{P} \subset \mathbb{R}^{3\times|J|}$ denotes the pose and J the joint set, see (9.1). We then view the pose $D(n)$ in the context of the ρ preceding frames and the ρ subsequent frames, yielding the $2\rho + 1$ frame numbers $[n - \rho : n + \rho]$. The 3D points corresponding to all of these poses form a point cloud $\{D^j(i) \mid i \in [n - \rho, n + \rho], j \in J\} \subset \mathbb{R}^3$ containing $K := |J|(2\rho + 1)$ points, which we denote by $P := (p_i)_{i\in[1:K]}$ with $p_i = (x_i, y_i, z_i)^\top \in \mathbb{R}^3$. As with all sliding window techniques, the boundary cases $n < \rho + 1$ and $n > T - \rho$ require special attention. Here, the motion needs to be padded by ρ additional frames in the front and in the back. Suitable strategies would be symmetric padding, i.e., padding with parts of a time-reversed version of the data stream or constant padding, i.e., replication of $D(1)$ and $D(T)$, respectively. Similarly, we transform $D(m)$ into $P' = (p'_i)_{i\in[1:K]}$ with $p'_i = (x'_i, y'_i, z'_i)^\top \in \mathbb{R}^3$. The comparison between the poses $D(n)$ and $D(m)$ will be performed on the basis of the corresponding point clouds P and P'.

To make the local distance measure invariant under certain global transformations, we allow the point cloud P' to move freely with respect to certain transformation parameters before being compared with P. Kovar and Gleicher [108] suggest to use the group of linear transformations consisting of all rotations about the y (vertical) axis and all translations in the xz plane (horizontal translations). Let $T_{\theta,x,z}$ denote the transformation that simultaneously rotates all 3D points of a point cloud about the y axis by an angle

$\theta \in [0, 2\pi)$ and then shifts the resulting points in the xz plane by an offset vector $(x, 0, z)^\top \in \mathbb{R}^3$. Similar to Sect. 10.1.1, we introduce some weights $w_i \in \mathbb{R}_{\geq 0}$ with $\sum_{i=1}^{K} w_i = 1$. These weights will be discussed later. Then, the *3D point cloud distance* c^{3D} between the poses $D(n)$ and $D(m)$ is defined as

$$c^{\mathrm{3D}}\Big(D(n), D(m)\Big) := \min_{\theta, x, z} \left(\sum_{i=1}^{K} w_i \| p_i - T_{\theta,x,z}(p_i') \|^2 \right), \tag{10.3}$$

where the minimum is taken over all possible transformation parameters $\theta \in [0, 2\pi)$ and $x, z \in \mathbb{R}$.

Opposed to the quaternion-based pose distance c^{Quat}, the 3D point cloud distance c^{3D} as proposed by Kovar and Gleicher is invariant only under a three-dimensional subgroup of the six-dimensional group of rigid 3D Euclidean transformations. Only admitting rotations about the y axis stems from the following observation: for a wide range of motion classes it is not semantically meaningful to consider two motions as equivalent that only differ with respect to a rotation about any other axis than the y axis. This is due to gravity usually pulling us downwards, in the $-y$ direction. For example, consider a jumping jack vs. a snow angel, which would be very similar if the full group of 3D Euclidean transformations were admitted. However, the state of "lying on the floor" has strong semantics, and the 3D point cloud distance does well in separating it from "standing upright." Furthermore, only admitting translations in the xz plane boils down to forbidding uneven terrain, which is realistic in many cases. Nevertheless, this restriction may have undesirable consequences in practical applications: sometimes, the ground level of motions changes between recording sessions, leading to a constant difference in the y coordinate – which leads to a large 3D point cloud distance even between otherwise identical motions. Similarly, the 3D point cloud distance as defined above is not invariant to scaling. However, such invariance could be incorporated by extending the group of admissible transformations.

Just as for the case of the quaternion distance, choosing suitable weights w_i, $i \in [1 : K]$, seems to be such a delicate issue that Kovar and Gleicher [106, 108] simply set $w_i := \frac{1}{K}$ for all i, which amounts to using no weights at all. They note that the weights could be used to stress or mask out certain parts of the body in the comparison, depending on the respective application. The weights could also be used to attribute different importance to different frames within the temporal context $[n - \rho : n + \rho]$, for example, to smoothly taper off the influence of the past and the future frames.

Our final remarks concern the minimization step that has to be performed in the computation of the 3D point cloud distance. Let $\hat{\theta}$, $\hat{x}$, and $\hat{z}$ denote the transformation parameters for which the minimum in (10.3) is assumed. It turns out that finding such optimal parameters constitutes a version of a so-called *Procrustes problem* and allows for an explicit closed-form solution [7]. We close this section by giving this explicit solution and refer to the literature for a proof.

Optimization Problem: POINTCLOUDDISTANCE

Input: Point clouds $P = (p_i)_{i\in[1:K]}$ with $p_i = (x_i, y_i, z_i)^\top \in \mathbb{R}^3$ and $P' = (p'_i)_{i\in[1:K]}$ with $p'_i = (x'_i, y'_i, z'_i)^\top \in \mathbb{R}^3$.
Weights $w_i \in \mathbb{R}_{\geq 0}$, $1 \leq i \leq K$, with $\sum_{i=1}^K w_i = 1$.

Output: Optimal parameters $\hat{\theta}$, $\hat{x}$, and $\hat{z}$ minimizing (10.3).

Solution:

$$\hat{\theta} = \arctan\left(\frac{\sum_{i=1}^K w_i(x_i z'_i - z_i x'_i) - (\bar{x}\bar{z'} - \bar{z}\bar{x'})}{\sum_{i=1}^K w_i(x_i x'_i + z_i z'_i) - (\bar{x}\bar{x'} + \bar{z}\bar{z'})}\right),$$

$$\hat{x} = \bar{x} - \bar{x'}\cos\hat{\theta} - \bar{z'}\sin\hat{\theta},$$

$$\hat{z} = \bar{z} + \bar{x'}\sin\hat{\theta} - \bar{z'}\cos\hat{\theta},$$

where $\bar{x} := \sum_i w_i x_i$.
The other barred terms are defined similarly.

10.1.3 Examples

We now discuss several examples to illustrate the properties of the local measures c^{Quat} and c^{3D}. In all examples of this section, a frame rate of 120 Hz has been used to record the mocap data. First, let us consider the two kicking motions indicated by Fig. 1.1. The first motion consists of 183 frames and represents a side kick, whereas the second motion consists of 250 frames and represents a front kick. Both kicks are performed with the right foot. The resulting cost matrix C with respect to $c = c^{\mathrm{Quat}}$ is shown in Fig. 10.2a, where the vertical axis corresponds to the first and the horizontal axis to the second motion. Recall from Sect. 4.1 that a tuple (n, m) is referred to as a *cell* corresponding to the nth frame of the first and the mth frame of the second motion. The entry $C(n, m)$ of a cost matrix C is then referred to as the *cost* of cell (n, m). Cells of low cost are indicated by blue, whereas cells of high cost are indicated by red.[1] Note that corresponding poses at the beginning and the end of the two kicking motions are similar to each other – actually, each of these poses is close to a standing pose – which is reflected by the cells of low costs in the four corners of C. However, corresponding poses during the kicking phases – even though similar from a semantic point of view – reveal significant differences in the respective angle configurations, which is reflected by the cells of high cost in the neighborhood of cell $(80, 100)$. Figure 10.2b shows the cost matrix with respect to $c = c^{\mathrm{3D}}$. Here the (relative) costs in a neighborhood of cell $(80, 100)$ are smaller than in the case $c = c^{\mathrm{Quat}}$. In other words, during the kicking phase, the spatial deviations of the 3D coordinates

[1] In the illustrations of the cost matrices, we use a relative color coding that depends on the minimal and maximal value appearing in C.

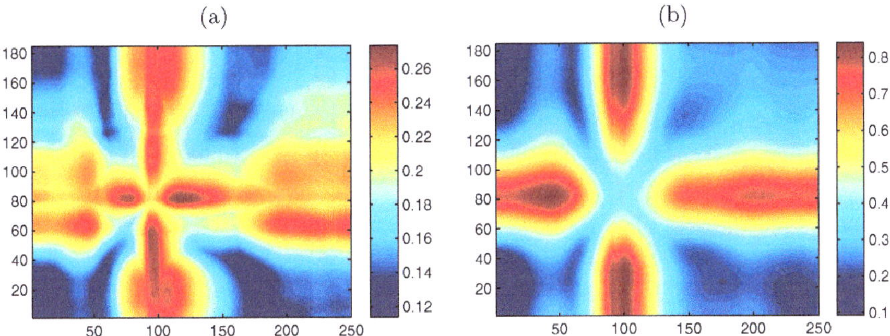

Fig. 10.2. Cost matrices with respect to **(a)** $c = c^{\mathrm{Quat}}$ and **(b)** $c = c^{\mathrm{3D}}$ of the two kicking motions indicated by Fig. 1.1. The vertical axis corresponds to the side kick, whereas the horizontal axis to the front kick

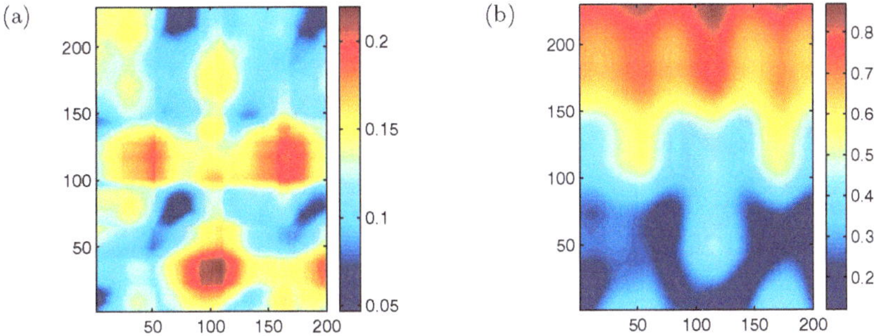

Fig. 10.3. Cost matrices with respect to (**a**) $c = c^{\mathrm{Quat}}$ and (**b**) $c = c^{\mathrm{3D}}$ of a climbing motion (vertical) and a walking motion (horizontal). Both motions consist of three steps starting with the right foot

measured by c^{3D} seem to be less significant than the deviations in the angle configurations measured by c^{Quat}.

In the next example, we compare a walking motion climbing up a staircase with a walking motion performed on the ground level. Both motions consist of three steps starting with the right foot. The cost matrix with respect to $c = c^{\mathrm{Quat}}$ is shown in Fig. 10.3. Note that c^{Quat} is invariant under all Euclidean motions in $\mathbb{R}^3$, including translations in y direction (vertical direction). Even though the actor in the first motion constantly moves upwards, the overall progression of angle configurations in the climbing motion (first motion) roughly coincides with the one in the usual walking motion (second motion). This is reflected by cells of relatively low costs along the main diagonal. On the other hand, the local cost measure c^{3D} is not invariant under translations in y-direction. This leads to high costs when comparing poses towards the end of the first motion (actor's body is well above ground level) with poses of the second motion (actor's body is at ground level), see Fig. 10.3b.

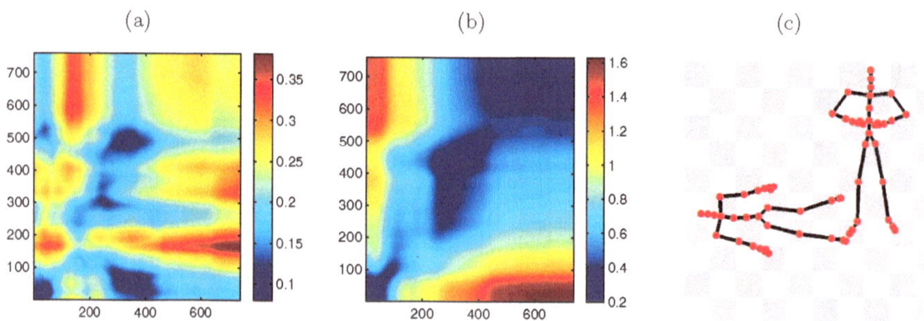

Fig. 10.4. Cost matrices with respect to (**a**) $c = c^{\mathrm{Quat}}$ and (**b**) $c = c^{\mathrm{3D}}$ of two motions, where the actors are lying down on the floor. The final poses of the motions are shown in (**c**)

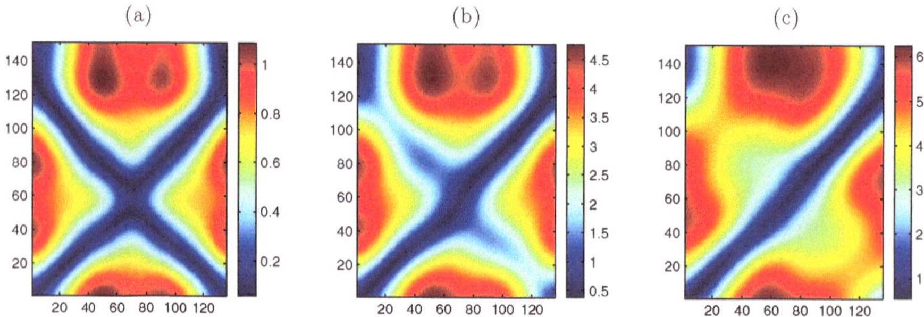

Fig. 10.5. Cost matrices with respect to $c = c^{\mathrm{3D}}$ for two jumping jacks using the parameters **(a)** $\rho = 0$, **(b)** $\rho = 10$, and **(c)** $\rho = 20$

In our third example, two different actors, starting from an upright standing pose, are lying down on the floor. The final poses of the two motions are indicated by Fig. 10.4c. The resulting cost matrices with respect to $c = c^{\mathrm{Quat}}$ and $c = c^{\mathrm{3D}}$ are shown in Fig. 10.4a, b, respectively. Note that the final poses of the two motions – even though similar from a semantic point of view – exhibit large differences particulary in the elbow angles, which results in significant c^{Quat}-costs as shown in the upper right corner (yellow/green) of Fig. 10.4a. This difference in the arm's position has a smaller overall effect on the c^{3D}-cost as illustrated by the upper right corner (blue) of Fig. 10.4b. A further interesting observation is that the c^{Quat}-costs between the standing poses at the beginning of one motion and lying poses at the end of the other motion are very small (see, e.g., the cells lying on the line $(500, 30)$ to $(750, 30)$). Conversely, the c^{3D}-costs of the same cells are extremely high. This is due to the fact that c^{Quat} is invariant under the full rotation group of $\mathbb{R}^3$, whereas c^{3D} is invariant only under rotations about the vertical axis.

The final example illustrates the effect on the cost matrix when increasing the parameter ρ in the local cost measure $c = c^{\mathrm{3D}}$, see Sect. 10.1.2. Recall that for the computation of $C(n,m)$ with respect to c^{3D}, one compares entire motion fragments of length $2\rho+1$ centered at the nth frame of the first motion and at the mth frame of the second motion, respectively. Figure 10.5 shows the cost matrices for the parameters $\rho = 0$, $\rho = 10$, and $\rho = 20$, where two different jumping jack motions are compared. For the case $\rho = 0$, no contextual information is used and the value $C(n,m)$ depends only on the nth frame of the first and the mth frame of the second motion. Note that in this case, C reveals low values not only along the main diagonal, but also along the "antidiagonal" (orthogonal to the main diagonal). In other words, the poses of the first jumping jack motion are similar to the corresponding poses of the time-reversed second jumping jack motion, see Fig. 10.5a. By increasing ρ, the temporal progression of the frames gains more and more influence, which leads to a smearing effect of the resulting cost matrix along the direction of the main diagonal, see Fig. 10.5b, c. For example, the value $\rho = 20$ corresponds to a window length of 51 frames, which corresponds to nearly half a second of motion (at a frame rate of 120 Hz). In such a temporal context, which also encodes local velocity information, the poses of the first jumping jack motion are no longer considered to be similar to corresponding poses of the time-reversed second jumping jack (where the poses within the temporal context are not reversed). In other multimedia domains, the strategy of incorporating contextual information into the local cost measure has been used to enhance certain structural properties of the resulting cost matrix, see, e.g., Müller and Kurth [137] and the references therein.

10.2 DTW-Based Motion Warping

Dynamic time warping constitutes an important tool widely used in the analysis and synthesis of motion data. In analysis applications, DTW allows for the design of semantically meaningful similarity measures that can cope with spatio-temporal variations in the motions to be compared. A further prevailing topic in computer animation is the synthesis of new, natural-looking motions from given prototype motions based on blending and morphing techniques, see, e.g., [22,75,107,217]. In the animation of virtual characters as used in video games, one starts with a small set of recorded prototypes motions to automatically synthesize new motions by taking suitable combinations. The synthesized motions then cover a large variety of possible movements that may be performed by a virtual football or basketball player or a martial artist, see Fig. 10.6. Modern locomotion synthesis systems use morphing techniques where the user supplies the system with some mocap data of a person walking. These data are then used to automatically adapt the walking style to different gaits [93].

A morphing transformation is generally obtained in a two-step process: first a domain transformation is performed to establish time correspondence between common features of the prototype objects, and then a range

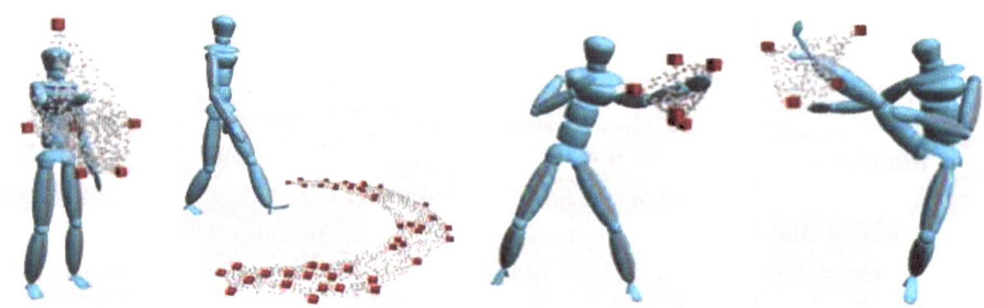

Fig. 10.6. Synthesis of new reaching, walking, punching, and kicking motions from given prototype motions based on morphing techniques, adopted from Kovar and Gleicher [107]. Red cubes show target locations of motion prototypes and grey cubes show the range of variation accessible through motion morphing

transformation is performed to blend or interpolate among the different mapped attribute values [22, 78]. In this approach, central questions concern the adequate definition of "correspondence" between different prototype motions as well as the warping procedure (domain transformation), in which the motions are locally expanded and compressed to run synchronously before the interpolation of the spatial motion attributes can be done on the warped motions.

For further details concerning morphing, we refer to the literature and discuss in the following the basic idea of DTW-based motion alignment and warping. Exemplarily, we use the quaternion distance measure c^{Quat} and do not discuss the influence of the specific distance measure at this point. First, let us consider the two walking motions shown in Fig. 9.4, both comprising five steps and both starting with the left leg. The first walking motion consists of roughly 260 frames, whereas the second walking motion, which is much slower being performed by an elderly person, consists of roughly 450 frames. To compare the two motions, we use dynamic time warping based on c^{Quat}. The resulting cost matrix is shown in Fig. 10.7, where the vertical axis corresponds to the first motion and the horizontal axis to the second motion. Without surprise, the cells in a neighborhood of the main diagonal are of low cost, which reflects the fact that the second motion is more or less a uniformly time-scaled version of the first motion. The optimal warping path, roughly running along the main diagonal, exhibits local horizontal path segments such as the one running from cell $(25, 50)$ to cell $(25, 60)$. The meaning of this path segments is that the frames 50–60 of the second motion are assigned to the single frame 25 of the first motion. A manual inspection of the motions shows that the horizontal path segments correspond to periods in time where in both motions both feet are in floor contact – these floor contacts, however, are much longer for the second motion. Cells with large costs (e.g, the cells around $(90, 55)$) correspond to periods with the following semantics: in the first motion the right foot has maximal extension to the front, whereas in the second motion

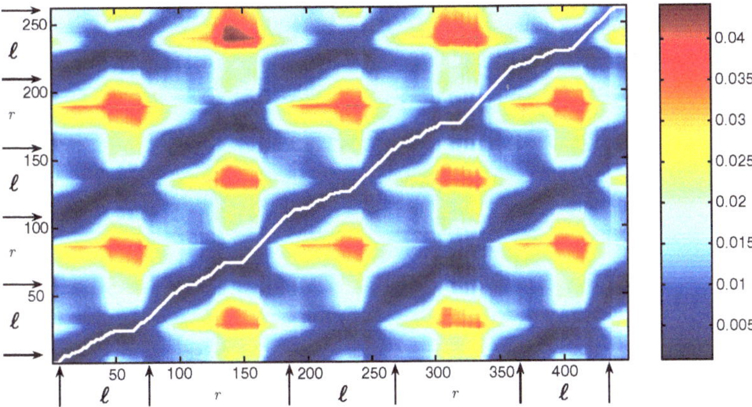

Fig. 10.7. Cost matrix and optimal warping path with respect to c^{Quat} of the walking motions shown in Fig. 9.4. The arrows indicate point in times where one foot passes the other one. The letters r/ℓ indicate periods in time, where the right/left foot lies in front of the left/right foot

the right foot has maximal extension to the back, or vice versa. In these cases the hip angles exhibit large relative differences, leading to a large overall pose distance. Furthermore, note that there are also cells in a neighborhood of secondary diagonals that exhibit low costs. Such a neighborhood occurs, for example, along the secondary diagonal from cell $(105, 1)$ to $(270, 260)$, which reflects the fact that the third to fifth step of the first motion is similar to the first three steps of the second motion. For a more thorough discussion of such phenomena we refer to Sect. 10.3.

As a further example, we compare two punching motions performed by two different actors. Both motions consist of a sequence of four punches towards the front of the body alternately executed by the left hand and right hand. The resulting cost matrix and optimal warping path is shown in Fig. 10.8. The second motion (horizontal axis) differs from the first motion (vertical axis) in that the second actor, after pushing his arm to the front, pauses for a short moment with stretched arm before pulling it back to the body again. This difference in the two motions is reflected by the horizontal path segments within the optimal warping path. The gradient of the warping path also reveals that the actual pushing and pulling motion is performed faster by the second actor than by the first one.

10.3 Motion Retrieval Based on Self-Similarity

Recall that the key idea of the DTW-based alignment between two sequences X and Y was to compute a cost matrix C with respect to a local cost measure and then to derive a cost minimizing warping path, which typically runs

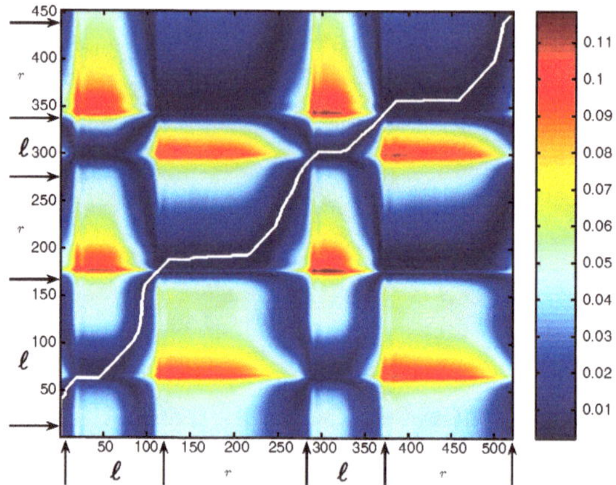

Fig. 10.8. Cost matrix and optimal warping path with respect to c^{Quat} for two punching motions, both consisting of two left–right punching combinations. The letters r/ℓ indicate periods in time, where the right/left hand is reached out in front of the body

through a "valley" of low cost in a neighborhood of the main diagonal of C. As was already mentioned in Sect. 10.2, the cost matrix C reveals further similarity relations between subsegments of X and Y, which are encoded by cost valleys running along secondary diagonals. This observation can be exploited for content-based motion retrieval based on self-similarity as will be explained in this section.

Instead of comparing two different motion streams, we now compare a motion data stream D to itself. In the following, let $X = (x_1, \ldots, x_N)$ denote the sequence of features $x_n \in \mathcal{F}$ extracted from the motion stream D and let $c : \mathcal{F} \times \mathcal{F} \rightarrow \mathbb{R}_{\geq 0}$ be a suitable local cost measure with $c(x, x) = 0$ for all features $x \in \mathcal{F}$. (For example, $\mathcal{F} = \mathcal{J}$, $x_n = D_{\mathcal{J}}(n)$ and $c = c^{\mathrm{Quat}}$ or $\mathcal{F} = \mathcal{P}$, $x_n = D(n)$ and $c = c^{\mathrm{3D}}$.) We then obtain a quadratic matrix $C \in \mathbb{R}^{N \times N}$ with $C(n, m) = c(x_n, x_m)$ for $n, m \in [1 : N]$, which is also referred to the *self-similarity matrix* of X. Obviously, the optimal warping path within C is just the main diagonal having a total cost of zero. Indeed, X is optimally aligned to itself by means of the identity.

We define a *partial warping path* to be a path $p = (p_1, \ldots, p_L)$ with $p_\ell = (n_\ell, m_\ell) \in [1 : N]^2$ for $\ell \in [1 : L]$ that fulfills the monotonicity condition and the step size condition but not necessarily the boundary condition, see Definition 4.1. The total cost of a partial warping path is defined as in the case of a warping path, see (4.2). The crucial observation is as follows: a partial warping path $p = (p_1, \ldots, p_L)$ of low total cost with $p_1 = (n_1, m_1)$ and $p_L = (n_L, m_L)$ indicates that the subsegment

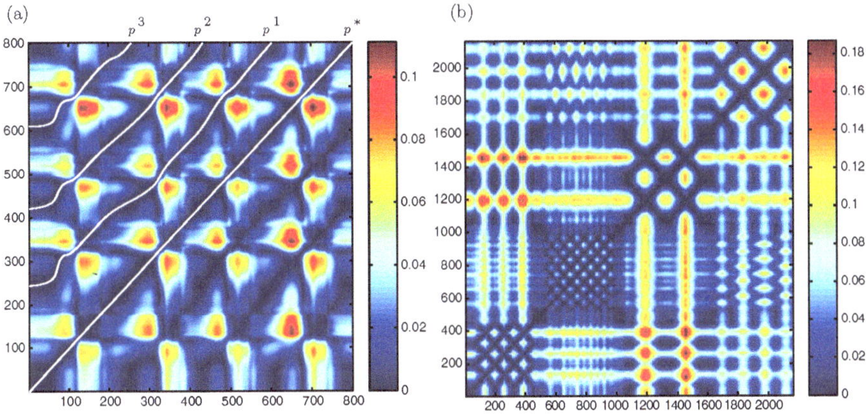

Fig. 10.9. **(a)** Self-similarity matrix C with respect to c^{Quat} for a motion consisting of four consecutive jumps with both legs as indicated by Fig. 12.9. The three partial warping paths p^1, p^2, and p^3 of low total cost (*white lines*) reflect the repetitive structure of the motion. **(b)** Self-similarity matrix C with respect to c^{Quat} for a database motion D consisting of four jumping jacks ($D(1 : 410)$), some running motion ($D(411 : 1070)$), two squats ($D(1071 : 1590)$), and two repetitions of an alternating elbow-to-knee motion ($D(1591\!:\!2150)$)

$(x_{n_1}, \ldots, x_{n_L})$ is similar to $(x_{m_1}, \ldots, x_{m_L})$. Furthermore, p encodes an alignment between these subsegments.

We illustrate this idea by means of a motion that consists of four consecutive jumps with both legs moving in parallel as indicated by Fig. 12.9. The corresponding feature sequence X consists of $N = 800$ features, each feature corresponding to one frame. The resulting self-similarity matrix C with respect to $c = c^{\text{Quat}}$ is shown in Fig. 10.9a. Since c is symmetric, the self-similarity C is symmetric as well so that we only have to consider the part above the main diagonal. The optimal warping path p^* is of course the main diagonal. There are several additional partial warping paths of low total cost. For example, the path p^1 that runs from cell $(240, 1)$ to cell $(800, 595)$ encodes that the motion fragment corresponding to the subsequence $X(240 : 800)$ is similar to the one corresponding to $X(1\!:\!595)$. This reflects the fact that the motion fragment consisting of the second to fourth jump is similar to the motion fragment consisting of the first to third jump. Similarly, the path p^2 reveals the similarity of the fragment consisting of the first two jumps and the fragment consisting of the last two jumps, and so on.

Each partial warping path of low cost corresponds to a pair of similar motion fragments. This principle can be exploited to identify similar motion fragments contained in some database. For simplicity, we suppose that the database consists of a single motion data stream, which is denoted by D.

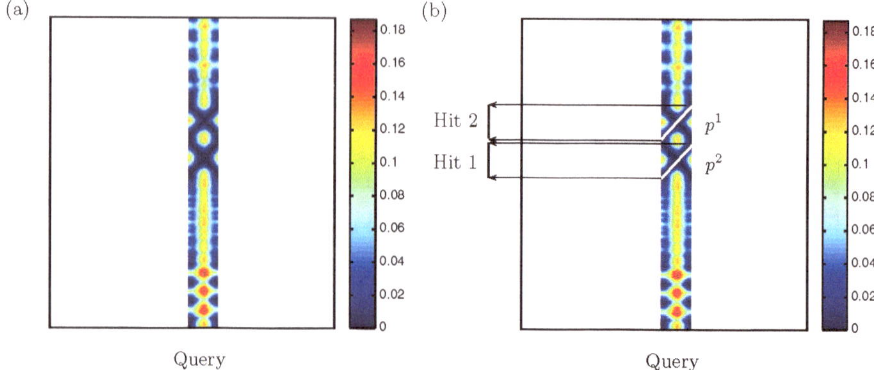

Fig. 10.10. (a) Part of the self-similarity matrix shown in Fig. 10.9 consisting of the "vertical stripe" above the query motion $D(1071:1320)$ (corresponding to the first squat). **(b)** The two partial warping paths of low cost (white lines) correspond to the two squats contained in the database

(This is no restriction since one can concatenate all database motions into a single data stream keeping track of document boundaries.) Then, given a query motion clip Q, the goal of content-based motion retrieval is to identify all motion fragments in D that are similar to Q. Such a motion fragment is also referred to as *hit*. We explain the idea of motion retrieval based on self similarity by means of an example. To this end, we consider a database motion sequence D that has a total length of $2,150$ frames and that consists of four jumping jacks ($D(1:410)$), some running motion ($D(411:1070)$), two squats ($D(1071:1590)$), and two repetitions of an alternating elbow-to-knee motion ($D(1591:2150)$). The self-similarity matrix of D with respect to c^{Quat} is shown in Fig. 10.9b. Suppose the query consists of the first squat ($D(1071:1320)$), then we consider the part of the self-similarity matrix consisting of the "vertical stripe" above the query fragment, see Fig. 10.10a. In the next step, all partial warping paths of low cost are identified within this stripe. Finally, the projection of each such path onto the vertical axis corresponds to a hit, see Fig. 10.10b. In our example, there are two partial warping paths p^1 and p^2 of low cost running from cell $(1071,1071)$ to $(1320,1320)$ and from cell $(1071,1330)$ to $(1320,1590)$, respectively. The projections of these paths result in the two hits $D(1071:1320)$ (the query itself) and $D(1330:1590)$ (the second squat). A much more elaborate and skillful method based on this general strategy has been applied to motion retrieval by Kovar and Gleicher [107].

There are several problems with a retrieval strategy based on a (precomputed) self-similarity matrix as described earlier. First, computing and storing the self-similarity is computationally expensive being quadratic in the number of frames. Hence, such a technique does not scale to large databases. Second, the features and local cost measures used to compute the self-similarity matrix are of numerical nature making it hard to identify semantically related motions, cf. Sect 9.2. And third, the suggested retrieval approach does not allow the user to incorporate a-priori knowledge of the motion. For example, if the user knows that the motion aspects of interest only regard the lower part of the body, he should be able to mask out irrelevant motion aspects such as the arm movements. In the subsequent sections, we introduce a relational-based approach to motion comparison and motion retrieval that will remedy some of these problems. Further strategies and related work are discussed in Sect. 10.4 and Sect. 12.4.2.

10.4 Further Notes

In this chapter, we have seen how DTW-based methods can be used to identify and align similar motions, which can then be further processed via blending and morphing techniques, see, e.g., [22, 75, 93, 106, 107, 217]. The curial point, however, is the particular choice of features and local cost measures, by which one can adjust the degree of invariance and the desired notion of similarity. As examples, we have introduced a quaternion-based pose distance using angle parameters and a 3D point cloud distance using 3D joint coordinates. The authors of [108] conjecture that measuring *perceptual* similarity between poses would require to obtain these point clouds from a downsampled version of the 3D mesh representation of the character's skin. Here, they argue that the skeleton-driven skin is all that is seen by an observer, and that the skeleton is only a means to an end. Such an approach is realistic in the context of professional animation systems, which provide that kind of skinning and mesh data. Hsu et al. [93] describe a DTW-based procedure for iterative motion warping to obtain more natural spatio-temporal alignments of two motion sequences. Only little work has been done to employ physics-based parameters in motion alignment [110, 130, 182]. In Chap. 14, we introduce a data-dependent adaptive local cost measure, by which irrelevant motion aspects are left unconsidered in the comparison [143].

Only recently, content-based motion retrieval has become an active research field, which is still gaining in importance. Most of the proposed retrieval strategies use a DTW variant at some stage [29, 70, 107, 123, 143, 220]. We discuss this literature in more detail in the motion retrieval context (Sect. 12.4.2).

In view of robustness and efficiency, several strategies have been suggested to replace or combine DTW-based techniques with methods based on local uniform scaling. For example, Keogh et al. [103] describe how to efficiently identify similar motion fragments that differ by some uniform scaling factor with respect to the time axis by adapting techniques from [100]. Similarly, Pullen and Bregler [169] use a simple resampling technique to compare motion fragments. In the music context, Müller et al. [140] use linear scaling techniques to retrieve similar audio excerpts form a database of CD recordings irrespective of the specific interpretation (Chap. 6). A combination of uniform scaling and dynamic time warping for general time series has been proposed in chee Fu [40], which can lead to more meaningful and robust alignments for certain applications. Other techniques for accelerating DTW computations are described in Sect. 4.5. In Chap. 11 we discuss how to avoid cost-intensive DTW computation by handling spatio-temporal deformation already at the feature level [146].

11

Relational Features and Adaptive Segmentation

Even though there is a rapidly growing corpus of motion capture data, there still is a lack of efficient motion retrieval systems that allow to identify and extract user-specified motions. Previous retrieval systems often require manually generated textual annotations, which roughly describe the motions in words. Since the manual generation of reliable and descriptive labels is infeasible for large datasets, one needs efficient *content-based* retrieval methods that only access the raw data itself. In this context, the *query-by-example* (QBE) paradigm has attracted a large amount of attention: given a query in form of a motion fragment, the task is to automatically retrieve all motion clips from the database containing parts or aspects similar to the query. The crucial point in such an approach is the notion of *similarity* used to compare the query with the database motions. For the motion scenario, two motions may be regarded as similar if they represent variations of the same action or sequence of actions. These variations may concern the spatial as well as the temporal domain. For example, the two jumps shown in Fig. 11.1 describe the same kind of motion, even though they differ considerably with respect to timing, intensity, and execution style (note, e.g., the arm swing). Similarly, the kicks shown in Fig. 1.1 describe the same kind of motion, even though they differ considerably with respect to direction and height of the kick. In other words, *semantically similar* motions need not be *numerically similar*, as is also pointed out in [107].

On the basis of a model in form of the kinematic chain, motion capture data has a much richer semantic content than, for example, pure video data of a motion, since the position and the meaning of all joints are known for every pose. We exploit this fact to bridge the semantic gap between numerical and semantic similarity by considering *relational features* that describe (boolean) geometric relations between specified points of a pose or short sequences of poses [146]. In Sect. 11.1, we will introduce several types of boolean relational features that encode spatial, velocity-based, as well as directional information. By applying feature-dependent temporal segmentations, the motions are then transformed into coarse sequences of binary vectors, which are invariant under

Fig. 11.1. *Top* 14 poses from a forceful jump. *Bottom* 14 poses from a weak jump

spatial as well as temporal deformations (Sect. 11.2). Motion comparison can then be performed efficiently on these vector sequences. In Sect. 11.3, we conclude this chapter and give references to the literature. The techniques introduced in this chapter have been first published in [146]. An accompanying video is available at [148].

In the following discussion, we need the notion of a *boolean feature*. Recall that the set of all poses is denoted by $\mathcal{P} \subset \mathbb{R}^{3\times|J|}$, with J being the joint set (Sect. 9.1). Then a boolean feature F is defined to be a function $F : \mathcal{P} \rightarrow \{0,1\}$, which only assumes the values 0 and 1. Obviously, any boolean expression of boolean functions (evaluated posewise) is a boolean function itself, examples being the conjunction $F_1 \wedge F_2$ and the disjunction $F_1 \vee F_2$ of boolean functions F_1 and F_2. Forming a vector of f boolean functions for some $f \geq 1$, one obtains a combined function

$$F : \mathcal{P} \rightarrow \{0,1\}^f. \tag{11.1}$$

From this point forward, F will be referred to as a *feature function* and the vector $F(P)$ as a *feature vector* or simply a *feature* of the pose $P \in \mathcal{P}$. Any feature function can be applied to a motion capture data stream $D : [1:T] \rightarrow \mathcal{P}$ in a posewise fashion, which is expressed by the composition $F \circ D$.

11.1 Relational Features

11.1.1 A Basic Example

As a basic example, we consider a relational feature that expresses whether the right foot lies in front of (feature value one) or behind (feature value zero) the plane spanned by the center of the hip (the root), the left hip joint, and the left foot for a fixed pose, cf. Fig. 11.2a. More generally, let $p_i \in \mathbb{R}^3$, $1 \leq i \leq 4$, be four 3D points, the first three of which are in general position. Let $\langle p_1, p_2, p_3 \rangle$ denote the oriented plane spanned by the first three points, where the orientation is determined by point order. Then define

$$B(p_1,p_2,p_3;p_4) := \begin{cases} 1, \text{ if } p_4 \text{ lies in front of or on } \langle p_1,p_2,p_3 \rangle, \\ 0, \text{ if } p_4 \text{ lies behind } \langle p_1,p_2,p_3 \rangle. \end{cases} \tag{11.2}$$

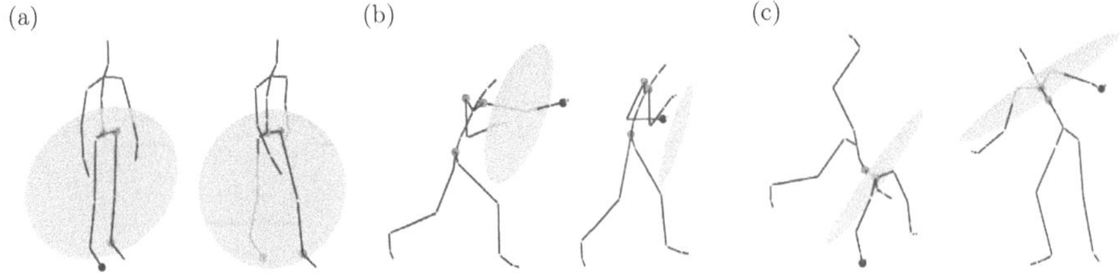

Fig. 11.2. Relational features describing geometric relations between the body points of a pose that are indicated by *red* and *black markers*. The respective features express whether **(a)** the right foot lies in front of or behind the body, **(b)** the left hand is reaching out to the front of the body or not, **(c)** the left hand is raised above neck height or not

From this we obtain a feature function $F_{\text{plane}}^{(j_1,j_2,j_3;j_4)} : \mathcal{P} \to \{0,1\}$ for any four distinct joints $j_i \in J$, $1 \leq i \leq 4$, by defining

$$F_{\text{plane}}^{(j_1,j_2,j_3;j_4)}(P) := B(P^{j_1}, P^{j_2}, P^{j_3}; P^{j_4}). \tag{11.3}$$

The concept of such relational features is simple but powerful, as we will illustrate by continuing the above example. Setting j_1 ="root," j_2 ="lankle," j_3 ="lhip," and j_4 ="rtoes," we denote the resulting feature by $F^r := F_{\text{plane}}^{(j_1,j_2,j_3;j_4)}$. The plane determined by j_1, j_2, and j_3 is indicated in Fig. 11.2a as a green disc. Obviously, the feature $F^r(P)$ is 1 for a pose P corresponding to a person standing upright. It assumes the value 0 when the right foot moves to the back or the left foot to the front, which is typical for locomotion such as walking or running. Interchanging corresponding left and right joints in the definition of F^r and flipping the orientation of the resulting plane, we obtain another feature function denoted by F^ℓ. Let us have a closer look at the feature function $F := F^r \wedge F^\ell$, which is 1 if and only if both, the right as well as the left toes, are in front of the respective planes. It turns out that F is very well suited to characterize any kind of walking or running movement. If a data stream $D : [1:T] \to \mathcal{P}$ describes such a locomotion, then $F \circ D$ exhibits exactly two peaks for any locomotion cycle, from which one can easily read off the speed of the motion (Fig. 11.3). On the other hand, the feature F is invariant under global orientation and position, the size of the skeleton, and various local spatial deviations such as sideways and vertical movements of the legs. Furthermore, F leaves any upper body movements unconsidered.

In the following, we will define feature functions purely in terms of geometric entities that are expressible by joint coordinates. Such relational features are invariant under global transforms (Euclidean motions, scalings) and are very coarse in the sense that they express only a single geometric aspect, masking out all other aspects of the respective pose. This makes relational features robust to variations in the motion capture data stream that are not correlated with the aspect of interest. Using suitable boolean expressions

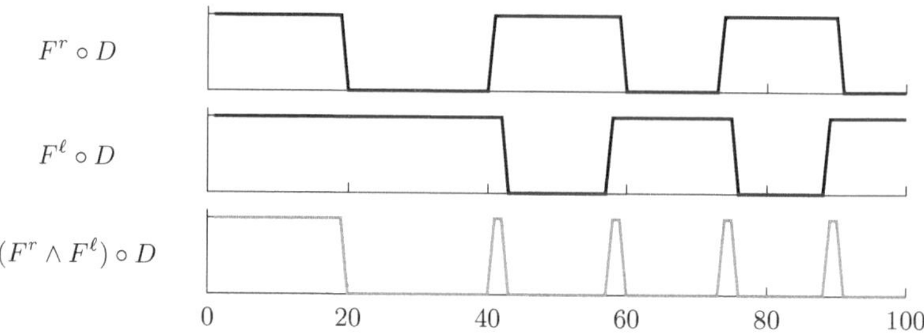

Fig. 11.3. Boolean features F^r, F^ℓ, and the conjunction $F^r \wedge F^\ell$ applied to the 100-frame walking motion $D = D_{\text{walk}}$ of Fig. 11.9

and combinations of several relational features then allows to focus on or to mask out certain aspects of the respective motion.

11.1.2 Generic Features

The four joints in $F_{\text{plane}}^{(j_1,j_2,j_3;j_4)}$ can be picked in various meaningful ways. For example, in the case $j_1 =$"root," $j_2 =$"lshoulder," $j_3 =$"rshoulder," and $j_4 =$"lwrist," the feature expresses whether the left hand is in front of or behind the body. Introducing a suitable offset, one can change the semantics of a feature. For the previous example, one can move the plane $\langle P^{j_1}, P^{j_2}, P^{j_3}\rangle$ to the front by one length of the skeleton's humerus. The resulting feature can then distinguish between a pose with a hand reaching out to the front and a pose with a hand kept close to the body, see Fig. 11.2b.

Generally, in the construction of relational features one can start with some *generic relational feature* that encodes certain joint constellations in 3D space and time. Such a generic feature depends on a set of joint variables, denoted by $j_1, j_2, \ldots$, as well as on a variable θ for a threshold value or threshold range. For example, the generic feature $F_{\text{plane}} = F_{\theta,\text{plane}}^{(j_1,j_2,j_3;j_4)}$ assumes the value one if joint j_4 has a signed distance greater than $\theta \in \mathbb{R}$ from the oriented plane spanned by the joints j_1, j_2, and j_3. Then each assignment to the joints $j_1, j_2, \ldots$ and the threshold θ leads to a boolean function $F : \mathcal{P} \to \{0,1\}$. For example, by setting $j_1 =$"root," $j_2 =$"lhip," $j_3 =$"ltoes," $j_4 =$"rankle," and $\theta = 0$ one obtains the (boolean) relational feature indicated by Fig. 11.2a.

Similarly, we obtain a generic relational feature $F_{\text{nplane}} = F_{\theta,\text{nplane}}^{(j_1,j_2,j_3;j_4)}$, where we define the plane in terms of a normal vector (given by j_1 and j_2), and fix it at j_3. For example, using the plane that is normal to the vector from the joint $j_1 =$"chest" to the joint $j_2 =$"neck" fixed at $j_3 =$"neck" and testing $j_4 =$"rwrist" against this plain with threshold $\theta = 0$, one obtains a feature that expresses whether the right hand is raised above neck height or not, cf. Fig. 11.2c.

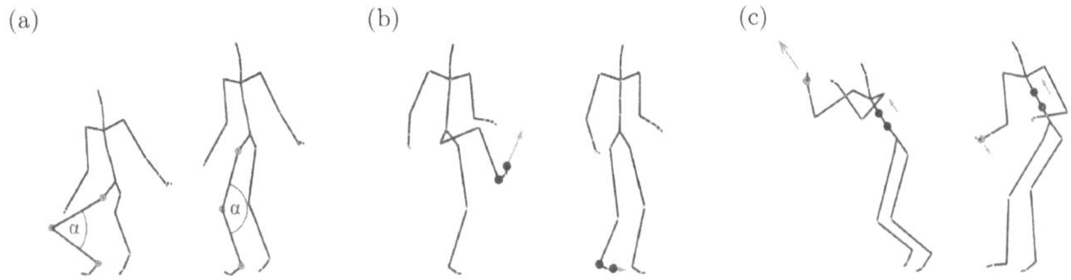

Fig. 11.4. Relational features that express whether **(a)** the right leg is bent or stretched, **(b)** the right foot is fast or not, **(c)** the right hand is moving upwards in the direction of the spine or not

Using another type of relational feature, one may check whether certain parts of the body such as the arms, the legs, or the torso are bent or stretched. To this end, we introduce the generic feature $F_{\text{angle}} = F^{(j_1,j_2;j_3,j_4)}_{\theta,\text{angle}}$, which assumes the value one if the angle between the directed segments determined by (j_1, j_2) and (j_3, j_4) is within the threshold range $\theta \subset \mathbb{R}$. For example, by setting $j_1 =$"rknee," $j_2 =$"rankle" $j_3 =$"rknee," $j_4 =$"rhip," and $\theta = [0, 120]$, one obtains a feature that checks whether the right leg is bent (angle of the knee is below 120°) or stretched (angle is above 120°), see Fig. 11.4a.

Other generic features may operate on velocity data that is approximated from the 3D joint trajectories of the input motion. An easy example is the generic feature $F_{\text{fast}} = F^{(j_1)}_{\theta,\text{fast}}$, which assumes the value one if joint j_1 has an absolute velocity above θ. Figure 11.4b illustrates the derived feature $F^{\text{rfootfast}} := F^{\text{rtoes}} \wedge F^{\text{rankle}}$, which is a movement detector for the right foot. $F^{\text{rfootfast}}$ checks whether the absolute velocity of both the right ankle (feature: F^{rankle}) and the right toes (feature: F^{rtoes}) exceeds a certain velocity threshold, θ_{fast}. If so, the feature assumes the value one, otherwise zero (Fig. 11.5). This feature is well suited to detect kinematic constraints such as footplants. The reason why we require both the ankle and the toes to be sufficiently fast is that we only want to consider the foot as being fast if all parts of the foot are moving. For example, during a typical walking motion, there are phases when the ankle is fast while the heel lifts off the ground, but the toes are firmly planted on the ground. Similarly, during heel strike, the ankle has zero velocity, while the toes are still rotating downwards with nonzero velocity. This feature illustrates one of our design principles for relational features. We construct and tune features so as to explicitly grasp the semantics of typical situations such as the occurrence of a footplant, yielding intuitive semantics for our relational features. However, while a footplant always leads to a feature value of zero for $F^{\text{rfootfast}}$, there is a large variety of other motions yielding the feature value zero (think of keeping the right leg lifted without moving). Here, the combination with other relational features is required to further classify the respective motions. In general, suitable combinations of our relational features prove to be very descriptive for full-body motions.

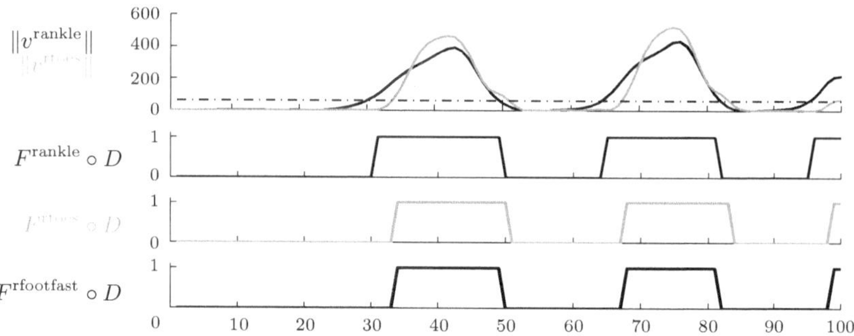

Fig. 11.5. *Top*: Absolute velocities in cm/s of the joints "rankle" ($\|v^{\mathrm{rankle}}\|$, *black*) and "rtoes" ($\|v^{\mathrm{rtoes}}\|$, *gray*) in the walking motion $D = D_{\mathrm{walk}}$ of Fig. 11.9. The *dashed line* at $\theta_{\mathrm{fast}} = 63$ cm/s indicates the velocity threshold. *Middle*: Thresholded velocity signals for "rankle" and "rtoes." *Bottom*: Feature values for $F^{\mathrm{rfootfast}} = F^{\mathrm{rtoes}} \wedge F^{\mathrm{rankle}}$

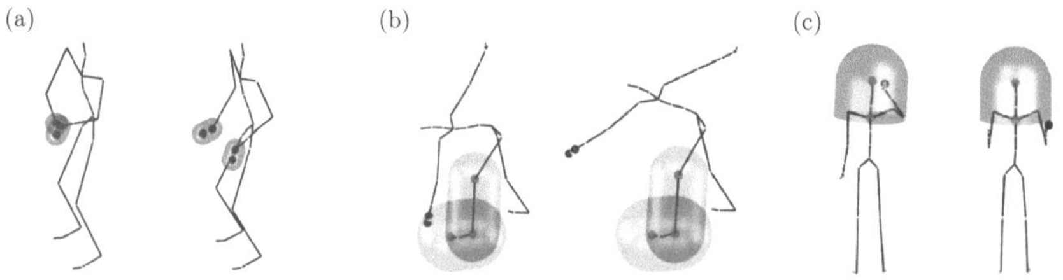

Fig. 11.6. Relational "touch" features that express whether **(a)** the two hands are close together or not, **(b)** the left hand is close to the leg or not, **(c)** the left hand is close to the head or not

Another velocity-based generic feature is denoted by $F_{\mathrm{move}} = F^{(j_1,j_2;j_3)}_{\theta,\mathrm{move}}$. This feature considers the velocity of joint j_3 relative to joint j_1 and assumes the value one if the component of this velocity in the direction determined by (j_1, j_2) is above θ. For example, setting $j_1 =$"belly," $j_2 =$"chest," $j_3 =$"rwrist," one obtains a feature that tests whether the right hand is moving upwards or not, see Fig. 11.4c. The generic feature $F^{(j_1,j_2,j_3;j_4)}_{\theta,\mathrm{nmove}}$ has similar semantics, but the direction is given by the normal vector of the oriented plane spanned by j_1, j_2, and j_3.

As a final example, we introduce generic features that check whether two joints, two body segments, or a joint and a body segment are within a θ-distance of each other or not. Here one may think of situations such as two hands touching each other, or a hand touching the head or a leg (Fig. 11.6). In the first case, this leads to a generic feature $F^{(j_1,j_2)}_{\theta,\mathrm{touch}}$, which checks whether the θ-neighborhoods of the joints j_1 and j_2 intersect or not. Similarly, one defines generic touch features for body segments.

11.1.3 Threshold Selection

Besides selecting appropriate generic features and suitable combinations of joints, the crucial point in designing relational features is to choose the respective threshold parameter θ in a semantically meaningful way. This is a delicate issue, since the specific choice of a threshold has a strong influence on the semantics of the resulting relational feature. For example, choosing $\theta = 0$ for the feature indicated by Fig. 11.2b results in a boolean function that checks whether the left hand is in front of or behind the body. By increasing θ, the resulting feature checks whether the left hand is reaching out to the front of the body. Similarly, a small threshold in a velocity-based feature such as $F^{(j_1)}_{\theta,\mathrm{fast}}$ leads to sensitive features that assume the value 1 even for small movements. Increasing θ results in features that only react for brisk movements. In general, there is no "correct" choice for the threshold θ – the specific choice will depend on the application in mind and is left to the designer of the desired feature set. In Sect. 11.1.4, we will specify a feature set that is suitable to compare the overall course of a full-body motion disregarding motion details.

To obtain a semantically meaningful value for the threshold θ in some automatic fashion, one can also apply supervised learning strategies. One possible strategy for this task is to use a training set $\mathcal{A}$ of "positive" motions that should yield the feature value one for most of its frames and a training set $\mathcal{B}$ of "negative" motions that should yield the feature value zero for most of its frames. The threshold θ can then be determined by a one-dimensional optimization algorithm, which iteratively maximizes the occurrences of the output one for the set $\mathcal{A}$ while maximizing the occurrences of the output zero for the set $\mathcal{B}$.

To make the relational features invariant under global scalings, the threshold parameter θ is specified relative to the respective skeleton size. For example, the value of θ may be given in terms of the length of the humerus, which scales quite well with the size of the skeleton. Such a choice handles differences in absolute skeleton sizes that are exhibited by different actors but may also result from different file formats for motion capture data.

Another problem arises from the simple quantization strategy based on the threshold θ to produce boolean features from the generic features. Such a strategy is prone to strong output fluctuations if the input value fluctuates slightly around the threshold. To alleviate this problem, we employ a robust quantization strategy using two thresholds: a stronger threshold θ_1 and a weaker threshold θ_2. As an example, consider a feature F^{sw} that checks whether the right leg is stretched sideways, see Fig. 11.7. Such a feature can be obtained from the generic feature $F^{(j_1,j_2,j_3;j_4)}_{\theta,\mathrm{nplane}}$, where the plane is given by the normal vector through j_1="lhip" and j_2="rhip" and is fixed at j_3="rhip." Then the feature assumes the value one if joint j_4="rankle" has a signed distance greater than θ from the oriented plane with a threshold $\theta = \theta_1 = 1.2$ measured in multiples of the hip width. As illustrated by Fig. 11.7a, the feature values may randomly fluctuate, switching between the numbers one and zero,

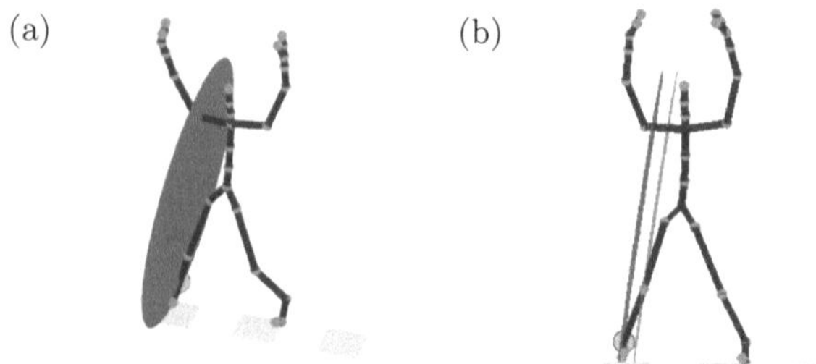

Fig. 11.7. Relational feature that expresses whether the right leg is stretched sideways or not. **(a)** The feature values may randomly fluctuate if the right ankle (*green dot*) is located on the decision boundary (*blue disc*). **(b)** Introducing a second "weaker" decision boundary (*red disc*) prevents the feature from fluctuations

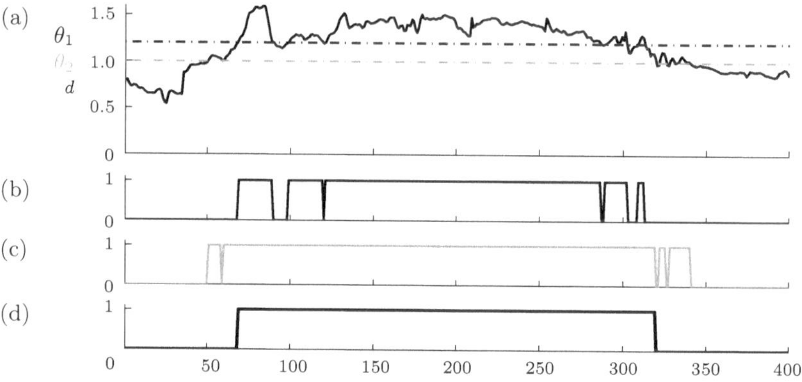

Fig. 11.8. **(a)** Distance d of the joint "rankle" to the plane that is parallel to the plane shown in Fig. 11.7a but passes through the joint "rhip," expressed in the relative length unit "hip width" (hw). The underlying motion is a Tai Chi move in which the actor is standing with slightly spread legs. The *dashed horizontal lines* at $\theta_2 = 1$ hw and $\theta_1 = 1.2$ hw, respectively, indicate the two thresholds, corresponding to the two planes of Fig. 11.7b. The remaining three curves show the binary feature values derived from d **(b)** using the threshold θ_1, **(c)** using the threshold θ_2, and **(d)** using the hysteresis thresholding strategy

if the right ankle is located on the decision boundary indicated by the blue disc. We therefore introduce a second decision boundary determined by a second, weaker, threshold $\theta_2 = 1.0$ indicated by the red disc in Fig. 11.7b. We then define a robust version $F^{\mathrm{sw}}_{\mathrm{robust}}$ of F^{sw} that assumes the value one as soon as the right ankle moves to the right of the stronger decision boundary (as before). But we only let $F^{\mathrm{sw}}_{\mathrm{robust}}$ return to the output value zero if the right ankle moves to the left of the weaker decision boundary. It turns out that this heuristic of *hysteresis thresholding* [64, Chapter 4] suppresses undesirable zero-one fluctuations in relational feature values very effectively, see Fig. 11.8.

11.1.4 Example for Some Feature Set

As an example, we now describe a set of $f = 39$ relational features that express intuitive and semantic qualities of a human pose. This feature set, which will be used in our experiments on motion retrieval and classification described in the subsequent chapters, has been specifically designed to focus on full-body motions. Therefore, we incorporated most parts of the body, in particular the end effectors, so as to create a well-balanced feature set. One guiding principle was to cover the space of possible end effector locations by means of a small set of pose-dependent space "octants" defined by three intersecting planes each (above/below, left of/right of, in front of/behind the body). Obviously, such a subdivision is only suitable to capture the rough course of a motion, since the feature function would often yield a constant output value for small-scaled motions. Here, the features of type $F_{\mathrm{move}}^{(j_1,j_2,j_3;j_4)}$ and $F_{\mathrm{nmove}}^{(j_1,j_2,j_3;j_4)}$ provide a finer view on motions by additionally considering directional information. Note that the proposed feature set may be replaced or modified as appropriate for the respective application.

The 39 relational features, given by Table 11.1, are divided into the three sets "upper," "lower,"and "mix," which are abbreviated as u, ℓ, and m, respectively. The features in the upper set express properties of the upper part of the body, mainly of the arms. Similarly, the features in the lower set express properties of the lower part of the body, mainly of the legs. Finally, the features in the mixed set express interactions of the upper and lower part or refer to the overall position of the body.

Features with two entries in the ID column exist in two versions pertaining to the right/left half of the body but are only described for the right half – the features for the left half can be easily derived by symmetry. The abbreviations "hl," "sw," and "hw" denote the relative length units "humerus length," "shoulder width," and "hip width," respectively, which are used to handle differences in absolute skeleton sizes. Absolute coordinates, as used in the definition of features such as F_{17}, F_{32}, or F_{33}, stand for virtual joints at constant 3D positions w.r.t. an (X, Y, Z) world system in which the Y axis points upwards. The symbols $Y_{\mathrm{min}}/Y_{\mathrm{max}}$ denote the minimum/maximum Y coordinates assumed by the joints of a pose that are not tested. Features such as F_{22} do not follow the same derivation scheme as the other features and are therefore described in words.

In general, all of our relational features are defined in such a way that the feature value zero corresponds to a neutral, standing pose. If a feature assumes the value one, something "extraordinary" that is related to the intended semantics of the feature is happening. To match this convention, the features F_{15}/F_{16} (features assume the value zero for a standing pose) have been given the opposite semantics of the features F^r/F^ℓ (features assume the value one for a standing pose) as introduced in the basic example (Sect. 11.1.1).

Table 11.1. A feature set consisting of 39 relational features

ID	Set	Type	j_1	j_2	j_3	j_4	θ_1	θ_2	Description
F_1/F_2	u	F_{nmove}	neck	rhip	lhip	rwrist	$1.8\,\mathrm{hl\,s}^{-1}$	$1.3\,\mathrm{hl\,s}^{-1}$	rhand moving forwards
F_3/F_4	u	F_{nplane}	chest	neck	neck	rwrist	$0.2\,\mathrm{hl}$	$0\,\mathrm{hl}$	rhand above neck
F_5/F_6	u	F_{move}	belly	chest	chest	rwrist	$1.8\,\mathrm{hl\,s}^{-1}$	$1.3\,\mathrm{hl\,s}^{-1}$	rhand moving upwards
F_7/F_8	u	F_{angle}	relbow	rshoulder	relbow	rwrist	$[0°, 110°]$	$[0°, 120°]$	relbow bent
F_9	u	F_{nplane}	lshoulder	rshoulder	lwrist	rwrist	$2.5\,\mathrm{sw}$	$2\,\mathrm{sw}$	hands far apart, sideways
F_{10}	u	F_{move}	lwrist	rwrist	rwrist	lwrist	$1.4\,\mathrm{hl\,s}^{-1}$	$1.2\,\mathrm{hl\,s}^{-1}$	hands approaching each other
F_{11}/F_{12}	u	F_{move}	rwrist	root	lwrist	root	$1.4\,\mathrm{hl\,s}^{-1}$	$1.2\,\mathrm{hl\,s}^{-1}$	rhand moving away from root
F_{13}/F_{14}	u	F_{fast}	rwrist				$2.5\,\mathrm{hl\,s}^{-1}$	$2\,\mathrm{hl\,s}^{-1}$	rhand fast
F_{15}/F_{16}	ℓ	F_{plane}	root	lhip	ltoes	rankle	$0.38\,\mathrm{hl}$	$0\,\mathrm{hl}$	rfoot behind lleg
F_{17}/F_{18}	ℓ	F_{nplane}	$(0,0,0)^\top$	$(0,1,0)^\top$	$(0, Y_{\mathrm{min}}, 0)^\top$	rankle	$1.2\,\mathrm{hl}$	$1\,\mathrm{hl}$	rfoot raised
F_{19}	ℓ	F_{nplane}	lhip	rhip	lankle	rankle	$2.1\,\mathrm{hw}$	$1.8\,\mathrm{hw}$	feet far apart, sideways
F_{20}/F_{21}	ℓ	F_{angle}	rknee	rhip	rknee	rankle	$[0°, 110°]$	$[0°, 120°]$	rknee bent
F_{22}	ℓ		Plane Π fixed at lhip, normal rhip→lhip. Test: rankle closer to Π than lankle?						feet crossed over
F_{23}	ℓ		Consider velocity v of rankle relative to lankle in rankle→lankle direction. Test: projection of v onto rhip→lhip line large?						feet moving towards each other, sideways
F_{24}	ℓ		Same as above, but use lankle→rankle instead of rankle→lankle direction.						feet moving apart, sideways
F_{25}/F_{26}	ℓ	F_{fast}	rankle				$2.5\,\mathrm{hl\,s}^{-1}$	$2\,\mathrm{hl\,s}^{-1}$	rfoot fast
F_{27}/F_{28}	m	F_{angle}	neck	root	rshoulder	relbow	$[25°, 180°]$	$[20°, 180°]$	rhumerus abducted
F_{29}/F_{30}	m	F_{angle}	neck	root	rhip	rknee	$[50°, 180°]$	$[45°, 180°]$	rfemur abducted
F_{31}	m	F_{plane}	rankle	neck	lankle	root	$0.5\,\mathrm{hl}$	$0.35\,\mathrm{hl}$	root behind frontal plane
F_{32}	m	F_{angle}	neck	root	$(0,0,0)^\top$	$(0,1,0)^\top$	$[70°, 110°]$	$[60°, 120°]$	spine horizontal
F_{33}/F_{34}	m	F_{nplane}	$(0,0,0)^\top$	$(0,-1,0)^\top$	$(0, Y_{\mathrm{min}}, 0)^\top$	rwrist	$-1.2\,\mathrm{hl}$	$-1.4\,\mathrm{hl}$	rhand lowered
F_{35}/F_{36}	m		Plane Π through rhip, lhip, neck. Test: rshoulder closer to Π than lshoulder?						shoulders rotated right
F_{37}	m		Test: Y_{min} and Y_{max} close together?						Y-extents of body small
F_{38}	m		Project all joints onto XZ-plane. Test: diameter of projected point set large?						XZ-extents of body large
F_{39}	m	F_{fast}	root				$2.3\,\mathrm{hl\,s}^{-1}$	$2\,\mathrm{hl\,s}^{-1}$	root fast

11.2 Adaptive Segmentation

As mentioned earlier, two semantically similar motions may exhibit considerable spatial as well as temporal deviations. One key idea is to absorb spatial variations by transforming a motion data stream D in a posewise fashion into a sequence $F \circ D$ of binary feature vectors using a suitable feature function F. In this section, we show how to achieve invariance toward local time deformations by introducing the concept of adaptive segmentation.

Let $F : \mathcal{P} \to \{0,1\}^f$ be a fixed feature function. Then, we say that two poses $P_1, P_2 \in \mathcal{P}$ are F*-equivalent* if the corresponding feature vectors $F(P_1)$ and $F(P_2)$ coincide, i.e., $F(P_1) = F(P_2)$. Then, an F*-run* of D is defined to

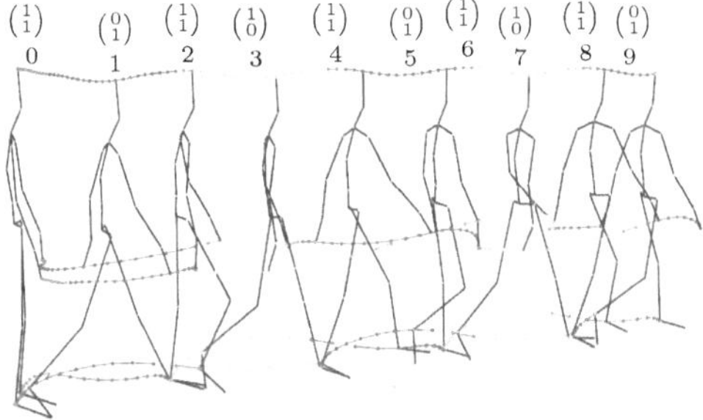

Fig. 11.9. F^2-segmentation of D_{walk}, where F^2-equivalent poses are indicated by uniformly colored trajectory segments. The trajectories of the joints "headtop," "rankle," "rfingers," and "lfingers" are shown

be a subsequence of D consisting of consecutive F-equivalent poses, and the F*-segments* of D are defined to be the nonextensible F-runs.

We illustrate these definitions by continuing the example from Sect. 11.1.1. Let $F^2 := (F^r, F^\ell) : \mathcal{P} \to \{0,1\}^2$ be the combined feature formed by F^r and F^ℓ so that the pose set $\mathcal{P}$ is partitioned into four F^2-equivalence classes. Applying F^2 to a walking motion D_{walk} results in the segmentation shown in Fig. 11.9, where the trajectories of selected joints have been plotted. F^2-equivalent poses are indicated by the same trajectory color: the red color represents the feature vector $(1,1)$, the blue color the vector $(0,1)$, and the green color the vector $(1,0)$. Note that no pose with feature vector $(0,0)$ appears in D_{walk}. Altogether, there are ten nonextensible runs constituting the F^2-segmentation of D_{walk}.

As a second example, we consider a motion consisting of a right foot kick followed by a left hand punch (Fig. 11.10). Here, we use a feature function $F^4 : \mathcal{P} \to \{0,1\}^2$ consisting of the four boolean functions, which describe the relations whether the right knee is bent or stretched (Fig. 11.4a), whether the left elbow is bent or stretched (F_8 of Table 11.1), whether the right foot is raised or not (F_{17} of Table 11.1), and whether the left hand is reaching out to the front of the body or not (Fig. 11.2b). Note that there are 16 different feature combinations with respect to F^4. Figure 11.10 shows the 11 segments of the resulting F^4-segmentation.

It is this feature-dependent segmentation that accounts for the postulated temporal invariance, the main idea being that motion capture data streams can now be compared at the segment level rather than at the frame level. To be more precise, let us start with the sequence of F-segments of a motion capture data stream D. Since each segment corresponds to a unique feature vector, the segments induce a sequence of feature vectors, which we simply refer to as the F*-feature sequence* of D and denote by $F[D]$. If $M+1$ is the number of F-segments of D and if $D(t_m)$ for $t_m \in [1:T]$, $0 \le m \le M$, is a

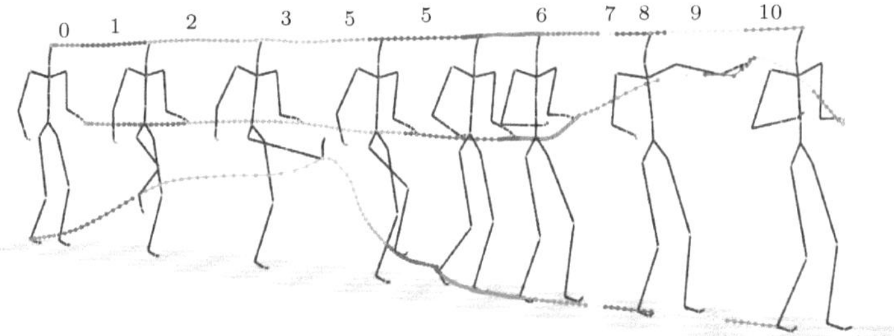

Fig. 11.10. Right foot kick followed by a left hand punch. The trajectories of the joints "headtop," "rankle," and "lfingers" are shown. The segments are induced by a 4-component feature function F^4 comprising the thresholded angles for "rknee" and "lelbow" as well as features describing the relations "right foot raised" and "left hand in front"

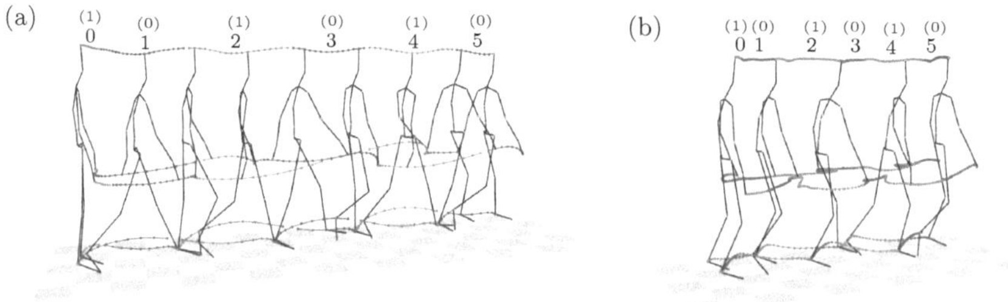

Fig. 11.11. (a) Restricting $F^2 = (F^r, F^\ell)$ to its first component results in an F^r-segmentation, which is coarser than the F^2-segmentation shown in Fig. 11.9. **(b)** Five steps of a slow walking motion performed by an elderly person resulting in exactly the same F^r-feature sequence as the much faster motion of (a)

pose of the m-th segment, then $F[D] = (F(D(t_0)), F(D(t_1)), \ldots, F(D(t_M)))$. For example, for the data stream D_{walk} and the feature function F^2 from Fig. 11.9, we obtain

$$F^2[D_{\text{walk}}] = \left(\tbinom{1}{1}, \tbinom{0}{1}, \tbinom{1}{1}, \tbinom{1}{0}, \tbinom{1}{1}, \tbinom{0}{1}, \tbinom{1}{1}, \tbinom{1}{0}, \tbinom{1}{1}, \tbinom{0}{1}\right). \quad (11.4)$$

Obviously, any two adjacent vectors of the sequence $F[D]$ are distinct. The crucial point is that time invariance is incorporated into the F-segments: two motions that differ by some deformation of the time axis will yield the same F-feature sequences. This fact is illustrated by Fig. 11.11. Another property is that the segmentation automatically adapts to the selected features, as a comparison of Figs. 11.9 and 11.11a shows. In general, fine features, i.e., feature functions with many components, induce segmentations with many short segments, whereas coarse features lead to a smaller number of long segments.

The main idea is that two motion capture data streams D_1 and D_2 can now be compared via their F-feature sequences $F[D_1]$ and $F[D_2]$ instead of comparing the data streams on a frame-to-frame basis. This has several advantages:

1. One can decide which aspects of the motions to focus on by picking a suitable feature function F.
2. Since spatial and temporal invariance are already incorporated in the features and segments, one can use efficient methods from (fault-tolerant) string matching to compare the data streams instead of applying cost-intensive techniques such as DTW at the frame level.
3. In general, the number of segments is much smaller than the number of frames, which accounts for efficient computations.

11.3 Further Notes

In this chapter, we have introduced the concept of relational features that encode the presence or absence of certain geometric or directional aspects among the body's joints. The crucial point is that relational features are invariant not only under global transformations but, in particular, to local spatial deformations. To account for temporal invariance, we labeled each individual frame by its corresponding bit vector with respect to a fixed function and segmented the data stream simply by merging adjacent frames with identical labels. It is the combination of relational features and induced temporal segmentations that accounts for spatio-temporal invariance, which is crucial for the identification of logically related motions. In the subsequent chapters, we will see how this concept can be used as a powerful tool for flexible, efficient, and automatic content-based motion retrieval and classification. For further explanation we also refer to our video available at [148].

To determine meaningful relational features, we used common sense and human knowledge about the typical range of motions in combination with the supervised learning technique to determine sensible threshold values (Sect. 11.1.3). As it turns out, the proposed feature set will suffice to prove the applicability of our concepts in the retrieval and classification scenarios of full-body motions. For the future, it would be desirable to back up the design process by expert knowledge from application fields such as computer animation, sports science, biomechanics, physical therapy, or cognitive psychology. Currently, however, these areas of science seem to have no theory about the role of certain body-defined planes or other relational features for human motion perception and understanding.

The idea of considering relational (combinatorial, geometric, qualitative) features instead of numerical (metrical, quantitative) features is not new and has already been applied by, e.g., Carlsson [30, 31] as well as Sullivan and Carlsson [198] in other domains such as visual object recognition in 2D and 3D, or action recognition and tracking. The following observations are of fundamental importance, see Carlsson [30]: first, relational structures are not only interesting for general recognition problems (due to their invariance properties) but also ideally suited for indexing (due to their discrete nature). Second, *relational* similarity of shapes correlates quite well with *perceptual (semantic)* similarity. These principles motivate the usage of progressions of geometric constellations to identify semantically similar movements. In Sect. 12.4, in the context of motion retrieval, we will continue our discussion of related work.

12

Index-Based Motion Retrieval

Given a database $\mathcal{D}$ of motion capture data streams, one typical retrieval task is to identify all motion clips contained in all documents $D \in \mathcal{D}$ that are semantically similar to a *query* example motion Q. Such motion clips are also referred to as *hits* with respect to the query Q. Several general questions arise at this point:

1. How should the data, the database as well as the query, be modeled?
2. How does a user specify a query?
3. What is the precise definition of a hit?
4. How should the data be organized to afford efficient retrieval of all hits with respect to a given query?

In Chap. 11, we gave an answer to the first question by introducing the concept of feature sequences, which represent motion capture data streams as coarse sequences of binary vectors. In Sect. 12.1, we will formally introduce the concepts of exact hits, fuzzy hits, and adaptive fuzzy hits. We then describe how one can compute such hits using an inverted file index. The proposed indexing and matching techniques can be put to use in a variety of query modes. Here, the possibilities range from isolated pose-based queries up to query-by-example (QBE), where the user supplies the system with a short query motion clip. In Sect. 12.2, we present a flexible and efficient QBE-based motion retrieval system and report on experimental results. Furthermore, we show how our relational approach to motion comparison can be used as a general tool for efficient motion preprocessing (Sect. 12.3). Finally, we discuss some problems and limitation of the presented index-based techniques and close with a discussion of related work (Sect. 12.4).

12.1 Indexing, Queries, and Hits

In this section, we use the following notations. The database consists of a collection $\mathcal{D} = (D_1, D_2, \ldots, D_I)$ of motion capture data streams or *documents*

D_i, $i \in [1:I]$. To simplify things, we may assume that $\mathcal{D}$ consists of one large document D by concatenating the documents $D_1, \ldots, D_I$, keeping track of document boundaries in a supplemental data structure. The query Q consists of an example motion clip. We fix a feature function $F : \mathcal{P} \rightarrow \{0,1\}^f$ and use the notation $F[D] = \mathbf{w} = (w_0, w_1, \ldots, w_M)$ and $F[Q] = \mathbf{v} = (v_0, v_1, \ldots, v_N)$ to denote the resulting F-feature sequences of D and Q, respectively. We then simply speak of the database $\mathbf{w}$ and the query $\mathbf{v}$.

12.1.1 Inverted File Index

To index our database $\mathcal{D}$ with respect to F, we use standard techniques as used in text retrieval [218]. For each feature vector $v \in \{0,1\}^f$, we store an *inverted list* (also referred to as *inverted file*) $L(v)$ that consists of all indices $m \in [0:M]$ of the sequence $\mathbf{w} = (w_0, w_1, \ldots, w_M)$ with $v = w_m$. $L(v)$ tells us which of the F-segments of D exhibit the feature vector v. As an example, let us consider the feature function $F^2 = (F^r, F^\ell)$ applied to the walking motion D as indicated by Fig. 11.9. From the resulting feature sequence

$$F^2[D] = \left(\left(\begin{smallmatrix}1\\1\end{smallmatrix}\right), \left(\begin{smallmatrix}0\\1\end{smallmatrix}\right), \left(\begin{smallmatrix}1\\1\end{smallmatrix}\right), \left(\begin{smallmatrix}1\\0\end{smallmatrix}\right), \left(\begin{smallmatrix}1\\1\end{smallmatrix}\right), \left(\begin{smallmatrix}0\\1\end{smallmatrix}\right), \left(\begin{smallmatrix}1\\1\end{smallmatrix}\right), \left(\begin{smallmatrix}1\\0\end{smallmatrix}\right), \left(\begin{smallmatrix}1\\1\end{smallmatrix}\right), \left(\begin{smallmatrix}0\\1\end{smallmatrix}\right)\right), \qquad (12.1)$$

we obtain the inverted lists $L\left(\left(\begin{smallmatrix}1\\1\end{smallmatrix}\right)\right) = \{0,2,4,6,8\}$, $L\left(\left(\begin{smallmatrix}0\\1\end{smallmatrix}\right)\right) = \{1,5,9\}$, $L\left(\left(\begin{smallmatrix}1\\0\end{smallmatrix}\right)\right) = \{3,7\}$, and $L\left(\left(\begin{smallmatrix}0\\0\end{smallmatrix}\right)\right) = \emptyset$. Inverted lists are sets represented as sorted, repetition-free sequences. Depending on the context, our notation for inverted lists and derived objects will switch between sequence and set notation.

The resulting index, which consists of the 2^f inverted lists $L(v)$ for $v \in \{0,1\}^f$, is denoted by $I_F^{\mathcal{D}}$ and referred to as *inverted file index* of the database $\mathcal{D}$. Since we store segment positions of the F-segmentation rather than frame positions in the inverted lists and since each segment position appears in exactly one inverted list, the index size is proportional to the number M of segments of D. Additionally, we store the segment lengths, so that the frame positions can be recovered. Practical issues on the indexing stage are discussed in Sect. 12.2.2.

In the subsequent sections, we will have to perform certain operations on inverted lists such as taking unions or intersections. By keeping the entries within each list $L(v)$ sorted, we facilitate an efficient computation of such operations using merge operations or binary search. More precisely, let A and B be sorted lists. Then $A \cap B$ and $A \cup B$ can be computed with $O(|A| + |B|)$ operations based on the merge techniques as used in sorting algorithms. In case the length $|A|$ of the list A is much shorter than the length $|B|$ of the other list B, one may apply binary search to speed up the computations. Here, it takes $O(\log(|B|)$ operations to search for one element of A within the list B, which leads to on overall complexity of $O(|A| \log(|B|))$ to compute $A \cap B$ and $A \cup B$. For example, if $|A| = 10$ and $|B| = 1\,000\,000$ the computation of

$A \cap B$ based on binary search requires roughly $10 \cdot 20 = 200$ operations opposed to $10 + 1\,000\,000 = 1\,000\,010$ operations, which are required in the merging strategy. Finally, note that the complement of a list A can be computed with $O(|A|)$ operations.

12.1.2 Exact Queries and Exact Hits

Recall that two motion clips are considered as similar (with respect to the selected feature function) if they exhibit the same feature sequence. Adapting concepts from [146], we introduce the following notions. Let $\mathbf{w} = (w_0, w_1, \ldots, w_M)$ and $\mathbf{v} = (v_0, v_1, \ldots, v_N)$ be the feature sequences of the database and the query, respectively. Then an *exact hit* is an element $k \in [0 : M]$ such that $\mathbf{v}$ is a subsequence of consecutive feature vectors in $\mathbf{w}$ starting from index k. In other words, writing in this case $\mathbf{v} \sqsubset_k \mathbf{w}$, one obtains

$$\mathbf{v} \sqsubset_k \mathbf{w} \quad :\Leftrightarrow \quad \forall n \in [0 : N] : \; v_n = w_{k+n}. \tag{12.2}$$

The set of all exact hits in the database $\mathcal{D}$ is defined as

$$H_{\mathcal{D}}(\mathbf{v}) := \{k \in [0 : M] \mid \mathbf{v} \sqsubset_k \mathbf{w}\}. \tag{12.3}$$

Then, the following equations show that $H_{\mathcal{D}}(\mathbf{v})$ can be evaluated efficiently by intersecting suitably shifted inverted lists:

$$\begin{aligned} H_{\mathcal{D}}(\mathbf{v}) &= \{k \in [0 : M] \mid (v_0 = w_k) \wedge (v_1 = w_{k+1}) \wedge \ldots \wedge (v_N = w_{k+N})\} \\ &= \bigcap_{n \in [0:N]} \{k \in [0 : M] \mid v_n = w_{k+n}\} \\ &= \bigcap_{n \in [0:N]} (\{\ell \in [0 : M] \mid v_n = w_\ell\} - n) \\ &= \bigcap_{n \in [0:N]} (L(v_n) - n), \end{aligned} \tag{12.4}$$

where the addition and subtraction of a list and a number is understood component-wise for every element in the list. The intersection in (12.4) can be computed iteratively, where we employ the following trick: instead of additively adjusting the input lists $L(v_n)$ (which are in general long) by $-n$, as suggested by (12.4), we adjust the lists appearing as intermediate results (which are in general much shorter) by $+1$ prior to each intersection step. One easily checks that after the final iteration, an adjustment of the resulting set by $-(N + 1)$ yields the set of exact hits. This idea gives rise to the following iterative algorithm (which will be extended in Sect. 12.1.4):

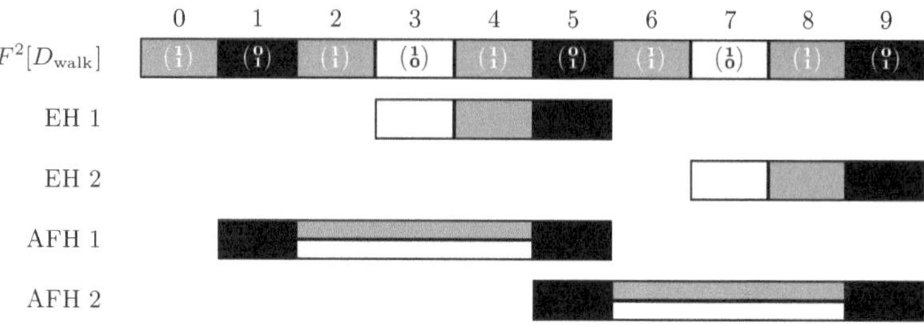

Fig. 12.1. Feature sequence $F^2[D_{\text{walk}}]$. The exact query $\mathbf{v} = (\binom{1}{0}, \binom{1}{1}, \binom{0}{1})$ results in two exact hits (EH). For the fuzzy query $\mathbf{V} = (V_0, V_1, V_2)$ with $V_0 = V_2 = \{\binom{0}{1}\}$ and $V_1 = \{\binom{1}{0}, \binom{1}{1}\}$, there are no fuzzy hits but two adaptive fuzzy hits (AFH)

Algorithm: ComputeExactHits		
Input:	$\mathbf{V}$	admissible fuzzy query
	$I_F^{\mathcal{D}}$	index consisting of the 2^f inverted lists $L(v)$, $v \in \{0,1\}^f$
Output:	$H_{\mathcal{D}}(\mathbf{v})$	list of exact hits.

(1) $L^0 := L(v_0) + 1$
(2) For $n = 0, \ldots, N-1$ compute $L^{n+1} := (L^n \cap L(v_{n+1})) + 1$
(3) $H_{\mathcal{D}}(\mathbf{v}) = L^N - (N+1)$

As an illustration, we apply this algorithm to our running example indicated by Fig. 11.9 and the query sequence $\mathbf{v} = (\binom{1}{0}, \binom{1}{1}, \binom{0}{1})$. Recall that in this case $L(\binom{1}{1}) = \{0,2,4,8\}$, $L(\binom{0}{1}) = \{1,5,9\}$, and $L(\binom{1}{0}) = \{3,7\}$.

(1) $L^0 := L(\binom{1}{0}) + 1 = \{4,8\}$
(2) $L^1 := (\{4,8\} \cap \{0,2,4,8\}) + 1 = \{5,9\}$
$L^2 := (\{5,9\} \cap \{1,5,9\}) + 1 = \{6,10\}$
(3) $H_{\mathcal{D}}(\mathbf{v}) = L^2 - 3 = \{3,7\}$

In other words, there are two exact hits for $\mathbf{v}$ starting with the fourth and eighth element of $\mathbf{w}$, respectively (Fig. 12.1).

12.1.3 Fuzzy Queries and Fuzzy Hits

In many situations, the user may be unsure about certain parts of the query and wants to leave certain parts of the query unspecified. Or, the user might want to mask out some of the f components of the feature function F to obtain a less restrictive search leading to more hits. To handle such situations, we introduce the concept of *fuzzy search*.

The idea is to allow at each position in the query sequence a whole set of possible, alternative feature vectors instead of a single one. This is modeled as follows. A *fuzzy set* is a nonempty subset $V \subset \{0,1\}^f$ with corresponding inverted list $L(V) := \bigcup_{v \in V} L(v)$. Then a *fuzzy query* is defined to be a sequence $\mathbf{V} = (V_0, V_1, \ldots, V_N)$ of fuzzy sets. Extending the definition in (12.2), a *fuzzy hit* is an element $k \in [0:M]$ such that $\mathbf{V} \sqsubset_k \mathbf{w}$, where

$$\mathbf{V} \sqsubset_k \mathbf{w} \quad :\Leftrightarrow \quad \forall n \in [0:N] : V_n \ni w_{k+n}. \tag{12.5}$$

Obviously, the case $V_n = \{v_n\}$ for $0 \le n \le N$ reduces to the case of an exact hit. Similar to (12.3), the set of all fuzzy hits is defined to be

$$H_{\mathcal{D}}(\mathbf{V}) := \{k \in [0:M] \mid \mathbf{V} \sqsubset_k \mathbf{w}\}. \tag{12.6}$$

In analogy to (12.4), the set $H_{\mathcal{D}}(\mathbf{V})$ can be computed via

$$H_{\mathcal{D}}(\mathbf{V}) = \bigcap_{n \in [0:N]} (L(V_n) - n). \tag{12.7}$$

This formula shows that the complexity of computing $H_{\mathcal{D}}(\mathbf{V})$ is proportional to $\sum_{n=0}^{N} |V_n|$, see also [42] for details.

12.1.4 Adaptive Fuzzy Hits

The concept of fuzzy hits as introduced above still lacks an important degree of flexibility. So far, the fuzziness only refers to the spatial domain (admitting alternative choices for the pose-based features) but does not take the temporal domain into account. More precisely, the segmentation of the document D is only determined by the feature function F, disregarding the fuzziness of the query $\mathbf{V}$. Here, it would be desirable if the matching strategy allowed for a fuzzy set to match with multiple successive elements of $\mathbf{w}$. For example, considering the fuzzy query $\mathbf{V} = (V_0, V_1, V_2)$ with $V_0 = V_2 = \{\binom{0}{1}\}$ and $V_1 = \{\binom{1}{0}, \binom{1}{1}\}$, one easily checks in Fig. 12.1 that the set of fuzzy hits, $H_{\mathcal{D}}(\mathbf{V})$ for $\mathcal{D} = D_{\text{walk}}$ is empty. By contrast, one obtains two hits for $\mathbf{V}$ using the concept of *adaptive fuzzy search*, see Fig. 12.1. Note that the fuzzy set V_1 is matched with the contiguous segment ranges 2–4 and 6–8, respectively. The two hits correspond to left/right/left step sequences.

The general strategy is to adjust the temporal segmentation of D to the fuzziness of the query *during the matching* as follows: supposing $w_k \in V_0$ with $w_{k-1} \notin V_0$ for some index $k_0 := k \in [0:M]$, we determine the maximal index $k_1 \ge k_0$ with $w_m \in V_0$ for all $m = k_0, k_0+1, \ldots, k_1 - 1$ and concatenate all segments corresponding to these w_m into one large segment. By construction,

$w_{k_1} \notin V_0$. Only if $w_{k_1} \in V_1$, we proceed in the same way, determining some maximal index $k_2 > k_1$ with $w_m \in V_1$ for all $m = k_1, k_1+1, \ldots, k_2-1$, and so on. In case we find a sequence of indices $k_0 < k_1 < \ldots < k_N$ constructed iteratively in this fashion we say that $k \in [0 : M]$ is an *adaptive fuzzy hit* and write $\mathbf{V} \sqsubset_k^{\mathrm{ad}} \mathbf{w}$. The set of all adaptive fuzzy hits for $\mathbf{V}$ in $\mathcal{D}$ is given by

$$H_{\mathcal{D}}^{\mathrm{ad}}(\mathbf{V}) := \{k \in [0 : M] \mid \mathbf{V} \sqsubset_k^{\mathrm{ad}} \mathbf{w}\}. \tag{12.8}$$

In view of this matching technique, we return to our running example with the above fuzzy query $\mathbf{V}$, see Fig. 12.1. We obtain $k_0 = 1$, $k_1 = 2$, $k_3 = 5$ for the first hit and $k_0 = 5$, $k_1 = 6$, $k_2 = 9$ for the second hit. In the case of the first hit, for example, this means that V_0 corresponds to segment 1 of $\mathbf{w}$, V_1 to segments 2–4, and V_2 to segment 5, amounting to a coarsened segmentation of D.

Not all fuzzy queries lead to sensible adaptive fuzzy hits. For example, the fuzzy query $\mathbf{V} = (V_0, V_1, V_2)$ with $V_0 = \{\binom{0}{0}, \binom{1}{0}\}$, $V_1 = \{\binom{1}{0}\}$, and $V_2 = \{\binom{0}{1}\}$ yields no adaptive fuzzy hits on the document $D = (w_0, w_1, w_2) = (\binom{0}{0}, \binom{1}{0}, \binom{0}{1})$, even though the three feature vectors of the document appear as elements of the three corresponding fuzzy sets. This is due to the greedy strategy of choosing the maximal index $k_1 \geq 0$ with $w_m \in V_0$ for all $m = 0, \ldots, k_1 - 1$; here, we obtain $k_1 = 2$, so an adaptive fuzzy hit for $\mathbf{V}$ would now have to satisfy $w_2 \in V_1$, which is not the case. The underlying problem is that the fuzzy sets V_0 and V_1 have a nonempty intersection, which makes the transition between these two sets during the matching process inherently ill-defined. We therefore only allow *admissible fuzzy queries* with $V_n \cap V_{n+1} = \emptyset$ for $n = 0, \ldots, N-1$. In case a query is not admissible, one possible strategy is to scan the query (from 0 to N) for fuzzy sets V_n that have nonempty intersections with their neighbors V_{n-1} and V_{n+1} (where $V_{-1} = V_{N+1} := \emptyset$). Such V_n are then replaced by $V_n \setminus (V_{n-1} \cup V_{n+1})$, leading to an admissible query.

Now, the important point is that $H_{\mathcal{D}}^{\mathrm{ad}}(\mathbf{V})$ can also be computed efficiently using the same index $I_F^{\mathcal{D}}$ as in the case of an exact hit. Before describing the algorithm, we need to introduce some more notations. Note that the list $L(V)$ for a fuzzy set V may contain consecutive segment indices (opposed to an inverted list $L(v)$). We consider nonextensible sequences of consecutive segment indices in $L(V)$. Suppose $L(V)$ consists of $K+1$ such sequences with starting segments $r_0 < r_1 < \ldots < r_K$ and lengths $t_0, t_1, \ldots, t_K$; then we define $R(V) := (r_0, r_1, \ldots, r_K)$ and $T(V) := (t_0, t_1, \ldots, t_K)$. For example, if $L(V) = (2, 4, 5, 6, 9, 10)$ then $R(V) = (2, 4, 9)$ and $T(V) = (1, 3, 2)$. Obviously, one can reconstruct $L(V)$ from $R(V)$ and $T(V)$. Note that by the maximality condition one has $r_k + t_k < r_{k+1}$ for $0 \leq k < K$. Extending the algorithm in Sect. 12.1.2, we can compute the set $H_{\mathcal{D}}^{\mathrm{ad}}(\mathbf{V})$ as follows:

Algorithm: COMPUTEADAPTIVEFUZZYHITS

Input:	$\mathbf{V}$	admissible fuzzy query
	$I_F^{\mathcal{D}}$	index consisting of the 2^f inverted lists $L(v)$, $v \in \{0,1\}^f$
Output:	$H_{\mathcal{D}}^{\mathrm{ad}}(\mathbf{V})$	list of adaptive fuzzy hits.

(1) $R^0 := R(V_0) + T(V_0)$, $T^0 := T(V_0)$

(2) For $n = 0, \ldots, N-1$ assume
$R^n = (p_0, \ldots, p_I)$, $T^n = (q_0, \ldots, q_I)$,
$R(V_{n+1}) = (r_0, \ldots, r_J)$, $T(V_{n+1}) = (t_0, \ldots, t_J)$,
$R^n \cap R(V_{n+1}) = (p_{i_0}, \ldots, p_{i_K}) = (r_{j_0}, \ldots, r_{j_K})$, for suitable indices $0 \le i_0 < \ldots < i_K \le I$ and $0 \le j_0 < \ldots < j_K \le J$. Then define
$R^{n+1} := (R^n \cap R(V_{n+1})) + (t_{j_0}, \ldots, t_{j_K})$,
$T^{n+1} := (q_{i_0}, \ldots, q_{i_K}) + (t_{j_0}, \ldots, t_{j_K})$.

(3) $H_{\mathcal{D}}^{\mathrm{ad}}(\mathbf{V}) = R^N - T^N$

The loop invariant for step (2) is that for $n = 0, \ldots, N$, we have $R^n = H_{\mathcal{D}}^{\mathrm{ad}}((V_0, \ldots, V_n)) + T^n$, where T^n holds the number of segments for each hit in $H_{\mathcal{D}}^{\mathrm{ad}}((V_0, \ldots, V_n))$. In other words, the entries of R^n denote hit candidates and always point to the beginning of the next (potentially matching) group of segments. To illustrate the algorithm, we continue our running example. From the lists $L(V_0) = L(V_2) = (1,5,9)$ and $L(V_1) = (0,2,3,4,6,7,8)$, we obtain $R(V_0) = R(V_2) = (1,5,9)$, $T(V_0) = T(V_2) = (1,1,1)$, and $R(V_1) = (0,2,6)$, $T(V_1) = (1,3,3)$. Then

(1) $R^0 := (1,5,9) + (1,1,1) = (2,6,10)$
$T^0 := (1,1,1)$

(2) $R^1 := ((2,6,10) \cap (0,2,6)) + (3,3) = (5,9)$
$T^1 := (1,1) + (3,3) = (4,4)$
$R^2 := ((5,9) \cap (1,5,9)) + (1,1) = (6,10)$
$T^2 := (4,4) + (1,1) = (5,5)$

(3) $H_{\mathcal{D}}^{\mathrm{ad}}(\mathbf{v}) = R^2 - T^2 = (1,5)$.

Finally, we note that fuzzy search can be complemented by the concept of *k-mismatch search*, which introduces another degree of fault tolerance. Here, one permits in the matching of a fuzzy query $\mathbf{V} = (V_0, V_1, \ldots, V_N)$ up to $k \in [0 : N]$ of the fuzzy sets to completely disagree with the database sequence $\mathbf{w}$. To maintain a certain degree of control over this mismatch mechanism, it is possible to restrict the positions within $\mathbf{V}$ where such mismatches may occur. Note that k-mismatch search allows for spatial variations via deviating feature values, whereas the temporal structure of the query must be matched exactly. Efficient dynamic programming algorithms for k-mismatch queries exist, see Clausen and Kurth [42] for further details.

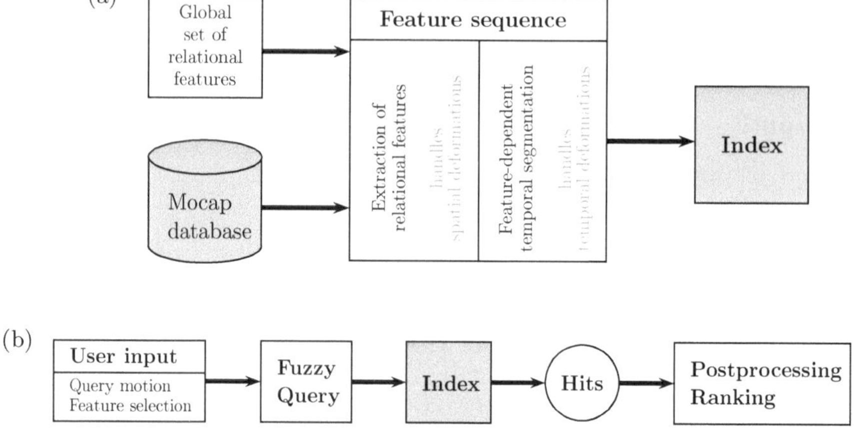

Fig. 12.2. **(a)** The preprocessing stage. **(b)** The query and retrieval stage

12.2 QBE Motion Retrieval

We now describe an efficient motion retrieval system based on the query-by-example (QBE) paradigm, which allows the user for intuitive browsing in a purely content-based fashion. A general query mechanism as well as the retrieval procedure is described in Sect. 12.2.1. We then deal with some practical aspects concerning the indexing stage (Sect. 12.2.2), report on our experimental results (Sect. 12.2.3), and discuss some ranking strategies (Sect. 12.2.4).

12.2.1 Query and Retrieval Mechanism

Our QBE retrieval system consists of several steps comprising database preprocessing, query formulation, retrieval, and hit postprocessing, see Fig. 12.2 for an overview.

In the *preprocessing step*, a global feature function F has to be designed that covers all possible query requirements and provides the user with an extensive set of semantically rich features. In other words, it is not imposed upon the user to specify such features (even though this is also possible). Having fixed a feature function F, an index $I_F^{\mathcal{D}}$ is constructed for a given database $\mathcal{D}$ and stored on disc (Sect. 12.2.2). For example, one may use the feature set comprising 39 relational features as described in Sect. 11.1.4. However, the described indexing and retrieval methods to be described are generic, and the proposed test feature set may be replaced as appropriate for the respective application.

In the QEB scenario, the *input* consists of a short query motion clip. Furthermore, the user should be able to incorporate additional knowledge about the query, e.g., by selecting or masking out certain body areas in the query. This is important to find, for example, all instances of "clapping ones hands" irrespective of any concurrent locomotion. To this end, the user selects relevant features from the given global feature set (i.e., components of F), where each feature expresses a motion aspect and refers to specific parts of the body. The query-dependent specification of motion aspects then determines the desired notion of similarity. In addition, parameters such as fault tolerance and the choice of a ranking or postprocessing strategy can be adjusted.

In the *retrieval stage*, the query motion is translated into a feature sequence, which can be thought of as a progression of geometric constellations. The user-specified feature selection is translated into a suitable fuzzy query, where the irrelevant features correspond to alternatives in the corresponding feature values. In the next step, the adaptive fuzzy hits are computed by the techniques as described in Sect. 12.1.4. Finally, the hits may be postprocessed by means of suitable ranking strategies as described in Sect. 12.2.4. In the subsequent sections, this procedure will be illustrated and discussed in more detail.

12.2.2 Indexing Stage

The retrieval algorithms for computing the hits are based on merging and intersecting operations of suitable inverted lists. In view of efficiency, it is important to have few lists (leading to few merging and intersecting operations), but at the same time it is important to have short lists (leading to fast merging and intersecting operations), which are mutually exclusive demands. Furthermore, recall that an index $I_F^{\mathcal{D}}$ consists of 2^f inverted lists, which becomes problematic in case f is large.

To remedy this problem, one can put an upper bound on the size of f by working with several indices in parallel. We explain this strategy by means of an example. For our experiments, we work with a feature function F with $f = 31$, where the 31 features are selected from the feature set of Table 11.1. Note that a single index for F would consists of 2^{31} inverted list, which is not manageable in practice. One strategy could be to store only the nonempty lists – actually only a small fraction of the inverted lists are nonempty since most feature combinations do not appear in real motion data. Another strategy is to suitably divide the feature set into several groups of features and build separate indexes for each of these groups. In our example, we divide the 31 features into three feature functions F_ℓ, F_u, and F_m consisting of 11, 12, and 8 components, respectively. The feature functions F_ℓ and F_u comprise boolean features expressing properties of the lower/upper part of the body (mainly of the legs/arms, respectively), whereas F_m consists of boolean features expressing interactions of the upper and lower part as well as global parameters such as velocities. The resulting indices $I_\ell^{\mathcal{D}}$, $I_\mathrm{u}^{\mathcal{D}}$, and $I_\mathrm{m}^{\mathcal{D}}$ consists of 2048, 4096, and 512 inverted lists, respectively. Queries involving features scattered across several indexes are then processed by querying each respective index to yield intermediary results, which are finally intersected to obtain the desired hits. However, the number of such additional operations is by far outweighed by the savings resulting from the significantly reduced overall number $(2^{11}+2^{12}+2^8)$ of inverted lists. Furthermore, list lengths are in practice well-balanced and medium-sized, allowing for fast merging and intersecting operations. A minor drawback is that the indexes $I_\ell^\mathcal{D}$, $I_\mathrm{u}^\mathcal{D}$, and $I_\mathrm{m}^\mathcal{D}$ require an amount of memory linear in the overall number of F_ℓ-, F_u-, and F_m-segments in $\mathcal{D}$, respectively. This effect, however, is attenuated by the fact that segment lengths with respect to F_ℓ, F_u, and F_m are generally larger compared with F-segment lengths, resulting in fewer segments.

We implemented our indexing and retrieval algorithms in MATLAB and tested them on a database $\mathcal{D}^{180}$ containing more than one million frames of motion capture data (180 min sampled at 120 Hz). The experiments were run on a 3.6 GHz Pentium 4 with 1 GB of main memory. The resulting indexes are denoted by I_ℓ^{180}, I_u^{180}, and I_m^{180}. In its columns, Table 12.1 shows the number f of feature components, the number 2^f of inverted lists, the number of nonempty inverted lists, the overall number of frames in the database, the overall number of segments of the corresponding segmentation, the index size in MB, the number of bytes per segment, and four running times t_r, t_f, t_i, and $\sum t$, measured in seconds. t_r is the portion of running time spent on data read-in, t_f is the feature extraction time, t_i is the inverted list build-up time, and $\sum t$ is the total running time. To demonstrate the scalability of our result, we quote analogous numbers for the indexes I_ℓ^{60}, I_u^{60}, and I_m^{60} built from a subset $\mathcal{D}^{60}$ of $\mathcal{D}^{180}$ corresponding to 60 min of motion capture data. The total size of $\mathcal{D}^{180}$ represented in the text-based AMC motion capture file format was 600 MB, a more compact binary double precision representation required about 370 MB. Typical index sizes ranged between 0.7 and 4.3 MB, documenting the drastic amount of data reduction our scheme achieves.

Table 12.1 shows that the number of segments (with respect to F_ℓ, F_u, and F_m) was only about 3 to 12% of the number of frames contained in the database. Observe that index sizes are proportional to the number of segments,

Table 12.1. Feature computation and index construction

Index	f	2^f	#(lists)	#(frames)	#(segs)	MB	$\frac{\text{bytes}}{\text{seg}}$	t_r	t_f	t_i	$\sum t$
I_ℓ^{60}	11	2 048	409	425 294	21 108	0.72	35.8	26	10	6	42
I_ℓ^{180}	11	2 048	550	1 288 846	41 587	1.41	35.5	71	26	13	110
I_u^{60}	12	4 096	642	425 294	53 036	1.71	33.8	26	13	10	49
I_u^{180}	12	4 096	877	1 288 846	135 742	4.33	33.4	71	33	25	129
I_m^{60}	8	256	55	425 294	19 067	0.60	33.0	26	20	3	49
I_m^{180}	8	256	75	1 288 846	55 526	1.80	34.0	71	54	12	137

The four running times t_r (data read-in), t_f (feature extraction), t_i (index construction), and $\sum t$ (total) are measured in seconds

Fig. 12.3. Selected frames from 16 query-by-example hits for a left hand punch. The query clip is highlighted. Query features: F_1, F_2, F_7, F_8 (Table 11.1)

the average number of bytes per segment is constant for all indexes. The total indexing time is *linear* in the number of frames. This fact is very well reflected in the table: for example, it took 42 s to build I_ℓ^{60}, which is roughly one third of the 110 s that were needed to build I_ℓ^{180}. Note that more than half of the total indexing time was spent on reading in the data, e.g., 71 s for the 180-min index. The scalability of our algorithms' running time and memory requirements permits us to use much larger databases than those treated by Kovar and Gleicher [107], where the preprocessing step to build up the index structure is quadratic in the number of frames (Sect. 12.4.2).

12.2.3 Experimental Results

As a first example, Fig. 12.3 shows all resulting 16 hits from a query for a punch (retrieval time: 12.5 ms), where only the four features F_1/F_2 (right/left hand moving forward) and F_7/F_8 (right/left elbow bent) have been selected, see Table 11.1. These four features induce an adaptive segmentation of the query consisting of six segments, which suffice to grasp the gist of the punching motion. Most of the retrieved hits constitute punching motions. However, there are also spurious hits such as a twist dancer to the far left of Fig. 12.3, who accidentally performs a punch-like motion. Here, note that the four selected features only refer to the upper part of the body leaving the motion of the legs unconsidered. Further reducing the number of features by selecting only F_2 and F_8 induces a 4-segment query sequence and results in 264 hits, comprising various kinds of punch-like motions involving both arms. Finally, increasing the number of selected features by adding F_5/F_6 induces an 8-segment query sequence resulting in a single hit.

As a second example, Fig. 12.4 depicts 9 hits out of the resulting 33 hits for a "squatting" motion (retrieval time: 18 ms) using the five features F_{19} (feet sideways far apart), F_{20} (right knee bent), F_{22} (feet crossed over), and F_{25}/F_{26} (right/left foot fast). The induced 5-segment query sequence is characteristic enough to retrieve 7 of the 11 "real" squatting motions contained in the database. Using a simple ranking strategy (e.g., a temporal

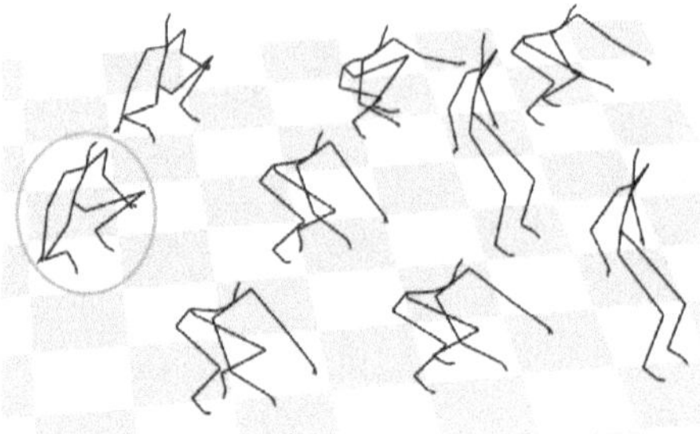

Fig. 12.4. Selected frames from 9 query-by-example hits for a squatting motion. The query clip is highlighted. Query features: F_{19}, F_{20}, F_{22}, F_{25}, and F_{26} (Table 11.1)

Fig. 12.5. Selected frames from 19 query-by-example hits for a right foot kick. The query clip is highlighted. Query features: F_{17}, F_{18}, F_{20}, and F_{21} (Table 11.1)

heuristic comparing the lengths of the matched segments of the query and a hit (Sect. 12.2.4)), these 7 hits appear as the top hits. The remaining 26 retrieved hits are false positives, two of which are shown to the right of Fig. 12.4 as the skeletons "sitting down" on a virtual table edge. One reason for this kind of false positives is that the relevant feature used in the query for the squatting motion thresholds the knee angle against a relatively high decision value of 120°. Hence, the knees of the sitting skeletons are just barely classified as "bent," leading to the confusion with a squatting motion. Also note the invariance of the knee angle feature to spatial deformations such as spreading the knees while squatting vs. keeping the thighs parallel. Omitting the velocity features F_{25} and F_{26} again results in an induced 5-segment query, this time, however, yielding 63 hits (containing the previous 33 hits with the same top 7 hits). Among the additional hits, one now also finds jumping and sneaking motions.

As a third example, Fig. 12.5 shows all 19 query results for a "kicking" motion (retrieval time: 5 ms) using F_{17}/F_{18} (right/left foot raised) and F_{20}/F_{21} (right/left knee bent). Out of these, 13 hits are actual martial arts kicks. The remaining six motions are ballet moves containing a kicking component.

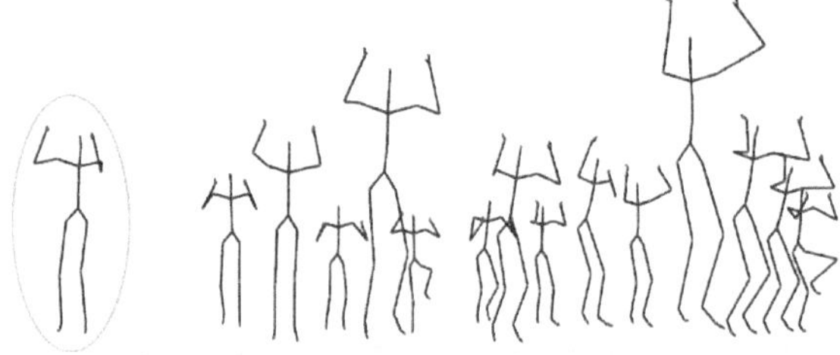

Fig. 12.6. Selected frames from the top 15 adaptive fuzzy hits for a jump query. Query features: F_5, F_6, F_{25}, and F_{26} (Table 11.1)

A manual inspection of $\mathcal{D}^{180}$ showed that there are no more than the 13 reported kicks in the database, demonstrating the high recall percentage our technique can achieve. Again, reducing the number of selected features leads to an increased number of hits. In general, a typical source of false positive hits is the choice of fuzzy alternatives in a query. For example, the ballet jumps in Fig. 12.5 were found as matches for a kicking motion because only the right leg was constrained by the query, leaving the left leg free to be stretched behind the body.

Finally, Fig. 12.6 shows the top 15 out of 133 hits for a very coarse adaptive fuzzy "jumping" query, which basically required the arms to move up above the shoulders and back down, while forcing the feet to lift off. Again, the hits were ranked according to the strategy based on a comparison of segment lengths. This example demonstrates how coarse queries can be applied to efficiently reduce the search space while retaining a superset of the desired hits.

The running time to process a query very much depends on the size of the database, the query length (the number of segments), the user-specified fuzziness of the query, as well as the number of resulting hits. In an experiment, we posed 10 000 random queries (guaranteed to yield at least one hit) for each of six query scenarios to the index I_{u}^{180}. Table 12.2 shows the average query times in milliseconds grouped by the hit count. For example, finding all exact F_{u}-hits for a query consisting of 5/10/20 segments takes on average 16–144/17–71/19–52 ms depending on the number of hits. Finding all adaptive fuzzy F_{u}-hits for a query consisting of 5/10/20 segments, where each fuzzy set of alternatives has a size of 64 elements, takes on average 23–291/28–281/42–294 ms.

12.2.4 Ranking Strategies

The feature function F is chosen to cover a broad spectrum of aspects appearing in all types of motions. Therefore, considering a specific motion class, many of the features are irrelevant for retrieval and should be masked out to

Table 12.2. Statistics on 10,000 random queries in the F_u-index for $\mathcal{D}^{180}$ for different query modes and lengths, grouped by the hit count, h

Type, #(segs)	1–9 hits			10–99 hits			≥100 hits		
	μ_h	σ_h	μ_t (ms)	μ_h	σ_h	μ_t (ms)	μ_h	σ_h	μ_t (ms)
Exact, $\|Q\| = 5$	3.0	2.4	16	44	28	20	649	567	144
Exact, $\|Q\| = 10$	1.7	1.6	17	34	22	26	239	147	71
Exact, $\|Q\| = 20$	1.1	0.6	19	32	26	36	130	5	52
Adaptive fuzzy, $\|Q\| = 5$	3.6	2.5	23	44	27	29	1,878	1,101	291
Adaptive fuzzy, $\|Q\| = 10$	2.4	2.1	28	40	26	35	1,814	1,149	281
Adaptive fuzzy, $\|Q\| = 20$	2.0	1.9	42	36	24	35	1,908	1,152	294

μ_h and σ_h are the average/standard deviation of h for the respective group, μ_t is the average query time in milliseconds. As regards the fuzzy queries, each fuzzy set contains 64 randomly chosen elements

avoid a large number of false negatives. Here, *false negatives* refer to database motion clips that are similar to the query but that are not retrieved by the system (e. g., due to over-specification). On the other hand, masking out a large number of features may lead to a larger percentage of *false positives*, i. e., hits that are not semantically related to the query but reveal the same relational aspects encoded by the user-specified features. One type of false positives arises from motions that incidentally fluctuate around the decision boundaries (thresholds) of our relational features (e. g., motion clips where the elbows are incidentally bent with approximately the chosen threshold angles, producing random fluctuations in the "elbow bent" features). Short queries, in particular, may thus yield a large number of arbitrary hits, which, however, typically exhibit some very short matching segments.

One important observation is that even though many useless hits might have been retrieved, one can easily eliminate many of the false positives in a postprocessing step. A first strategy is to rank the hits by comparing the proportions of the segment lengths of the respective hit with the corresponding proportions of the query motion. Here note that the segment lengths, even though they were left unconsidered in the retrieval process, often encode characteristic timing information. Considering such timing information, many of the false positives (which often contain very short matching segments) can be found at the end of the ranking list. This strategy was very successful in our experiments, where the ranking reflected the subjective (semantic) notion of similarity to a high degree – the aforementioned "arbitrary hits" were clearly separated from the "correct hits."

A second strategy is to postprocess the hits by means of more refined DTW-based methods using local distance measures as described in Chap. 10. Here, we propose a DTW-based ranking strategy that uses the Hamming distance between the binary vectors of the features sequences. Depending on the user needs, the Hamming distance could only consider the user-specified features selected for the query or could also be based on all features of the global feature set. Let $c(n, m)$ denote the Hamming distance between the nth vector in the feature progression of the query and the mth vector in the feature progression of the hit. Then the hit's ranking value is determined by the cost of the DTW-path in the cost matrix $C = (c(n, m))$ (cf. Sect. 4.1). Here, the important point is that even when working with a very coarse feature set (possibly leading to hundreds of hits), the data are still significantly (and very efficiently) reduced from a couple of hours (database) to a couple of minutes (hits), which is then well within reach of applying computationally expensive DTW-techniques.

12.3 Further Applications

We have seen that the concept of relational features facilitates automatic and efficient motion segmentation, indexing, and retrieval. In this section, we give further applications that illustrate how this concept can be used as a general tool for efficient motion preprocessing. In Sect. 12.3.1, we sketch a two-stage approach for accelerating DTW-based motion alignment. We then show how to efficiently cut down the search space in retrieval applications using a simple keyframe-based preselection strategy (Sect. 12.3.2). Finally, we propose a general strategy for handling low-level descriptive queries in an automatic way without using manually generated annotations (Sect. 12.3.3).

12.3.1 Accelerating DTW-Based Motion Alignment

The concept of relational features may be used to speed up the computation of DTW-based alignments (Chap. 4). Recall that two motion clips are considered as similar if they possess (more or less) the same progression of relational features. Matching two such progressions obtained from similar motions can be regarded as a time alignment of the underlying motion data streams. Even though such alignments may be too coarse in view of applications such as morphing or blending, they are quite accurate with respect to the overall course of motion. For the time alignment of motion data streams, we therefore suggest the following two-stage procedure: first compute (in linear time) a coarse segment-wise alignment based on our index. Then refine this alignment resorting to classical DTW-based techniques. The important point is that once a coarse alignment is known, the DTW step can be done efficiently since the underlying cost matrix needs only be computed within an area corresponding to the frames of the matched segments. This is also illustrated by Fig. 12.7, where the two walking motions of Fig. 11.11 are aligned. To avoid alignment artifacts enforced by segment boundaries, the restricted area is slightly enlarged as indicated by the gray area. A similar idea has been applied in the multiscale DTW approach (Sect. 4.3).

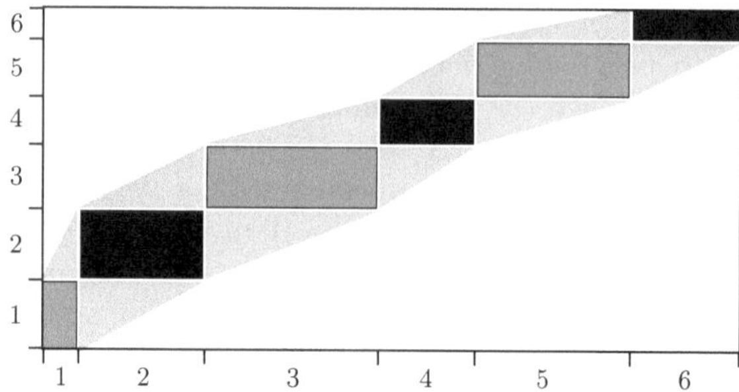

Fig. 12.7. The segment-wise time alignment of the two walking motions shown in Fig. 11.11a (corresponding to the vertical axis) and Fig. 11.11b (corresponding to the horizontal axis) can be refined via DTW-based methods restricted to the *gray area*

12.3.2 Keyframe-Based Preselection

Often certain types of motions exhibit characteristic aspects that already discriminate these motions from most other types of motions. For example, a cartwheel motion can be distinguished from most other motions simply by checking if the body is upside down in the course of motion. As a second example, every jumping jack motion contains a pose with both hands raised above the head and the feet moving apart, followed by a pose with both hands dropped below the shoulders and the feet moving together. In other words, simply checking few characteristic poses, so-called *keyframes*, in the temporal context, one can exclude all motions in the database that do not share the characteristic progression of relations. By this strategy one can often cut down the search space very efficiently from several hours to a couple of minutes of mocap data. The reduced dataset can then by analyzed and processed by more refined techniques.

Instead of using the matching concept based on adaptive fuzzy hits, we propose a crude but simple strategy based on *keyframe search*, which may be applied in a preprocessing step to efficiently cut down the set of candidate motions prior to computing actual hits. We label a small number of characteristic keyframes in the query motion and select for each such frame a small number of relational features to be of interest. This can be done manually or in some automatic way as will be sketched in Sect. 14.4. Then, we extract those parts from the unknown motion database that exhibit feature vectors matching the feature vectors of the specified keyframes in the correct order within suitable time bounds. This preselection can be computed efficiently using the same inverted file index as described in Sect. 12.1.1.

More precisely, a *keyframe query* consists of a sequence of keyframes, where each keyframe is specified by a pose as well as a subset of features with respect to a fixed f-dimensional feature function F. Furthermore, let

$\Theta \in \mathbb{N}$ denote a *time bound* parameter that specifies the maximum admissible distance between the occurrence of the first and the last keyframe. The sequence of keyframes can be represented by a sequence $(V_1, \ldots, V_J)$ of fuzzy sets $V_j \subseteq \{0,1\}^f$, $j \in [1 : J]$. Similar to Sect. 12.1.4, the keyframe query is called *admissible* if $V_j \cap V_{j+1} = \emptyset$ for $j \in [1 : J-1]$.

As before, we suppose that the mocap database is represented by a single document $D : [1 : T] \to \mathcal{P}$. Then a *keyframe hit* bounded by Θ is defined to be a pair $(r, s) \in [1 : T]^2$ such that there exists an ascending sequence of frame numbers $r = t_1 < t_2 < \cdots < t_J = s$ satisfying

$$\forall j \in [1 : J] : \; F(D(t_j)) \in V_j \quad \text{and} \quad t_J - t_1 + 1 \leq \Theta. \tag{12.9}$$

A keyframe hit (r, s) describes the start and end frames of some *matching interval*
$[r : s] \subseteq [1 : T]$ such that the resulting motion clip $D([r : s])$ contains all keyframes in the specified order. Note that two different keyframe hits may correspond to strongly overlapping matching intervals basically describing the same motion clip. To eliminate some of this redundance, we introduce the notion of a *reduced* keyframe hit. Here, a keyframe hit (r, s) is called reduced if $F(D(r+1)) \notin V_1$ and if s is minimal with respect to all ascending sequences starting with r and satisfying (12.9).

Using the inverted file index of Sect. 12.1.1, we now describe an algorithm to compute all reduced keyframe hits for a given keyframe query and time bound parameter Θ. Let $L_j := L(V_j)$ be the sorted union of inverted lists corresponding to V_j (Sect. 12.1.3). Since we are interested in frame numbers rather than segment numbers (as stored in the inverted lists), we use the additional information about segment boundaries stored along with the index to convert the inverted lists L_j into suitable lists of frame numbers. For any segment number h pertaining to the F-segmentation of D, let $s(h)$ and $e(h)$ denote the corresponding start and end frames, respectively. Writing the jth inverted list as $L_j = (h_j^1, \ldots, h_j^{\ell_j})$, we define the list

$$\Lambda_1 := (e(h_1^1), \ldots, e(h_1^{\ell_1})), \tag{12.10}$$

containing the *end frames* of the segments corresponding to L_1. The remaining lists Λ_j for $2 \leq j \leq J$ are defined as

$$\Lambda_j := (s(h_j^1), \ldots, s(h_j^{\ell_j})), \tag{12.11}$$

containing the *start frames* of the segments corresponding to L_j. We make this distinction for the following reason: recall that we are interested in finding keyframe hits (r, s) with corresponding sequences $r = t_1 < t_2 < \cdots < t_J = s$ satisfying the property (12.9). For any reduced keyframe hit, t_1 will be the end frame of a segment contained in V_1, and k_J will be the start frame of a segment contained in V_J. Hence, it is sufficient to restrict our search to such frames.

Next, we perform a linear, simultaneous sweep through the lists $\Lambda_1, \ldots, \Lambda_J$ using the pointers $p_1, \ldots, p_J$, which are initialized to 1. The first candidate for a start index is $r = t_1 := \Lambda_1(p_1)$, since, by construction, $F(D(t_1)) \in V_1$. Incrementing the list pointer p_2, we then search the second inverted list for a candidate position $t_2 := \Lambda_2(p_2)$ with $t_2 > t_1$ satisfying $|t_2 - t_1| \leq \Theta$. If such a t_2 exists, we continue with the third list by incrementing p_3 to find a $t_3 := \Lambda_3(p_3)$ with $t_3 > t_2$ satisfying $|t_3 - t_1| \leq \Theta$, and so on. In case a sequence of indices $t_1 < t_2 < \ldots < t_J$ can be constructed in this way, we report a hit $(r, s) = (t_1, t_J)$. We then increment p_1 (leaving the other pointers at their current position) and then restart the procedure at Λ_1. If at any point during the procedure a suitable t_j cannot be found, we know that the current t_1 cannot be the starting position r of a keyframe hit (r, s). The search is, therefore, restarted with an incremented p_1. If at any point a list pointer passes the end of its list, the procedure is terminated. Obviously, in this algorithm one has to perform *one* single sweep through the lists L_j, $j \in [1 : J]$, to retrieve *all* reduced keyframe hits. Therefore, running time as well as memory requirements linearly depend on the sum $\sum_{j=1}^{J} \ell_j$ of the lengths $\ell_j = |L_j|$.

The matching intervals obtained by the reduced keyframe hits may not be disjoint. In particular, this can be observed in periodic motions such as walking. To further reduce the amount of data that has to be processed by subsequent, more refined matching strategies (Sect. 14.4), one may perform another linear postprocessing sweep through the matching intervals, conjoining overlapping intervals.

Table 12.3 summarizes some of our experimental results conducted on a database consisting of 210 min of systematically recorded mocap data. For example, the database contains 24 cartwheel motions performed by five different actors. Using only two keyframes, our algorithm reduces the dataset from 210 to 2.8 min of motion data (corresponding to 1.3%) within 20 ms. The reduced dataset still contains 20 of the originally 24 cartwheels. Similarly, for a jumping jack motion, the preselected dataset, computed in 90 ms, consists of 0.7% of the original data still containing 50 of the original 52 jumping jacks. Selecting unsuitable keyframes and too restrictive features, however, may discard a large number of hit candidates. This is illustrated by the elbow-to-knee example, where only 16 out of 29 original performances survived in the reduced dataset. Further experimental results as well as a more detailed discussion of the keyframe-based search strategy in combination with a DTW-based postprocessing can be found in Sect. 14.4.

12.3.3 Toward Scene Descriptions

Often, the user will only have a sketchy idea of which kind of movement to look for in the motion capture database, for example a "right foot kick followed by a left hand punch." That is, the query may not be given as an example motion but only be specified by a vague textual description. In processing such queries, one so far has to revert to annotations such as manually generated descriptive

Table 12.3. Keyframe-based preselection for various types of motions from a database consisting of 210 min of mocap data

	Cartwheel	Jumping jack	Elbow to knee	Kick with right foot	Sit down on chair	Walk feet cross over	Hop with right leg
#(kf)	2	2	2	2	3	5	2
sel (min)	2.8	1.5	0.8	2.2	3.8	4.9	1.7
sel (%)	1.3	0.7	0.4	1.0	1.8	2.3	0.8
t_{sel} (ms)	20	90	30	30	170	140	250
N	24	52	29	30	20	16	75
N_{sel}	20	50	16	30	16	16	64

The first row (#(kf)) indicates the number of keyframes used for data reduction. The second and third row show the size of the reduced dataset in minutes and percentage, respectively. The forth row (t_{sel}) indicates the time in milliseconds needed for the preselection step. Finally, the last two rows give the numbers N and N_{sel} of motions of the respective type contained in the database and in the reduced dataset, respectively

labels attached to the motion data. Our indexing strategy, on the other hand, is purely content-based and thus facilitates fully automated preprocessing. Furthermore, it provides an efficient way of focusing on different aspects of the motion and selecting the desired search granularity *after* the preprocessing stage (and not *before* as for manual annotations). First experiments to implement sketchy queries such as the one above with our retrieval technique are based on the following observation: vaguely specified motions such as kicking, punching, or clapping can often be specified by a very small set of basic geometric constellations. For example, a kick of the right foot can be quite well characterized by the concurrent progression of the constellations "rknee straight/bent/straight/bent/straight" and "rfoot lowered/raised/lowered." In addition, the intermediary constellations "rknee straight" and "rfoot raised" should overlap at some point in time. Figure 12.8 shows a *scene description* for the motion "right foot kick followed by a left hand punch," where overlapping constellations are indicated by an arrow and unspecified constellations, realizable by our fuzzy concept, are indicated by the symbol "*". In our future work, we plan to generalize our keyframe-based retrieval strategy to efficiently process such scene descriptions. Another interesting research problem would be to learn complex scene descriptions from example motions to fully automate queries at higher semantic levels. In Chap. 13, we will introduce a concept that allows for learning a compact motion representation for a given class of semantically related motions.

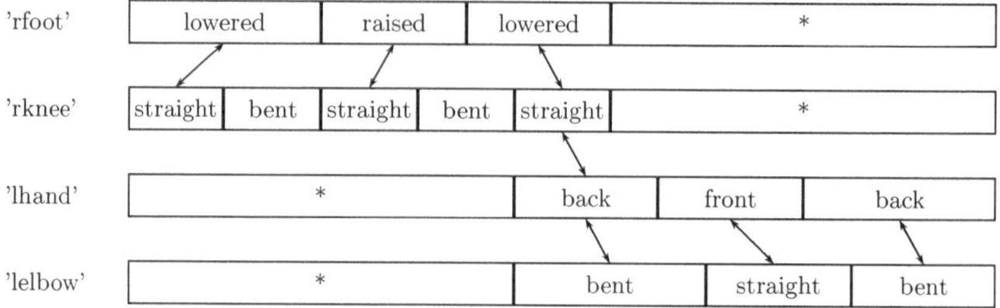

Fig. 12.8. Scene description for the movement "right foot kick followed by a left hand punch" as shown in Fig. 11.10. The symbol "*" indicates that a constellation is unspecified. *Arrows* indicate which constellations are to occur simultaneously

12.4 Further Notes

In this chapter, we have presented automated methods for efficient indexing and content-based retrieval of motion capture data. Adapting the notion of fuzzy queries and fuzzy hits, we introduced the concept of adaptive temporal segmentation, by which segment lengths are not only adjusted to the granularity of the feature function but also to the fuzziness of the query. Using an inverted file index, one can compute the motion hits very efficiently by intersecting and merging inverted lists. One decisive advantage of our index structure is that the time as well as the space to construct and store the index is linear in the size of the database. This solves the problem of scalability emerging in DTW-based approaches. See also [146] and the accompanying video available at [148].

12.4.1 Problems and Limitations

We now discuss some problems and limitations attached to the retrieval strategy as presented so far. It turns out that using all features at the same time in the retrieval process is far too restrictive – even in combination with fault tolerance strategies such as fuzzy or mismatch search – possibly leading to a large number of false negatives. Therefore, for each query, the user has to adequately select a small subset of features reflecting the important aspects of the query motion. On the one hand, the manual specification enables the user to incorporate prior knowledge about the query, thus providing a great deal of flexibility; on the other hand, the manual selection, in general, requires some experience as well as parameter tuning. Not only can this be a tedious process, but also prohibits batch processing as required for motion reuse techniques such as [6, 35, 169]. For example, in the blending scenario described by Pullen and Bregeler [169], a fully automatic feature selection would be important to identify similar motions in a large database for many different motion

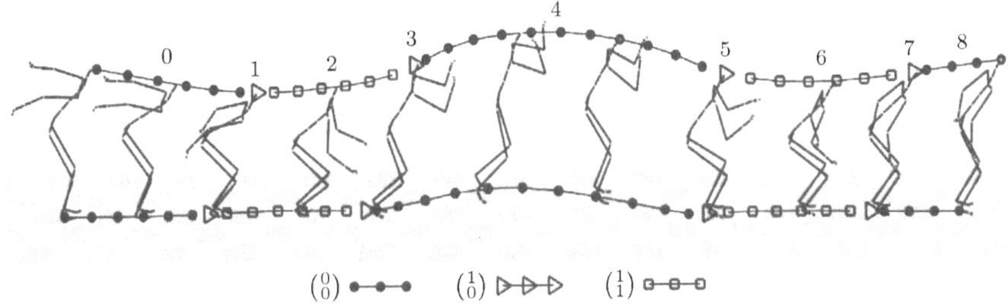

Fig. 12.9. Segmentation of a parallel leg jumping motion with respect to a combination of the feature F_{20} (*right knee bent*) and the feature F_{21} (*left knee bent*). Here the nine segments correspond to the sequence of feature values. Poses assuming the same feature values are indicated by identically marked trajectory segments. The trajectories of the joints "headtop" and "rankle" are shown

clips without manual intervention. In Chap. 13, we will discuss an approach for learning class representations in the form of so-called motion templates, which can be used to locally and globally adjust the feature selection.

We next discuss a problem of the index-based retrieval method that is due to the existence of co-called *transition segments*. Recall that the key idea of using relational features for motion retrieval is based on the observation that most human motions are characterized by the progression of few characteristic geometric constellations. Even though certain constellations may be characteristic for an entire class of semantically related motions, the *transition* from one constellation to the subsequent characteristic constellation may not be consistent throughout the motions of the same type. This fact is problematic in view of the relative strict retrieval techniques described so far. We substantiate this problem by means of a concrete example. Let us consider a parallel leg jumping motion as illustrated by Fig. 12.9. In the following discussion, we focus on the movement of the legs. Both legs are kept parallel throughout the jump sequence and are basically stretched during the initial phase of the arm-swing (segment 0). The legs are then bent into a half-squatting position (segments 1–2), preparing the following push-off (starting shortly before segment 3), during which the legs are stretched once more. In the landing phase (starting shortly before segment 5), the legs absorb the energy of the jump by bending as deep as before push-off. The jump sequence is concluded by stretching the legs into a normal standing position (segments 7–8). The resulting feature sequence with respect to the features F_{20} (right knee bent) and F_{21} (left knee bent) is given by

$$\left(\binom{0}{0}, \binom{1}{0}, \binom{1}{1}, \binom{1}{0}, \binom{0}{0}, \binom{1}{0}, \binom{1}{1}, \binom{1}{0}, \binom{0}{0}\right). \tag{12.12}$$

This alternating constellations of bending and stretching the knees is characteristic for many kinds of parallel-leg jumping motions, which is reflected by the feature combinations $\binom{1}{1}$ (legs bent) and $\binom{0}{0}$ (legs stretched), respectively.

The feature vectors $\binom{1}{0}$, however, only arise because the actor does not bend or stretch both legs at the same time. Instead, he has a tendency to bent the right knee a bit earlier than the left knee in the transition from $\binom{0}{0}$ to $\binom{1}{1}$, and to keep the right leg bent a bit longer in the transition from $\binom{1}{1}$ to $\binom{0}{0}$. Considering jumping motions by a different actor may lead to other transitions such as $\binom{0}{0} \rightarrow \binom{0}{1} \rightarrow \binom{1}{1}$ or $\binom{0}{0} \rightarrow \binom{1}{1}$. Each of the aforementioned transitions encodes the motion "bending both legs," which is all that matters to our query. To some extent, this uncertainty in the transition segments can be absorbed by our fuzzy concept. For example, using the sequence $V = \{V_0, V_1, V_2\}$ with $V_0 = \{\binom{0}{0}\}$, $V_1 = \{\binom{0}{1}, \binom{1}{0}\}$, and $V_0 = \{\binom{1}{1}\}$ one can retrieve the two transitions $\binom{0}{0} \rightarrow \binom{0}{1} \rightarrow \binom{1}{1}$ as well as $\binom{0}{0} \rightarrow \binom{1}{0} \rightarrow \binom{1}{1}$. However, the direct transition $\binom{0}{0} \rightarrow \binom{1}{1}$, where both features values switch at the same time, has still not been covered. This is due to the inability of our retrieval methods to express *insertions* and *deletions* within a sequence: the fuzzy set handling the transitory segment *must* be matched and may not be omitted in case it is not needed. Recall that the less efficient technique of DTW can handle such insertions and deletions: at this point, the tradeoff between efficiency and retrieval quality, corresponding to the choice between index-based retrieval and DTW, becomes apparent. To remedy the problem of simultaneous changes of feature values (which occurs very rarely at a sampling rate of 120 Hz), we can modify our retrieval algorithm so as to handle "dummy segments" within queries for such transitions [56].

12.4.2 Related Work

We close this chapter with a discussion of related work. In view of massively growing multimedia databases of various types and formats, efficient methods for indexing and content-based retrieval have become an important issue. Vast literature exists on indexing and retrieval in text, image, and video data, see, e. g., Witten et al. [218] and Sebe et al. [187] and references therein. For the music scenario, Clausen and Kurth [42] give a unified approach to content-based retrieval; their group theoretical concepts generalize to other domains as well. The problem of indexing large time series databases has also attracted great interest in the database community, where DTW-based and LCS-based retrieval approaches are pursued, see [100, 117, 208] and Sect. 4.5.

Because of possible spatio-temporal variations, the difficult task of identifying similar motion clips still bears open problems. Many of the previous approaches to motion comparison are based on DTW in conjunction with features that are semantically close to the raw data, using 3D positions, 3D point clouds, joint angle representations, or PCA-reduced versions thereof, see [29, 70, 93, 103, 107, 123, 169, 183, 220, 222]. In the following, we discuss some of these approaches in more detail.

In their motion blending application, Pullen and Bregler [168, 169] identify similar motion clips within a motion database, by suitably decomposing the database motions into fragments. The fragments are uniformly scaled to make them the same length as the given query fragment before calculating the Euclidean distance. In their QBE retrieval approach, Wu et al. [220] proceed in two stages: they first identify start and end frames of possible candidate clips utilizing a pose-based index and then compute the actual distance from the query via DTW. Cardle et al. [29] sketch how to use DTW-based indexing techniques to accomplish the retrieval task. Adapting techniques from [100], Keogh et al. [103] describe a lower-bounding indexing technique to efficiently identify similar motion fragments that differ by some uniform scaling factor with respect to the time axis. In the approach of Kovar and Gleicher [107], numerically similar motions are identified by means of a DTW-based index structure termed *match web*. A multistep search spawning new queries from previously retrieved motions allows for the identification of semantically similar motions using numerically similar motions as intermediaries. The authors report good retrieval results, which are particularly suitable for their blending application.

Conceptually similar to our approach, Liu et al. [123] handle spatio-temporal deformations by absorbing a certain degree of motion variations at the feature level. They use principle component analysis (PCA) and k-means clustering to transform motions into sequences of cluster centroids. However, they provide no details on their retrieval method except that exact as well as approximate string matching (a variant of DTW) have been successfully applied to compare a query's cluster transition signature against that of a large subset of the CMU mocap database [44]. Forbes and Fiume [70] also use PCA-based techniques to reduce the dimensionality of the pose space. They claim that the Euclidean distance in the PCA space is a good measure of similarity for individual poses. Under this assumption, they pursue the strategy of first extracting certain characteristic poses from the PCA-reduced query clip, which are then used to identify cluster of similar poses (so-called seed points) within the database. The motion fragments around these seed points are then compared with the query by a DTW-based similarity measure applied to the PCA-reduced trajectories. Some results for mocap databases with lengths ranging from 70 s to 11 min (120 Hz sampling rate) are reported, where query times are on the order of half a second. Yu et al. [222] perform motion comparison by means of *Labanotation* [88, 210], a versatile movement notation language developed in the dance community. Here, the query as well as the database motions have to be transcribed into Labanotation [91]. Retrieval is then performed on the Laban sequences via DTW. However, querying a 19 000-frame mocap database for a short motion clip comprising 150 frames takes 139 s using their method.

Our approach to content-based motion retrieval differs fundamentally from previous approaches: instead of using numerical features and DTW-based techniques that rely on numerical local cost measures to compare individual frames of data streams, we handle spatio-temporal deformation already at the feature level (transforming the motions into feature sequences). This allows for applying exact matching algorithms at the segment level, which can be performed very efficiently supported by an inverted file index. One crucial consequence is that the time and space complexity of building and storing our index structure is *linear*, $O(N)$, in the number N of frames of the underlying database opposed to DTW-based strategies such as the web match, which are *quadratic*, $O(N^2)$, in N. For example, to compute and store the match web for a database containing 37 000 frames requires, as reported in [107], roughly 50 min and 76 MB, whereas our index (consisting of I_ℓ^{180}, I_u^{180}, and I_m^{180}, see Sect. 12.2.2), for a database containing 1 000 000 frames needs roughly 6 min and less than 5 MB.

Using relational features, our retrieval system is based on a rather coarse-grained notion of similarity that (intentionally) blends out a large degree of motion details. This allows us to efficiently cut down the search space to a small subset that still contains most of the relevant motion clips. Here, one strength of our method is that it works purely content-based without needing an ontology of annotations manually applied to the respective collection of motion data [6].

However, to assess the retrieval quality as well as to further process the motions of the reduced dataset, one needs a finer-grained notion of similarity, which will crucially depend on the respective application. As has also been pointed out by Gleicher [76], the similarity measure should be *operational* in the sense that it closely reflects the semantics and requirements of the specific application. For example, in motion synthesis applications, two motions may be considered as similar if they can be combined to yield new motions with the desired properties and predetermined constraints. Such an operational approach to motion similarity has been used by Kovar et al. [106–108] for their motion morphing and blending applications. As another example, a physical therapist who is studying gait anomalies may be interested in some metric that can capture slight differences in locomotion styles. Here, most of the motion details that are of interest for the therapist are factored out in our present retrieval scenario. For the future, it seems promising to combine and integrate various similarity measures and retrieval methods to obtain a system that, on the one hand, facilitates scalable search of large databases and, on the other hand, yields the operational metric needed for the specific application.

13

Motion Templates

In this chapter, we introduce a method for capturing the spatio-temporal characteristics of an entire motion class of semantically related motions in a compact and explicit matrix representation called a *motion template* (MT). Motion templates, which can be regarded as generalized boolean feature matrices, are formally introduced in Sect. 13.1. Employing an iterative warping and averaging strategy, we then describe an efficient algorithm that automatically derives a motion template from a class of training motions. We summarize the main ideas of this algorithm in Sect. 13.2 before giving the technical details in Sect. 13.3. In Sect. 13.4, we report on our experiments on template learning and discuss a number of illustrative examples to demonstrate the descriptive power of motion templates. Finally, in Sect. 13.5, we close with some general remarks on the multiple alignment problem underlying our learning procedure. In this and the following chapter, we close follow Müller and Röder [143]. An accompanying video is available at [144].

13.1 Basic Definitions and Properties

For the remainder of this chapter, we fix a feature function $F : \mathcal{P} \to \{0,1\}^f$ consisting of f relational features. Applying F to a motion data stream $D : [1 : K] \to \mathcal{P}$ of length K in a posewise fashion yields a binary matrix $X \in \{0,1\}^{f \times K}$, which we refer to as *feature matrix* of D. By definition, the kth column of X, denoted by $X(k)$, equals the feature vector $F(D(k))$, $k \in [1 : K]$.

Feature matrices reveal specific motion characteristics with a direct semantic interpretation. As an example, we consider a jogging motion D of length K and a walking motion D' of length K' each consisting of four steps (Fig. 13.1). At first sight, the trajectories and corresponding poses seem to be

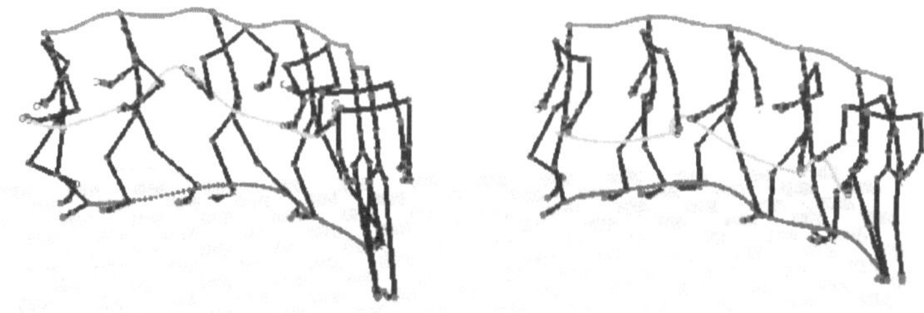

Fig. 13.1. *Left:* four steps of a curved jogging motion. *Right:* four steps of a curved walking motion. Both motions start from a standing pose. The trajectories of the joints "headtop," "rankle," "lankle," "rfingers," and "lfingers" are shown

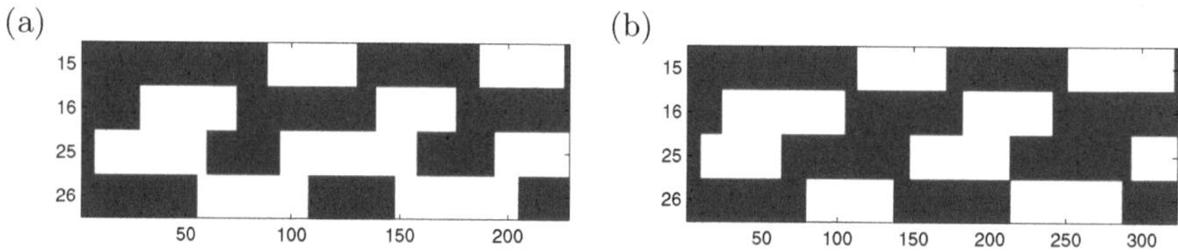

Fig. 13.2. Relational feature matrices for **(a)** "jogging" and **(b)** "walking" using $f = 4$ features. The matrices are color coded as black (0) and white (1), and the numbers in front of the rows refer to Table 11.1. Time (measured in frames) runs along the horizontal axis

rather similar and, depending on the application, one could consider these motions to be of the same type because both of them are instances of locomotion. To expose common characteristics of the two locomotions as well as subtle difference between jogging and walking, we consider the feature function F^4 consisting of the $f = 4$ features F_{15}/F_{16} ("r/lfoot behind l/rleg") and F_{25}/F_{26} ("r/lfoot fast"), see Table 11.1. The resulting feature matrices $X \in \{0,1\}^{f \times K}$ and $X' \in \{0,1\}^{f \times K'}$ with respect to F^4 are shown in Fig. 13.2, where the value one is encoded by black and the value zero by white. A comparison of Fig. 13.2a,b reveals that the first two rows of the feature matrices are identical in structure for the two motions. This corresponds to the fact that both walking and jogging motions are dominated by an alternation of the condition "left/right foot in front" (as has already been discussed in Sect. 11.1.1). The last two rows, however, reveal a subtle but crucial difference between jogging and walking: opposed to the walking motion, the jogging motion exhibits characteristic poses where both features F_{25} and F_{26} simultaneously assume the value one, see the neighborhoods of frames 54, 100, 150, and 200 of Fig. 13.2a. These are exactly the air phases of the jogging motion, where both feet are

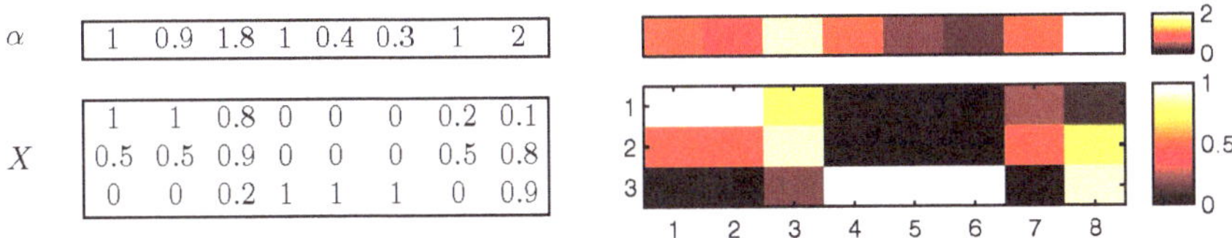

α	1	0.9	1.8	1	0.4	0.3	1	2
X	1	1	0.8	0	0	0	0.2	0.1
	0.5	0.5	0.9	0	0	0	0.5	0.8
	0	0	0.2	1	1	1	0	0.9

Fig. 13.3. *Left:* WMT (X, α) of dimension $f = 3$ and length $K = 8$. The first row corresponds to the weight vector α and the last three rows to the motion template X. *Right:* Color-coded representation of (X, α). For the MT X, the value one is encoded by *white*, the value zero by *black*, and the intermediary values by shades of *red* and *yellow*

simultaneously fast. Concluding this example, note that even though feature matrix representations discard a lot of detail contained in the raw motion data, important information regarding the overall course of motion as well as subtle characteristics are made explicit.

Generalizing boolean feature matrices, we now introduce the concept of *motion templates* (MTs), which is suited to express the essence of an entire class of motions. A motion template of *dimension* f and *length* K is a real-valued matrix $X \in [0,1]^{f \times K}$. As for feature matrices, each row of an MT corresponds to one relational feature, and time (in frames) runs along the columns. In the following, we assume that all MTs under consideration have the same fixed dimension f. Intuitively, an MT can be thought of as a "fuzzified" version of a feature matrix. For the proper interpretation of the matrix entries, we refer to Sect. 13.2, where we describe how to learn an MT from training motions by a combination of warping and averaging operations.

During the learning procedure, a *weight vector* $\alpha \in \mathbb{R}^K_{>0}$ is associated with an MT, where the total weight $\bar{\alpha} := \sum_{k=1}^{K} \alpha(k)$ is at least one. We say that the kth column $X(k)$ of X has weight $\alpha(k)$. These weights are used to keep track of the time warping operations. Initially, each column of an MT corresponds to the real time duration of one frame, which we express by setting all weights $\alpha(k)$ to one. Subsequent time warping may change the amount of time that is allotted to an MT column. The respective weights are then modified so as to reflect the new time duration. Hence, the weights allow us to unwarp an MT back to real time. For short, the pair (X, α) is also referred to as *weighted motion template* (WMT). For an example, we refer to Fig. 13.3. From this point forward, we will consider a feature matrix as a WMT by attaching a weight vector α initialized to $\alpha(k) = 1$ for all $k \in [1 : K]$ (we then also write $\alpha \equiv 1$).

13.2 MT Learning Procedure: Overview

Given a set of N example motion clips for a specific motion class, such as the four cartwheels shown in Fig. 13.4a, our goal is to automatically learn a

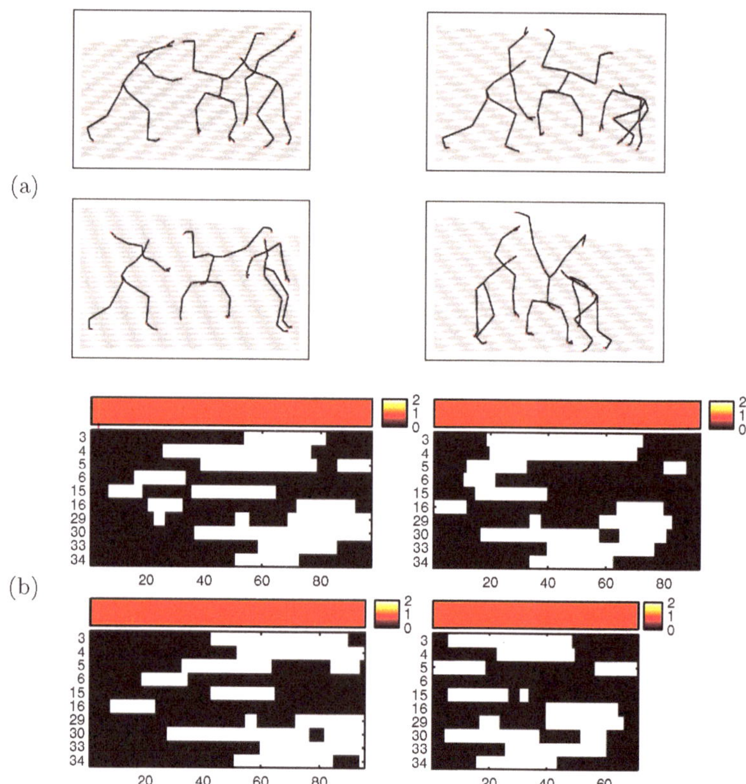

Fig. 13.4. **(a)** Selected frames from four different cartwheel motions. **(b)** Relational feature matrices resulting from the motions in (a) for selected features with associated weight vectors. The rows are numbered in accordance with the relational features defined in Table 11.1. The weight vectors are displayed as color-coded horizontal bars

meaningful MT that grasps the essence of the class. We start by computing the feature matrices for a fixed set of features, as shown in Fig. 13.4b, where, for the sake of clarity, we only display a subset comprising 10 features from our test feature set.

Now, the goal is to compute a semantically meaningful average over the N input MTs, which would simply be the arithmetic mean of the feature matrices if all of the motions agreed in length and temporal structure. However, our MTs typically differ in length and reflect the temporal variations that were present in the original motions. This fact necessitates some kind of temporal alignment prior to averaging (Sect. 13.3.1). We do this by choosing one of the input MTs as the *reference MT*, say X with weight vector α initialized to $\alpha \equiv 1$. We then apply dynamic time warping to compute

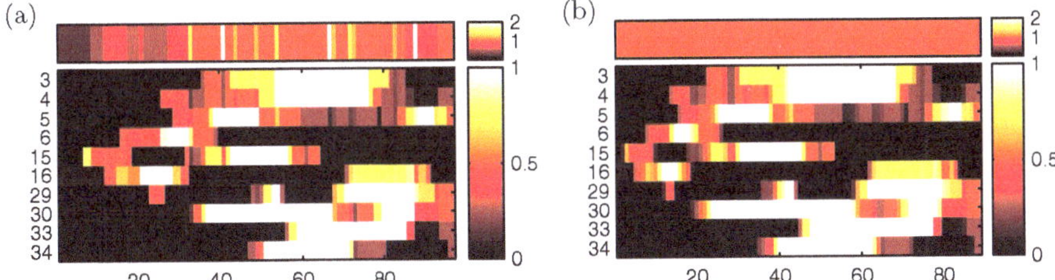

Fig. 13.5. (a) Average MT and average weights computed from the MTs in Fig. 13.4b after all four MTs have been aligned with the top left MT, which acted as the reference. **(b)** Unwarped version of the MT in (a)

optimal alignments of the remaining MTs with X. Here, we measure local distances of feature vectors by the Manhattan distance, which coincides with the Hamming distance for boolean feature vectors. Let (Y, β) be one of the WMTs to be aligned with (X, α). According to the optimal alignment with X, the MT Y is locally contracted and stretched to yield an MT Z, where time contractions are resolved by forming a weighted average of the columns in question while time stretching is simulated by duplicating MT columns. Similarly, the weight β is transformed into a weight γ as follows: in case column $X(k)$ of the reference was matched to multiple columns of Y, the new weight $\gamma(k)$ is the sum of the weights of the matching columns in Y. In case n columns $X(k), \ldots, X(k+n-1)$ of the reference are matched to a single column $Y(\ell)$, the new weights $\gamma(k+i)$ are set to $\frac{1}{n}\beta(\ell)$ for $i = 0, \ldots, n-1$, i.e., the weight $\beta(\ell)$ is equally distributed among the n duplicated columns. For details, we refer to Sect. 13.3.1.

Now that all MTs and associated weight vectors have the same length as the reference MT, we compute the weighted average over all MTs in a column-wise fashion as well as the arithmetic mean vector, again denoted by γ, of all weight vectors (Sect. 13.3.3). Note that the total weight, $\bar{\gamma}$, equals the average length of the input motions. Figure 13.5a shows the results for our cartwheel example, where the top left MT in Fig. 13.4 acted as the reference. Finally, we unwarp the average MT according to the weight vector (Sect. 13.3.2). Column ranges with $\gamma(k) < 1$ are unwarped by contracting the respective MT columns into one average column (e.g., $k = 1, \ldots, 6$ in Fig. 13.5a), while columns with $\gamma(k) > 1$ are unwarped by duplicating the respective column (e.g., $k = 42$). Since, in general, columns will not have integer or reciprocal integer weights, we additionally perform suitable partial averaging between adjacent columns such that all weights but the last are one in the resulting unwarped MT, see Fig. 13.5b. Note that the total weight, $\bar{\gamma}$, is preserved by the unwarping procedure. The average MT now constitutes a combined representation of all the input motions, but it is still biased by the influence of the reference MT, to which all of the other MTs have been aligned. Our experiments show that it is possible to eliminate this bias by the following strategy (Sect. 13.3.4). We let each of the original MTs act as the reference and perform for each reference the entire averaging and unwarping procedure as described earlier. This yields N averaged MTs corresponding to the different references. Then, we use these N MTs as the input to a second pass of mutual warping, averaging, and unwarping, and so on. The procedure is iterated until no major changes occur (Fig. 13.8).

A motion template learned from training motions belonging to a specific motion class $\mathcal{C}$ is referred to as the *class motion template* $X_{\mathcal{C}}$. A class MT has the following interpretation: black and white regions indicate periods in time (horizontal axis) where certain features (vertical axis) consistently assume the same values zero and one in all training motions, respectively. By contrast, red to yellow regions indicate inconsistencies mainly resulting from variations in the training motions (and partly from inappropriate alignments). Intuitively, each row of such a class template can be viewed as an estimate of the temporal distribution of the corresponding feature's values, which is highly class-dependent and thus an essential characteristic for $\mathcal{C}$. The interpretation of a single matrix entry $x_{ik} := X_{\mathcal{C}}(k)_i \in [0,1]$ is that on average, a proportion of x_{ik} of the training motions assumed the feature value one for feature i at frame number k. Here, the semantics of a frame number k along a class template's time axis is that of an average time position, which makes sense because corresponding columns of the MTs have been aligned. Some illustrative examples will be discussed in Sect. 13.4.3.

13.3 MT Learning Procedure: Technical Details

We now give the technical details for the iterative warping and averaging procedure outlined in Sect. 13.2. We first describe the warping step foradjusting the length of an arbitrary WMT with respect to some reference WMT (Sect. 13.3.1), which is followed by an averaging (Sect. 13.3.3) and unwarping (Sect. 13.3.2) step. By iteratively repeating these steps with alternating references, we finally obtain a reference-free class motion template (Sect. 13.3.4).

13.3.1 Reference-Based WMT Warping

Let (X,α) be a WMT of length K. We define a *warping operator* $\Omega^{\mathrm{W}}_{(X,\alpha)}$, which transforms an arbitrary WMT (Y,β) of length L into a WMT of length K using (X,α) as a reference. To compare two WMTs, we use classical dynamic time warping (Sect. 4.1) based on a weighted Manhattan distance. More precisely, for $k \in [1:K]$ and $\ell \in [1:L]$, we define the local cost measure $c(k,\ell)$ between the vectors $X(k)$ and $Y(\ell)$ by

$$c(k,\ell) := \frac{\alpha(k)+\beta(\ell)}{2\cdot f}\sum_{i=1}^{f}|X(k)_i - Y(\ell)_i|, \tag{13.1}$$

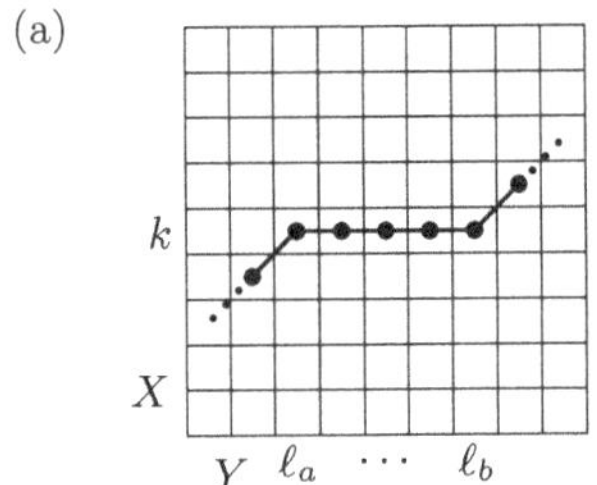

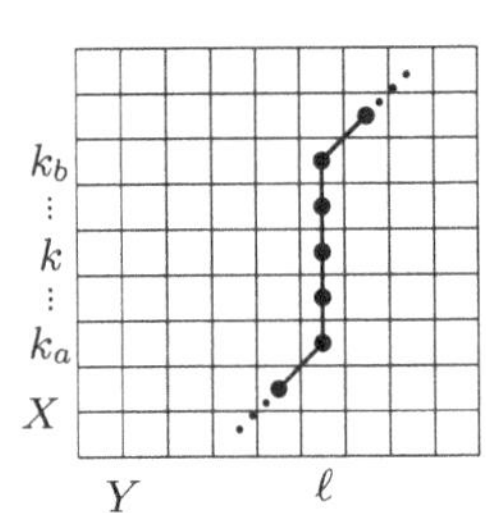

Fig. 13.6. Local sections of an optimal warping path aligning the MTs X and Y. **(a)** The column $X(k)$ is assigned to the columns $Y(\ell_a)$ to $Y(\ell_b)$. **(b)** The columns $X(k_a)$ to $X(k_b)$ including column $X(k)$ are assigned to column $Y(\ell)$

where $X(k)_i$ and $Y(\ell)_i$ denote the vector entries. Let $p^* = (p_1, \ldots, p_M)$ with $p_m = (k_m, \ell_m)$, $m \in [1 : M]$, denote the optimal warping path (of smallest length-lexicographical order) between (X, α) and (Y, β). We define a new WMT (Z, γ) of length K as follows. For $k \in [1 : K]$, there are two cases, as illustrated by Fig. 13.6. In the first case, k appears more than once in the list $(k_1, \ldots, k_M)$. Let a denote the minimum index with $p_a = (k, \ell_a)$ and b the maximum index with $p_b = (k, \ell_b)$. Note that the step size condition enforces $\ell_{a+i} = \ell_a + i$ for $i \in [0 : b - a]$. We then set

$$\gamma(k) := \sum_{i=a}^{b} \beta(\ell_i), \qquad Z(k) := \frac{1}{\gamma(k)} \sum_{i=a}^{b} \beta(\ell_i) Y(\ell_i). \tag{13.2}$$

In the second case, k appears exactly once in the list $(k_1, \ldots, k_M)$, say $k = k_m$. Setting $\ell := \ell_m$, let a denote the minimum index with $p_a = (k_a, \ell)$ and b the maximum index with $p_b = (k_b, \ell)$. Note that the step size condition enforces $k_{a+i} = k_a + i$ for $i \in [0 : b - a]$. We then set

$$\gamma(k) := \frac{1}{(b - a + 1)} \beta(\ell), \qquad Z(k) := Y(\ell). \tag{13.3}$$

Intuitively, we average in the first case over all Y-columns that match to the same column $X(k)$ and replicate in the second case the column $Y(\ell)$ by the number of X-columns being matched to $Y(\ell)$. The warping operator $\Omega^{\mathrm{W}}_{(X,\alpha)}$ with respect to (X, α) is defined by

$$\Omega^{\mathrm{W}}_{(X,\alpha)}(Y, \beta) := (Z, \gamma). \tag{13.4}$$

Obviously, $\sum_{k=1}^{K} \gamma(k) = \sum_{\ell=1}^{L} \beta(\ell)$ and $\Omega^{\mathrm{W}}_{(X,\alpha)}(X, \alpha) = (X, \alpha)$.

13.3.2 WMT Unwarping

We now introduce further operators for the manipulation of WMTs. We start with the *uniqueness operator* Ω^{U}, which cancels consecutive repetitions in X.

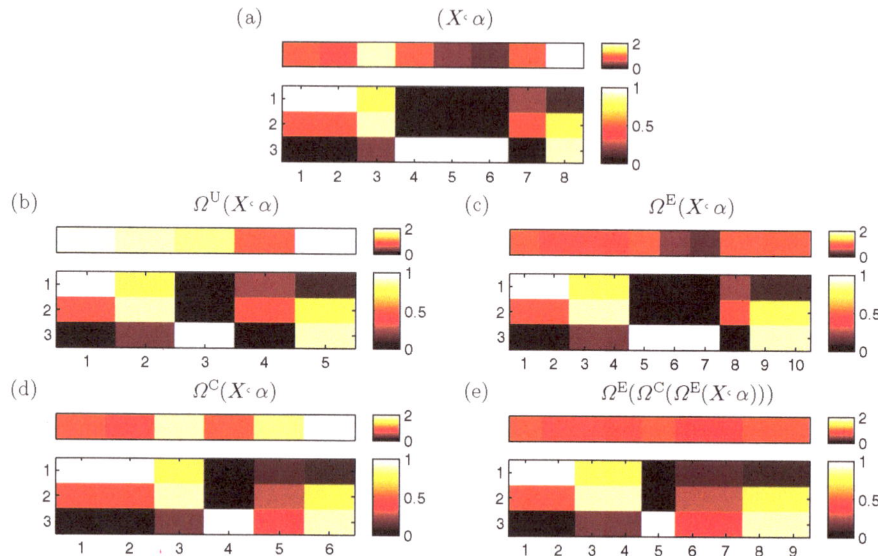

Fig. 13.7. **(a)** WMT (X, α) of dimension $f = 3$ and length $K = 8$. **(b)–(e)** Resulting WMTs after applying various operators

Let (X, α) be an arbitrary WMT of length K. In case there is some $k \in [1 : K-1]$ with $X(k) = X(k+1)$, we replace $\alpha(k)$ by the sum $\alpha(k)+\alpha(k+1)$ and delete $X(k+1)$ and its assigned weight $\alpha(k+1)$. This procedure is repeated until all adjacent vectors are different. The resulting WMT will be denoted by $\Omega^{\mathrm{U}}(X, \alpha)$ (Fig. 13.7b). The operator Ω^{U} is motivated by the following observation: note that a DTW distance $\mathrm{DTW}((X, \alpha), (Y, \beta)) = 0$ between two arbitrary WMTs (X, α) and (Y, β) does not necessarily imply $(X, \alpha) = (Y, \beta)$. However, it is not hard to see that in this case the underlying matrices of $\Omega^{\mathrm{U}}(X, \alpha)$ and $\Omega^{\mathrm{U}}(Y, \beta)$ coincide.

We next define an *expansion operator* Ω^{E}, which locally expands a WMT in case a weight exceeds a threshold of 1.5. To this end, we consider all $k \in [1 : K]$ satisfying $\alpha(k) \geq 1.5$. Let $[\alpha(k)]$ denote the rounded integer value of $\alpha(k)$. We then replace the vector $X(k)$ by a sequence of $[\alpha(k)]$ copies of $X(k)$ and assign to each such copy the weight $\alpha(k)/[\alpha(k)]$. Note that $0.75 \leq \alpha(k)/[\alpha(k)] < 1.5$. The resulting WMT will be denoted by $\Omega^{\mathrm{E}}(X, \alpha)$ (Fig. 13.7c).

Similarly, we define a *contraction operator* Ω^{C}, which contracts a WMT in case a weight falls below a threshold of 0.75. For $k \in [1 : K]$ with $\alpha(k) < 0.75$ and $k' := \mathrm{argmin}\{\alpha(k-1), \alpha(k+1)\}$, we replace the two weights $\alpha(k)$ and $\alpha(k')$ by the single weight $a := \alpha(k) + \alpha(k')$ and the two vectors $X(k)$ and $X(k')$ by the single vectors $\big(\alpha(k)X(k) + \alpha(k')X(k')\big)/a$. This procedure is repeated until all weights are above the threshold 0.75. (This always works, since $\sum_{k=1}^{K} \alpha(k) \geq 1$ by definition.) The resulting WMT will be denoted by $\Omega^{\mathrm{C}}(X,\alpha)$ (Fig. 13.7d).

Note that the operators Ω^{U}, Ω^{E}, and Ω^{C} do not change the total weight of a WMT. Furthermore, one can easily check that all weights of the WMT $\Omega^{\mathrm{E}}(\Omega^{\mathrm{C}}(\Omega^{\mathrm{E}}(X,\alpha)))$ lie between 0.75 and 1.5 (Fig. 13.7e). This procedure will be referred to as *unwarping*. By partial averaging between adjacent columns, as described in Sect. 13.2, one can even achieve that all weights (except of possibly the last weight) are one in the resulting unwarped WMT. As our experiments showed, the additional partial averaging step has no significant impact on the final class MT. Hence, in the subsequent sections, we work with the unwarping operator $\Omega^{\mathrm{E}} \circ \Omega^{\mathrm{C}} \circ \Omega^{\mathrm{E}}$.

13.3.3 Reference-Based WMT Averaging

We next describe how to average over an entire set $\mathcal{Y} = \{(Y_1, \beta_1), \ldots, (Y_N, \beta_N)\}$ of WMTs of arbitrary lengths with respect to some reference template (X,α) of length K. In a first step, we adjust the WMTs of $\mathcal{Y}$ by computing $(Z_n, \gamma_n) := \Omega^{\mathrm{W}}_{(X,\alpha)}(Y_n, \beta_n)$ for $n \in [1:N]$. Then, a WMT (Z,γ) of length K is defined by

$$\gamma(k) := \frac{1}{N+1}\Big(\alpha(k) + \sum_{n=1}^{N} \gamma_n(k)\Big), \tag{13.5}$$

$$Z(k) := \frac{1}{(N+1)\cdot\gamma(k)}\Big(\alpha(k)X(k) + \sum_{n=1}^{N} \gamma_n(k) Z_n(k)\Big) \tag{13.6}$$

for $k \in [1:K]$. Finally, we unwarp the weights as described in Sect. 13.3.2. The averaging operator $\Omega^{\mathrm{A}}_{(X,\alpha)}$ with respect to (X,α) is thus defined by

$$\Omega^{\mathrm{A}}_{(X,\alpha)}(\mathcal{Y}) := \Omega^{\mathrm{E}}(\Omega^{\mathrm{C}}(\Omega^{\mathrm{E}}(Z,\gamma))). \tag{13.7}$$

Note that the total weights of (Z,γ) and of $\Omega^{\mathrm{A}}_{(X,\alpha)}(\mathcal{Y})$ equal the average of the total weights of (X,α) and all (Y_n, β_n), $n \in [1:N]$.

13.3.4 Reference-Free WMT Averaging

Now, we introduce an iterative averaging procedure resulting in a WMT that does not depend on a reference WMT. Let $\mathcal{X}^0 = \{(X_1^0, \alpha_1^0), \ldots, (X_N^0, \alpha_N^0)\}$ be a set of WMTs constituting the input of our algorithm. In each iteration step $m \in \mathbb{N}$, a new set $\mathcal{X}^m = \{(X_1^m, \alpha_1^m), \ldots, (X_N^m, \alpha_N^m)\}$ of WMTs is computed from the set $\mathcal{X}^{m-1}$ by setting

$$(X_n^m, \alpha_n^m) := \Omega^{\mathrm{A}}_{(X_n^{m-1}, \alpha_n^{m-1})}\Big(\mathcal{X}^{m-1} \setminus \{(X_n^{m-1}, \alpha_n^{m-1})\}\Big) \tag{13.8}$$

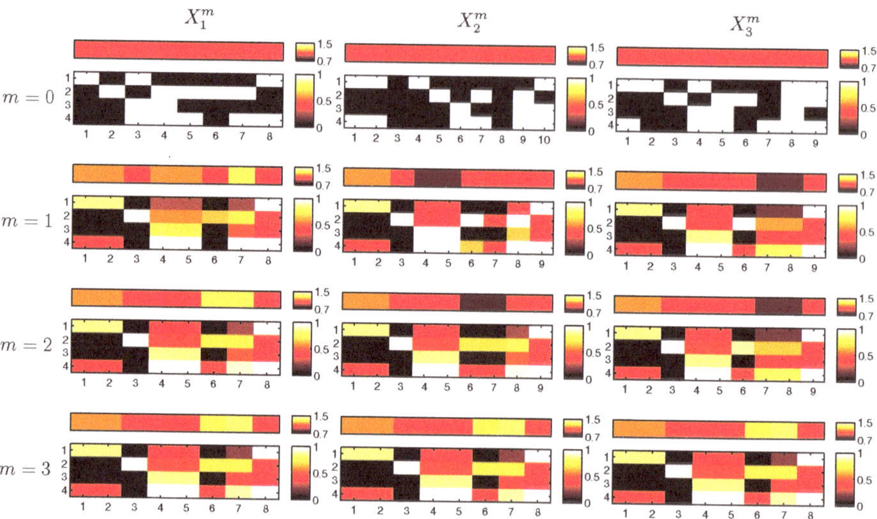

Fig. 13.8. Averaging procedure starting with the set $\mathcal{X}^0$ (*first row*) consisting of $N = 3$ WMTs. Each iteration corresponds to one row. After $M = 3$ iterations, all WMTs of $\mathcal{X}^M$ (*last row*) are identical and $\Omega^{\mathrm{A}}(\mathcal{X}^0) = (X_1^3, \alpha_1^3)$

for $n \in [1 : N]$. In other words, each $(X_n^{m-1}, \alpha_n^{m-1}) \in \mathcal{X}^{m-1}$ is taken as a reference WMT to perform the alignment and averaging procedure. The resulting set $\mathcal{X}^m$ again consists of N WMTs. Now, the crucial point is that, due to the averaging process, the WMTs in $\mathcal{X}^m$ are more similar to each other than the WMTs in $\mathcal{X}^{m-1}$. It turns out that there exists an unwarped WMT $(X^\infty, \alpha^\infty)$ and some iteration parameter $M \in \mathbb{N}$ such that $(X^\infty, \alpha^\infty) = (X_n^m, \alpha_n^m)$ for all $n \in [1 : N]$ and $m \geq M$. The reference-free averaging operator Ω^{A} is then defined by

$$\Omega^{\mathrm{A}}(\mathcal{X}^0) := (X^\infty, \alpha^\infty). \tag{13.9}$$

We refer to Fig. 13.8 for an illustration of our iterative procedure and conclude this section with some remarks. Let $\bar{\omega} \in \mathbb{R}$ denote the average over the total weights of the WMTs in $\mathcal{X}^0$. Then it follows from the above discussion that all WMTs (X_n^m, α_n^m) for $m > 0$ as well as the limit WMT $(X^\infty, \alpha^\infty)$ exhibit the same total weight $\bar{\omega}$. Furthermore, the unwarping step (13.7) prevents the computed WMTs from degeneration in the sense that the length L of any such WMT is kept within the bounds $\frac{2}{3}\bar{\omega} \leq L \leq \frac{4}{3}\bar{\omega}$. In particular, this holds for the WMT $(X^\infty, \alpha^\infty)$. Finally, in all of our conducted experiments, the iterative averaging procedure converged in an exponential-like fashion, assuming the limit $(X^\infty, \alpha^\infty)$ in very few steps, see also Fig. 13.9.

13.4 Experimental Results

13.4.1 Motion Class Database

For our experiments, we systematically recorded several hours of motion capture data containing a number of well-specified motion sequences, which were executed several times and performed by five different actors. Using this data, we built up the database $\mathcal{D}_{210}$ consisting of roughly 210 min of motion data. Then, from $\mathcal{D}_{210}$, we manually cut out suitable motion clips and arranged them into 64 different classes. Each such motion class (MC) contains 10–50 different realizations of the same type of motion, covering a broad spectrum of semantically meaningful variations. For example, the motion class "CartwheelLeft" contains 21 variations of a cartwheel motion, all starting with the left hand. The resulting *motion class database* $\mathcal{D}^{\mathrm{MC}}$ contains 1,457 motion clips of a total length corresponding to roughly 50 min of motion data. Table 13.1 gives an overview of some of the motion classes contained in $\mathcal{D}^{\mathrm{MC}}$. We split up $\mathcal{D}^{\mathrm{MC}}$ into two disjoint databases $\mathcal{D}^{\mathrm{MCT}}$ and $\mathcal{D}^{\mathrm{MCE}}$, each consisting of roughly half the motions of each motion class. The database $\mathcal{D}^{\mathrm{MCT}}$ serves as the *training database* to derive the motion templates, whereas $\mathcal{D}^{\mathrm{MCE}}$ is used as a training-independent *evaluation database*. We preprocessed all databases by computing and storing the feature matrices along with the motion capture data.

13.4.2 WMT Computation and Time Complexity

For each of the 64 motion classes, we determine a WMT as follows. Let $\mathcal{C} \subset \mathcal{D}^{\mathrm{MCT}}$ be the set of motions contained in a particular motion class and (X_D, α_D) denote the feature matrix of $D \in \mathcal{C}$ regarded as a WMT

Table 13.1. The MC-database $\mathcal{D}^{\mathrm{MC}}$ contains 10–50 different variations for each of its 64 motion classes

Motion Class $\mathcal{C}$	Comment	Size	N	$\bar{\omega}$	M	$t(\mathcal{C})$
CartwheelLeft	Left hand first on floor	21	11	105.3	6	17.0
ElbowToKnee	Start: relbow/lknee	27	14	36.6	5	4.9
JumpingJack	Start, end: neutral pose	52	26	35.5	6	19.3
KickFrontRightFoot	1 kick	30	15	53.3	5	9.4
KickSideRightFoot	1 kick	30	15	48.9	6	10.1
LieDownFloor	Start: standing pose	20	10	165.0	5	25.6
RotateRightArmBackward3	3 times	16	8	80.8	4	3.8
RotateRightArmForward3	3 times	16	8	83.6	4	3.9
Squat	Arms stretched out in front	52	26	47.3	5	24.6
WalkCrossoverRight3	3 steps, left over right	13	7	137.3	4	6.8
WalkSidewaysRight3	3 steps	16	8	123.0	3	5.5

This table shows eleven of the motion classes, along with their respective size, the size N of the training subset, the average total weight $\bar{\omega}$, as well as the number M of iterations and the running time $t(\mathcal{C})$ in seconds needed to compute $X_{\mathcal{C}}^{\infty}$.

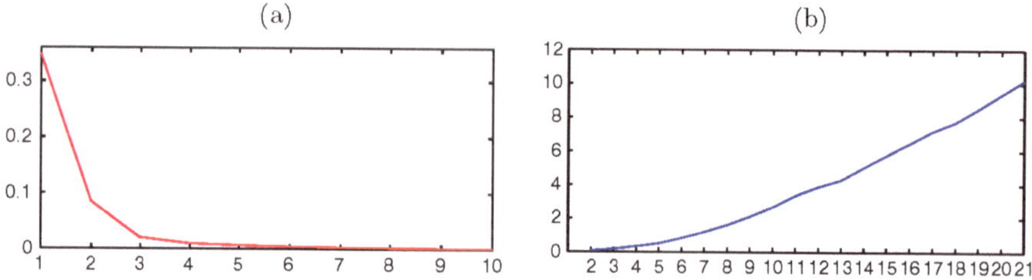

Fig. 13.9. **(a)** The maximum DTW distance $\mu(\mathcal{X}_{\mathcal{C}}^m)$ for $m = 1, \ldots, 10$ for the motion class "CartwheelLeft" with 11 training motions. $M = 6$ iterations are needed to fall below the error threshold $\varepsilon = 0.005$. **(b)** Average running time of one iteration step against the number N of training motions for "CartwheelLeft"

(with $\alpha_D \equiv 1$). Then the set $\mathcal{X}_{\mathcal{C}}^0 := \{(X_D, \alpha_D) \,|\, D \in \mathcal{C}\}$ defines a WMT $(X_{\mathcal{C}}^\infty, \alpha_{\mathcal{C}}^\infty) := \Omega^{\mathrm{A}}(\mathcal{X}_{\mathcal{C}}^0)$. The matrix $X_{\mathcal{C}}^\infty$ will be referred to as the *class MT* with respect to $\mathcal{C}$. To speed up computation time, we compute the class MT only up to some small approximation error. To this end, we define the maximum DTW distance $\mu(\mathcal{X}_{\mathcal{C}}^m)$ as

$$\mu(\mathcal{X}_{\mathcal{C}}^m) := \max\left\{\mathrm{DTW}\big((X,\alpha),(Y,\beta)\big) \,\big|\, (X,\alpha),(Y,\beta) \in \mathcal{X}_{\mathcal{C}}^m\right\}. \tag{13.10}$$

Then the iterative averaging process is performed for $m = 1, 2, 3 \ldots$ until $\mu(\mathcal{X}_{\mathcal{C}}^m)$ falls below some fixed error threshold ε. Letting M denote the minimum iteration number such that $\mu(\mathcal{X}_{\mathcal{C}}^M) \leq \varepsilon$, we set $(X_{\mathcal{C}}^\infty, \alpha_{\mathcal{C}}^\infty) := (X^M, \alpha^M)$ for some $(X^M, \alpha^M) \in \mathcal{X}_{\mathcal{C}}^M$. In our experiments, we chose $\varepsilon = 0.005$, resulting in approximations of very high accuracy. Only 3–7 iterations were required to fall below this error threshold, depending on the respective motion class, see also Fig. 13.9a and Table 13.1.

We now comment on the running time behavior of our algorithm. Let $\mathcal{C}$ be a class containing N training motions. Then one has to compute $N(N-1)/2$ DTWs for each iteration step. Since the computational cost for DTW is quadratic in the length of the WMTs, the total running time $t(\mathcal{C})$ to compute the class MT basically depends on the size N and the average length $\bar{\omega}$ of the motion clips contained in $\mathcal{C}$ in a quadratic fashion as well as on the number M of iterations in a linear fashion:

$$t(\mathcal{C}) = O(N^2 \cdot \bar{\omega}^2 \cdot M). \tag{13.11}$$

This is also illustrated by Fig. 13.9b. For our experiments, we used a sampling rate of 30 Hz, which turned out to be sufficient in view of WMT quality. The duration of the training motion clips ranged from half a second (for short motions such as clapping ones hands or arm rotations) up to 10 s (for more complex motion sequences such as lying down on the floor), leading to WMT lengths between 15 and 300. The number of training motions used for each class ranged from 7 to 26. Using 3–7 iterations, it took on average 7.5 s to compute a class MT on a 3.6 GHz Pentium 4 with 1 GB of main memory, see

Table 13.1. For example, for the class "RotateRightArmForward3," the total computation time was $t(\mathcal{C}) = 3.9\,\mathrm{s}$ with $\bar{\omega} = 83.6$, $N = 8$, and $M = 4$, whereas for the class "CartwheelLeft," it took $t(\mathcal{C}) = 17.0\,\mathrm{s}$ with $\bar{\omega} = 105.3$, $N = 11$, and $M = 6$.

Note that, opposed to most statistical learning procedures, our iterative algorithm is completely deterministic and produces a unique class MT. If the training set contains a sufficient degree of variations, the derived class representation generalizes well to motions that are not contained in the training set (Sect. 14.3).

13.4.3 Examples

We now discuss some representative examples in detail, illustrating the descriptive power of our MT concept. More results can be found at our web site [144]. As first example, we consider the class "CartwheelLeft," which exhibits significant variations between different realizations. Figure 13.11a shows the class MT learned from 11 example motions, which form a superset of the motions shown in Fig. 13.4. Recall that black and white regions in a class MT correspond to consistent aspects of the training motions, while colored regions correspond to variable aspects. The following observations demonstrate that the essence of the cartwheel motion is captured by our class MT. Considering the regions marked by boxes in Fig. 13.10 (which only shows some selected rows of the class MT), the white region (a) reflects that during the initial phase of a cartwheel, the right hand moves to the top (F_5 in Table 11.1). Furthermore, region (b) shows that the right foot moves behind the left leg (F_{15}). This can also be observed in the first poses of Fig. 13.4. Then, both hands are above the shoulders (F_3, F_4), as indicated by region (c), and the actor's body is upside down (F_{33}, F_{34}), see region (d) and the second poses in

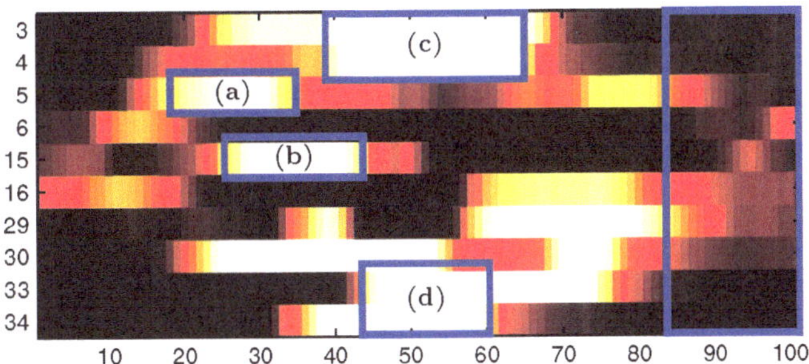

Fig. 13.10. Selected rows of the class MT for "CartwheelLeft" based on $N = 11$ training motions, where the row number correspond to the features listed in Table 11.1. The framed regions are discussed in the text. For an illustration of the complete class MT, we refer to Fig. 13.11a

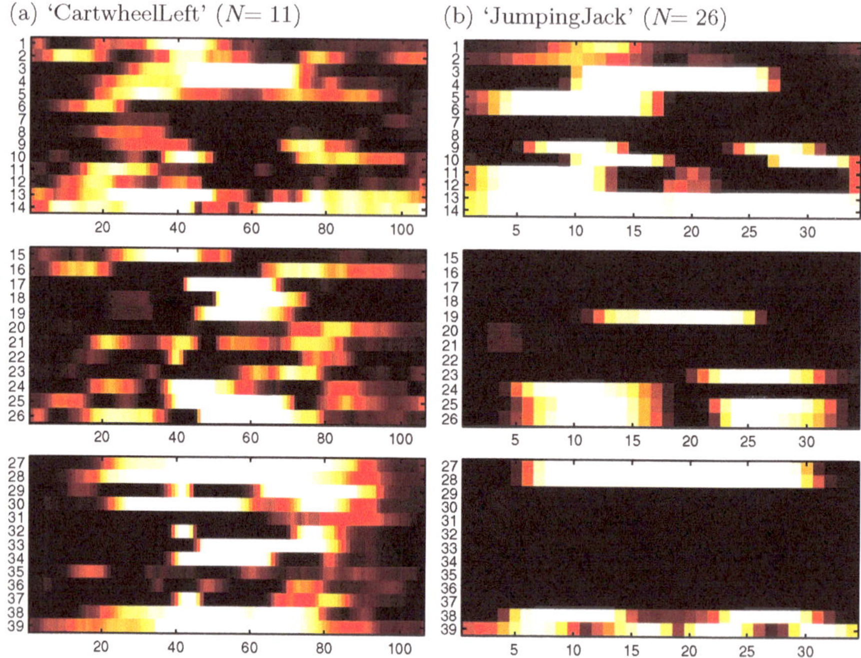

Fig. 13.11. Class MT for **(a)** the class "CartwheelLeft" based on $N = 11$ and **(b)** the class "JumpingJack" based on $N = 26$ training motions, respectively. For the sake of clarity, the rows corresponding to the 39 features are grouped according to the lower, upper, and mixed feature set, see Table 11.1

Fig. 13.4. The landing phase exhibits large variations between different realizations, which is reflected by numerous intermediary values within the region (e). For example, some actors lost their balance in this phase, resulting in rather chaotic movements, compare the third poses in Fig. 13.4.

Next, we discuss the relatively homogeneous class "JumpingJack," which contains motions of one jumping cycle starting and ending in the neutral standing pose. The homogeneity of the training motions is reflected by the absence of any major colored regions in the resulting class template (Fig. 13.11b), which was derived from 26 different training motions. For example, in all motions, the right and the left hand are first moved to the top (F_5 and F_6 assume value one), then remain raised (F_3 and F_4 assume value one), before they drop again at the end. The left and right elbows are not bent throughout the entire movement (F_7 and F_8 only assume the value zero). Furthermore, there are two phases where the hands are apart relative to the shoulders (F_9) and where the hands move together (F_{10}). Also very consistent is the movement of the lower part of the body. For example, the feet first move apart (F_{24}), then are apart (F_{19}), and finally move together (F_{23}). The two phases where both feet are in the air are reflected by the values of the velocity features F_{25} and F_{26}. The colored regions in the class MT are due to two different phenomena. First, as explained before, colored regions result from variations of certain body movements within the same motion class. For example, in "JumpingJack" some of the actors moved their hands slightly to the front before their hands touched above their head, resulting in a colored region in the rows corresponding to F_1 and F_2. Second, colored regions in the MT appear when a feature flips from one value to the other. Such transitions are hard to be matched consistently by the DTW procedure, in particular in the case that several features flip within a short period of time in a more or less random order.

The motion classes "WalkSidewaysRight3" and "WalkCrossoverRight3" are closely related, and it may depend on the respective application whether or not one considers them as semantically similar. The respective class MTs

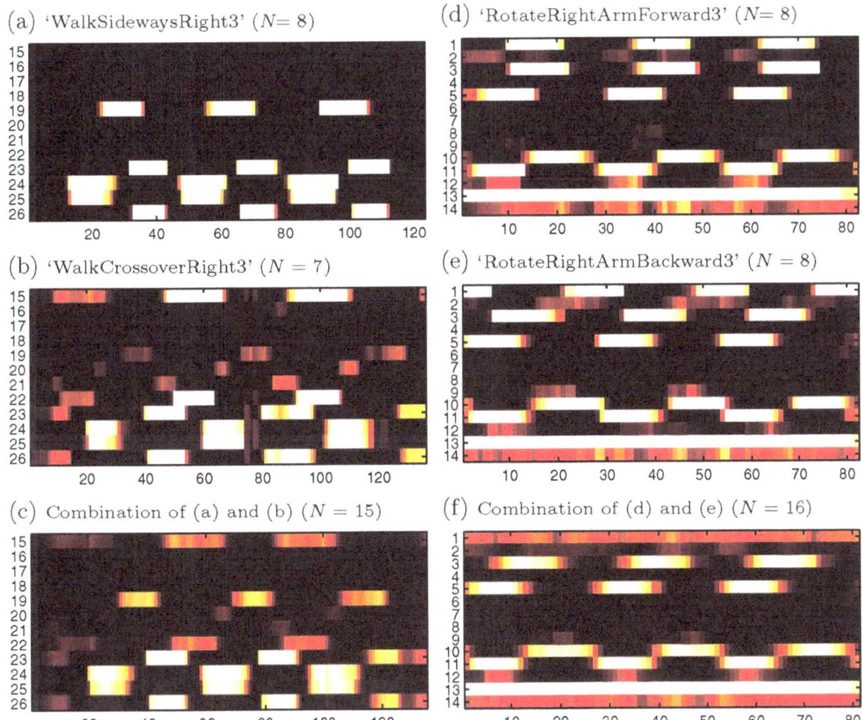

Fig. 13.12. *Left part:* class MTs (shown only for the lower feature set) for **(a)** "WalkSidewaysRight3" with $N = 8$, **(b)** "WalkCrossoverRight3" with $N = 7$, and **(c)** the combination of the two classes with $N = 15$ training motions. *Right part:* class MTs (shown only for the upper feature set) for **(d)** "RotateRightArm-Forward3," **(e)** "RotateRightArmBackward3," and **(f)** its combination

are shown for the lower feature set in Fig. 13.12a, b, respectively. For both classes, the movement starts with a sideways step to the right. Hence the right foot assumes a high velocity (F_{25}) and the feet move apart (F_{24}). In the second phase, the left foot moves with a high velocity (F_{26}) towards the right foot (F_{23}). However, for the first motion class, the feet are kept parallel throughout the movement, whereas for the second motion class, the left foot moves across in front of the right foot. This difference is reflected by features F_{15} (right foot behind left foot) and F_{22} (feet crossed). Here, either of the features is sufficient to distinguish the two motion classes. Finally, we used the training motions of both classes to derive a single, combined MT (Fig. 13.12c). Indeed, the resulting MT very well reflects the common characteristics as well as the disagreements of the two involved classes.

As a final example, consider the motion classes "RotateRightArmForward3" and "RotateRightArmBackward3." Even though the two corresponding class MTs – the parts of the upper feature set are shown in Fig. 13.12d, e, respectively – exhibit a similar zero-one distribution, there is one characteristic difference: in the forward movement, the right arm moves forwards (F_1) exactly when it is raised above the shoulder (F_3 is one). In contrast, in the backward movement the right arm moves forwards (F_1) exactly when it is below the shoulder (F_3 is zero). The combined WMT is shown in Fig. 13.12f.

13.5 Further Notes

In this chapter, we introduced a new concept for capturing the spatio-temporal characteristics of an entire motion class in one explicit, compact matrix representation. As one major advantage, motion templates have a direct, semantic interpretation; they can be easily be edited, extended, restricted, as well as mixed and combined with other MTs, thus providing a great deal of flexibility [143]. Furthermore, we described an iterative DTW-based warping and averaging algorithm to automatically learn a class MT from example motions. Here, one main problem was to simultaneously align a large number of motion sequences. Multiple alignments of $N > 2$ time series can be computed by a straightforward generalization of DTW with space and time complexities that are exponential in N. In fact, it has been shown that the multiple alignment problem is NP-complete [212]. Polynomial time approximation algorithms rely on a fixed reference time series [89]. For the multiple alignment of several motions in a morphing application, Kovar and Gleicher [106] also revert to a fixed reference motion. In contrast, our idea was to cancel out the influence of a reference motion by some iterative process. In all of our experiments, our procedure converged very quickly to a unique motion template. This template reflected the averaged temporal feature progression of the training motions without any noticeable bias. Also, changing parameters in the warping and unwarping stages (e.g., using different unwarping strategies as described in Sect. 13.2 and Sect. 13.3.2) did not have any significant impact on the resulting class MT demonstrating the robustness of our overall training procedure. As a theoretical issue, the convergence of our DTW-based warping and averaging algorithm for MT computation has so far not been proven. Also, it would be interesting to assess the suitability of this algorithm for other multiple alignment problems.

The concept of class motion templates, as we will describe in Chap. 14, can be used for robust and automatic annotation, classification, and retrieval of motion data. In this context, we will continue our discussion of related work and future research directions (Sect. 14.6).

14

MT-Based Motion Annotation and Retrieval

Summary. Given a class of semantically related motions, we have derived a class motion template that captures the consistent as well as the inconsistent aspects of all motions in the class. The application of MTs to automatic motion annotation and retrieval, which is the content of this chapter, is based on the following interpretation: the consistent aspects of a class MT represent the class characteristics that are shared by all motions, whereas the inconsistent aspects represent the class variations that are due to different realizations. The key idea in designing a distance measure for comparing a class MT with unknown motion data is to mask out the inconsistent aspects – a kind of class-dependent adaptive feature selection – so that related motions can be identified even in the presence of significant spatio-temporal variations. In Sect. 14.1, we define such a distance measure, which is based on a subsequence variant of DTW. Our concepts of MT-based annotation and retrieval are then described in Sect. 14.2 and Sect. 14.3, respectively, where we also report on our extensive experiments [143, 144]. To substantially speed up the annotation and retrieval process, we introduce an index-based (the index being independent of the class MTs) preprocessing step to cut down the set of candidate motions by using suitable keyframes (Sect. 14.4). In Sect. 14.5, we compare MT-based matching to several baseline methods (based on numerical features) as well as to adaptive fuzzy querying. Finally, related work and future research directions are discussed Sect. 14.6.

14.1 MT-Based Matching

To compare a class MT with the feature matrix resulting from an unknown motion data stream, we use a subsequence variant of DTW. The crucial point of our matching strategy is the local cost measure, which disregards the inconsistencies encoded in the class MT. To this end, we introduce a *quantized MT*, which has an entry 0.5 at all positions where the class MT indicates inconsistencies between different executions of a training motion within the same class. More precisely, let $\delta \in [0, 0.5)$, be a suitable threshold. Then for an MT $X \in [0,1]^{f \times K}$, we define the quantized MT by replacing each entry

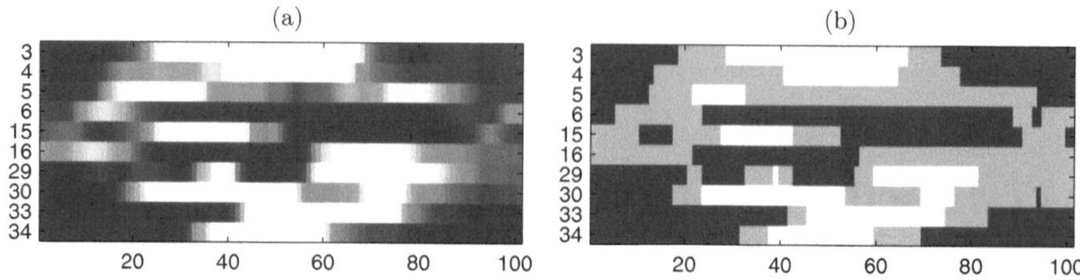

Fig. 14.1. (a) Class MT for "CartwheelLeft" from Fig. 13.10. **(b)** The corresponding quantized MT

of X that is below δ by zero, each entry that is above $1-\delta$ by one, and all remaining entries by 0.5. In our experiments, we used the threshold $\delta = 0.1$. Figure 14.1 shows an example of a quantized MT.

Now, let D be a motion data stream. The goal is to identify subsegments of D that are similar to a given motion class $\mathcal{C}$. Let X be a quantized class MT of length K and Y the feature matrix of D of length L. We define for $k \in [1:K]$ and $\ell \in [1:L]$ a local cost measure $c^{\mathrm{Q}}(k,\ell)$ between the vectors $X(k)$ and $Y(\ell)$. Let $I(k) := \{i \in [1:f] \mid X(k)_i \neq 0.5\}$, where $X(k)_i$ denotes a matrix entry of X. Then, if $|I(k)| > 0$, we set

$$c^{\mathrm{Q}}(k,\ell) := c^{\mathrm{Q}}(X(k),Y(\ell)) := \frac{1}{|I(k)|} \sum_{i \in I(k)} |X(k)_i - Y(\ell)_i|, \tag{14.1}$$

otherwise we set $c^{\mathrm{Q}}(k,\ell) = 0$. In other words, $c^{\mathrm{Q}}(k,\ell)$ only accounts for the consistent entries of X with $X(k)_i \in \{0,1\}$ and leaves the other entries unconsidered. Furthermore, to avoid degenerations in the DTW alignment, we use the modified step size condition $p_{m+1} - p_m \in \{(2,1),(1,2),(1,1)\}$, cf. Sect. 4.2.1. This forces the slope of the warping path to assume values between $\frac{1}{2}$ and 2. Then, the *distance function* $\Delta_{\mathcal{C}} : [1:L] \to \mathbb{R} \cup \{\infty\}$ is defined by

$$\Delta_{\mathcal{C}}(\ell) := \frac{1}{K} \min_{a \in [1:\ell]} \Big(\mathrm{DTW}\big(X\,,\,Y(a:\ell)\big)\Big), \tag{14.2}$$

where $Y(a:\ell)$ denotes the submatrix of Y consisting of columns a through $\ell \in [1:L]$. (Because of the modified step size condition, some of the DTW distances in (14.2) may not exist, which are then set to ∞.) Note that the function $\Delta_{\mathcal{C}}$ can be computed by a standard subsequence DTW (Sect. 4.4). Furthermore, one can derive from the resulting DTW matrix for each $\ell \in [1:L]$ the index $a_\ell \in [1:\ell]$ that minimizes (14.2). The interpretation of $\Delta_{\mathcal{C}}$ is as follows: a small value $\Delta_{\mathcal{C}}(\ell)$ for some $\ell \in [1:L]$ indicates that the motion subsegment of D starting at frame a_ℓ and ending at frame ℓ is similar to the motions of the class $\mathcal{C}$. Note that using the local cost function c^{Q} of (14.1) based on the quantized MT (instead of simply using the Manhattan distance c) is of crucial importance, as illustrated by Fig. 14.2.

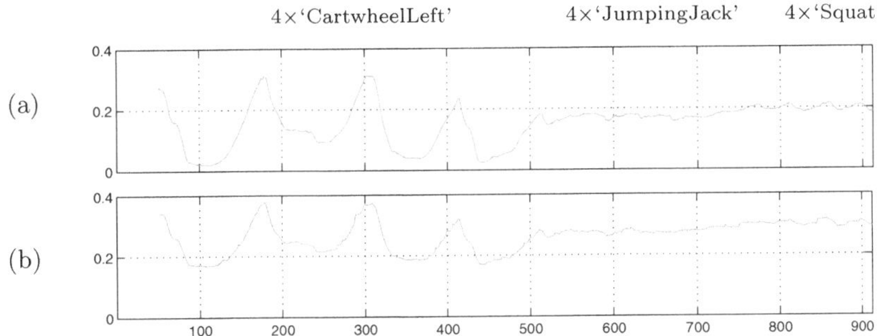

Fig. 14.2. **(a)** Distance function $\Delta_{\mathcal{C}}$ based on c^{Q} of (14.1) for the quantized class MT "CartwheelLeft" and a motion sequence D consisting of four cartwheel (reflected by the four local minima close to zero), four jumping jacks, and four squats. **(b)** Corresponding distance function based on the Manhattan distance without MT quantization, leading to a much poorer result

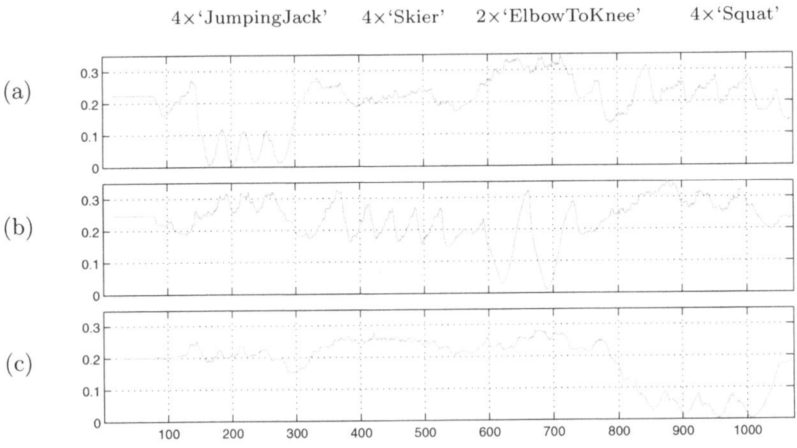

Fig. 14.3. Resulting distance functions for a 35-s gymnastics sequence consisting of four jumping jacks, four repetitions of a skiing coordination exercise, two repetitions of an alternating elbow-to-knee motion, and four squats with respect to the class MTs for **(a)** "JumpingJack," **(b)** "ElbowToKnee," and **(c)** "Squat"

14.2 MT-Based Annotation

In the *annotation* scenario, we are given an unknown motion data stream D for which the presence of certain motion classes $\mathcal{C}_1, \ldots, \mathcal{C}_P$ at certain times is to be detected. These motion classes are identified with their respective class MTs $X_1, \ldots, X_P$, which are assumed to have been precomputed from suitable training data. Now, the idea is to match the input motion D with each of the

X_p, $p = 1, \ldots, P$, yielding the distance functions $\Delta_p := \Delta_{\mathcal{C}_p}$. Then, every local minimum of Δ_p close to zero indicates a motion subsegment of D that is similar to the motions in $\mathcal{C}_p$. As an example, we consider the distance functions for a 35-s gymnastics motion sequence with respect to the motion classes $\mathcal{C}_1$ = "JumpingJack," $\mathcal{C}_2$ = "ElbowToKnee," and $\mathcal{C}_3$ = "Squat" (Fig. 14.3). For $\mathcal{C}_1$, there are four local minima between frames 100 and 300, which match the template with a cost of nearly zero and exactly correspond to the four jumping jacks contained in D, see Fig. 14.3a. Note that the remaining portion of D is clearly separated by Δ_1, yielding a value far above 0.1. Analogously, the two local minima in Fig. 14.3b and the four local minima in Fig. 14.3c correspond to the two repetitions of the elbow-to-knee exercise and the four squats, respectively. The choice of suitable quality thresholds for Δ_p as well as an evaluation of our experiments will be discussed in the next sections.

14.3 MT-Based Retrieval

The goal of content-based motion retrieval is to automatically extract all semantically similar motions of some specified type scattered in a motion capture database $\mathcal{D}$. By concatenating all documents of $\mathcal{D}$, we may assume that the database is represented by one single motion data stream D. To retrieve all motions represented by a class $\mathcal{C}$, we compute the distance function $\Delta_{\mathcal{C}}$ with respect to the precomputed class MT. Then, each local minimum of $\Delta_{\mathcal{C}}$ below some quality threshold $\tau > 0$ indicates a hit. To determine a suitable threshold τ and to measure the retrieval quality, we conducted extensive experiments based on several databases. We start with the evaluation database $\mathcal{D}^{\mathrm{MCE}}$, which consists of 718 motion clips corresponding to 24 min of motion data, see Sect. 13.4.1. Recall that $\mathcal{D}^{\mathrm{MCE}}$ is disjoint to the training database $\mathcal{D}^{\mathrm{MCT}}$, from which the class MTs were derived. Fixing a quality threshold τ, we computed a set H_τ of hits for each of the 64 class MTs in a fully automated batch mode. Based on a manually generated annotation of $\mathcal{D}^{\mathrm{MCE}}$ used as the ground truth, we then determined the subset $H_\tau^+ \subseteq H_\tau$ of relevant hits corresponding to motion clips of the respective class. At this point we want to emphasize that our retrieval strategy does not require any previous segmentation of the motion data to be searched through. The only reason for using the trimmed data of $\mathcal{D}^{\mathrm{MCE}}$ in this retrieval experiment is that having the ground truth for this data facilitates fully automated evaluation (Table 14.1). In the retrieval experiment to be discussed in Sect. 14.4 we use the uncut and unlabeled databases $\mathcal{D}_{210}$ and $\mathcal{D}_{180}^{\mathrm{CMU}}$, where the evaluation had to be done manually.

Table 14.1 shows some representative retrieval results for six different choices of τ. For example, for the motion class "ClapHandsAboveHead" and the quality threshold $\tau = 0.02$, all of the seven resulting hits are relevant – only one clapping motion is missing. Increasing the quality threshold to $\tau = 0.04$, one obtains 16 hits containing all of the eight relevant hits. However, one also

Table 14.1. Retrieval results for the evaluation database $\mathcal{D}^{\mathrm{MCE}}$ for various class MTs

Motion Class $\mathcal{C}$	N	$\lvert H_\tau \rvert$ / $\lvert H_\tau^+ \rvert$						K	$t(\Delta_\mathcal{C})$
CartwheelLeft	10	1	4	6	8	9	10	106	12.97
		1	4	6	8	9	10		
ClapHandsAboveHead	8	5	7	16	39	61	81	25	4.14
		5	7	8	8	8	8		
DepositFloorR	16	4	7	15	31	48	70	73	9.59
		0	1	4	12	13	15		
DepositHighR	14	11	21	47	60	103	197	67	11.39
		8	11	13	13	13	13		
DepositLowR	14	16	22	33	58	104	174	68	10.56
		8	10	12	13	13	13		
DepositMiddleR	14	21	52	122	180	236	295	74	14.61
		7	10	13	13	13	13		
ElbowToKnee	13	8	11	12	13	13	22	36	4.19
		8	11	12	13	13	13		
GrabFloorRHand	8	6	8	11	20	41	75	61	8.36
		5	7	7	8	8	8		
GrabHighRHand	14	15	22	49	58	115	201	68	11.39
		7	9	12	13	14	14		
GrabLowR	14	14	20	33	55	88	150	73	10.95
		8	10	13	14	14	14		
GrabMiddleR	14	20	71	150	227	323	408	59	14.77
		7	9	11	12	12	13		
HitRHandHead	6	29	70	152	257	343	432	55	16.83
		5	5	6	6	6	6		
HopBothLegs	18	13	19	22	32	126	334	24	6.56
		13	17	17	18	18	18		
HopLLeg	20	15	19	23	51	96	174	19	3.83
		15	19	20	20	20	20		
HopRLeg	21	17	19	22	35	66	107	18	3.08
		17	19	21	21	21	21		
JogLeftCircleRFootStart4	8	7	20	39	43	63	84	60	8.00
		5	7	8	8	8	8		
JogRightCircleRFootStart4	8	6	13	37	41	53	74	59	7.73
		4	8	8	8	8	8		
JumpDown	7	3	3	6	6	10	46	66	8.28
		3	3	6	6	7	7		
JumpingJack	26	25	26	26	26	33	40	34	4.11
		25	26	26	26	26	26		
KickFrontLFoot	14	4	8	23	126	328	465	54	13.64
		4	5	11	14	14	14		
KickFrontRFoot	15	3	6	26	90	239	385	54	13.70
		3	5	12	12	13	14		
KickSideLFoot	13	2	6	16	26	87	299	55	11.61
		1	3	8	12	13	13		
KickSideRFoot	15	6	13	27	48	163	359	51	12.88
		5	9	14	14	15	15		
LieDownFloor	10	4	6	8	8	9	11	172	28.05
		4	6	8	8	9	9		
PunchFrontLHand	15	1	5	22	42	173	414	48	12.77
		1	4	11	13	14	15		
PunchFrontRHand	15	4	6	27	92	289	442	54	13.70
		4	5	8	10	12	12		
PunchSideLHand	15	7	13	24	48	229	500	41	13.05
		5	8	11	13	14	15		
PunchSideRHand	14	2	9	40	104	355	500	41	12.69
		2	8	10	14	14	14		
RotateBothArmsBwd1	8	7	7	8	8	11	30	29	3.47
		7	7	8	8	8	8		
RotateBothArmsFwd1	8	8	8	8	8	15	47	30	3.69
		8	8	8	8	8	8		
RotateLArmBwd1	8	5	6	10	14	51	157	29	5.34
		5	6	8	8	8	8		
RotateLArmFwd1	8	5	6	9	21	55	176	28	5.67
		5	6	8	8	8	8		
RotateRArmBwd1	8	6	6	7	34	70	151	27	5.16
		6	6	7	8	8	8		
RotateRArmFwd1	8	6	6	7	39	77	186	28	6.13
		6	6	7	8	8	8		
RotateRArm(Bwd&Fwd)1	16	12	12	39	101	235	453	26	9.95
		12	12	13	15	16	16		
Shuffle2StepsLStart	6	11	20	57	122	255	383	48	13.22
		5	6	6	6	6	6		

Motion Class $\mathcal{C}$	N	$\lvert H_\tau \rvert$ / $\lvert H_\tau^+ \rvert$	K	$t(\Delta_\mathcal{C})$
Shuffle2StepsRStart	6	15 25 67 152 228 295 5 5 6 6 6 6	62	13.66
SitDownChair	10	4 9 17 29 53 70 4 8 10 10 10 10	83	12.92
SitDownFloor	10	9 15 25 34 48 61 4 6 9 10 10 10	106	15.78
SitDownKneelTieShoes	8	5 7 8 9 13 20 5 7 8 8 8 8	166	22.16
SitDownTable	10	19 31 80 125 184 260 9 10 10 10 10 10	69	15.22
Skier	15	12 13 15 16 25 56 12 13 15 15 15 15	36	4.80
Squat	26	23 24 26 26 26 27 23 24 26 26 26 26	48	5.69
StaircaseUp3	14	9 11 18 32 59 143 9 11 13 13 14 14	77	11.64
StandUpKneelToStand	8	4 6 15 26 53 89 4 5 7 7 7 7	49	6.86
StandUpLieFloor	10	3 7 10 14 18 23 3 7 8 8 10 10	129	16.47
StandUpSitChair	10	7 15 18 32 50 74 7 10 10 10 10 10	70	9.33
StandUpSitFloor	10	9 10 18 25 39 58 6 6 8 10 10 10	94	12.44
StandUpSitTable	10	29 53 104 191 266 345 10 10 10 10 10 10	59	15.50
ThrowBasketball	7	3 5 5 16 38 68 3 5 5 7 7 7	94	12.33
ThrowFarR	7	3 8 11 20 62 92 3 7 7 7 7 7	128	17.61
ThrowSittingHighR	7	3 7 11 14 19 29 3 6 7 7 7 7	72	9.19
ThrowSittingLowR	7	3 7 11 14 31 53 3 7 7 7 7 7	66	9.27
ThrowStandingHighR	7	4 5 22 48 129 210 4 5 6 7 7 7	78	13.39
ThrowStandingLowR	7	5 7 18 128 247 340 5 6 7 7 7 7	81	15.59
TurnLeft	15	37 72 195 310 397 489 12 15 15 15 15 15	49	16.50
TurnRight	15	39 81 174 261 344 427 11 13 14 15 15 15	54	16.38
WalkFwdRFootStart4	8	17 21 25 44 69 131 7 7 8 8 8 8	82	11.42
WalkBwdRFootStart4	7	6 7 7 24 72 101 6 7 7 7 7 7	97	13.41
WalkCrossoverRight3	6	6 7 14 24 32 56 6 6 6 6 6 6	136	17.87
WalkSidewaysRight3	8	7 8 11 14 28 49 7 8 8 8 8 8	123	16.08
WalkLeftCircle4	9	13 21 24 41 73 118 7 8 9 9 9 9	83	11.61
WalkRightCircle4	7	14 18 22 41 67 129 5 6 7 7 7 7	84	11.78
Walk(Crossover & Sideways)Right3	14	13 14 18 29 49 69 13 14 14 14 14 14	135	17.95

Note that $\mathcal{D}^{\mathrm{MCE}}$ is disjoint to the training database $\mathcal{D}^{\mathrm{MCT}}$, from which the class MTs were derived. N denotes the number of relevant motions contained in $\mathcal{D}^{\mathrm{MCE}}$. $|H_\tau|$ (first rows) denotes the number of hits and $|H_\tau^+|$ (second rows) the number of relevant hits with respect to the quality thresholds $\tau = 0.01, 0.02, 0.04, 0.06, 0.08,$ and 0.1. Finally, K denotes the length of the class MT and $t(\Delta_\mathcal{C})$ the running time in seconds required to compute the respective distance function $\Delta_\mathcal{C}$.

obtains eight false positives, mainly coming from the jumping jack class, which contains a similar arm movement. Generally, the precision and recall values are very good for whole-body motions such as "JumpingJack," "Cartwheel," or "LieDownFloor" – even in the presence of large variations within a class. Short motions with few characteristic aspects such as the class "GrabHighRHand" are more problematic. For $\tau = 0.04$, one obtains 49 hits containing 12 of the 14 relevant movements. Confusion arises mainly with similar classes such as "DepositHighRHand" or "GrabMiddleRHand" and with subsegments of more complex motions containing a grabbing-like component such as the beginning of a cartwheel. Even from a semantical point of view, it is hard to distinguish such motions. Similar confusion arises with increasing values of τ for the kicking, walking/jogging, rotation, or sitting classes. However, most of the relevant hits could be found among the top ranked hits in all cases.

For the classes "RotateRArmFwd1" and "RotateRArmBwd1" all relevant movements could be correctly identified. Using a combined MT (cf. Fig. 13.12f) the two classes could not be distinguished any longer – the characteristics that had separated the two classes were now regarded as mere variations and therefore masked out in the retrieval process.

The above experiments imply that the quality threshold $\tau = 0.06$ constitutes a good trade-off between precision and recall. Since the distance function Δ_C yields a ranking of the retrieved hits in a natural way, our strategy is to admit some false positives rather than to have too many false negatives. Furthermore, note that the running time of MT-based retrieval depends linearly on the size of the database, where the bottleneck is the computation of the distance function Δ_C. For example, in case of the 24-min database $\mathcal{D}^{\mathrm{MCE}}$ it took 4–28 s to process one query – depending on the respective MT length and the number of hits, see Table 14.1.

14.4 Keyframe-Based Preprocessing

To speed up the retrieval on large databases, we introduce a keyframe-based preselection step, as described in Sect. 12.3.2, in order to cut down the set of candidate motions prior to computing Δ_C. To this end, we label a small number of characteristic columns of each class MT as *keyframes*. The underlying assumption is that most instances of a motion belonging to the respective motion class will exhibit the feature vectors pertaining to the keyframes in the same order as specified by the keyframe positions. In our experiments, the keyframes were picked automatically using a simple heuristic: we basically chose two to five columns from the quantized MT that had many "white" entries (i.e., entries close to one, indicating some consistent action) and few "gray" entries (i.e., entries indicating inconsistencies).

Let $X \in \{0, 0.5, 1\}^{f \times K}$ denote the quantized class MT and assume that the keyframes are given by $v_j := (X(k_j)) \in \{0, 0.5, 1\}^f$, $j \in [1:J]$, for indices $1 \leq k_1 < k_2 < \ldots < k_J \leq K$. Each keyframe vector v_j induces a fuzzy set V_j via

$$V_j := \{v \in \{0,1\}^f \mid c^Q(v_j, v) = 0\} \tag{14.3}$$

with the cost measure c^Q as defined in (14.1). Intuitively, V_j contains all vectors v that coincide with v_j at positions corresponding to "black and white" entries while ignoring the "gray" entries. The resulting sequence $(V_1, V_2, \ldots, V_J)$ is then used as input for the keyframe-search algorithm of Sect. 12.3.2. Here, the admissibility condition has to be ensured in the automatic keyframe selection described earlier. Furthermore, we set the time bound parameter to $\Theta := 2(k_J - k_1 + 1)$, which accounts for motions that are slower than the template by up to a factor of two.

As suggested in Sect. 12.3.2, we further process the keyframe hits by conjoining overlapping matching intervals. Furthermore, to account for possible global differences in motion speed between the class MT and the unknown motion, we extend the resulting intervals by a certain number of frames K_ℓ and K_r to the left and to the right, respectively. In our experiments, we chose $K_\ell := 3k_1$ and $K_r := 3(K - k_J + 1)$.

Once a set of suitable motion segments has been precomputed in this way, we compute the distance function Δ_C only on those motion segments. This strategy has been applied to our (unlabeled) 210-min database $\mathcal{D}_{210}$, which was introduced in Sect. 13.4.1. Some retrieval results as well as running times are summarized in Table 14.2a. To assess retrieval quality, we manually inspected the set $H_{0.06}$ of hits as well as the database $\mathcal{D}_{210}$ for each class to determine the set $H^+_{0.06}$ of relevant hits. For example, the database $\mathcal{D}_{210}$ contains 24 left cartwheels. Using two automatically determined keyframes, it took 20 ms to reduce the data from 210 to 2.8 min – 1.3% of the original data. Then, MT retrieval was performed on the preselected 2.8 min of motion, which resulted in 21 hits and took 0.83 s. These hits contained 20 of the 24 cartwheels.

Even though keyframes are a powerful tool to significantly cut down the search space, there is also an attached risk: one single inappropriate keyframe may suffice to produce a large number of false negatives. For example, this happened for the classes "RotateRArmFwd1" or "SitDownChair." Here, using more appropriate manually selected keyframes significantly improved the retrieval result, compare (a) and (b) of Table 14.2. For example, in case of "SitDownChair" the procedure based on automatic keyframe selection resulted in only $|H^+_{0.06}| = 4$ (out of 20 possible) correct hits, whereas manual keyframe selection produced 16 correct hits. A further benefit of the keyframe approach is that the large number of false positives, as typical for short and unspecific motions, may be cut down by adding a single keyframe. For example, using two automatically selected keyframes the number of false positives amounted to $|H_{0.06}| - |H^+_{0.06}| = 128 - 30 = 98$ for the motion class "GrabHighRHand" (Table 14.2a). By using one more manually selected keyframe, the number of false positives could be reduced to $|H_{0.06}| - |H^+_{0.06}| = 59 - 30 = 29$, while still obtaining all of the 30 correct hits (Table 14.2b). For future work, we plan

Table 14.2. Retrieval results for **(a)** the database $\mathcal{D}_{210}$ and $\tau = 0.06$ based on automatic keyframe selection and **(b)** manually selected keyframes

(a)

Motion Class $\mathcal{C}$	#(kf)	sel (m)	sel (%)	t(kf)	N	$\|H_{0.06}\|$	$\|H_{0.06}^{+}\|$	$t(\Delta_\mathcal{C})$
CartwheelLeft	2	2.8	1.3	0.02	24	21	20	0.83
ElbowToKnee	2	0.8	0.4	0.03	29	16	16	0.13
GrabHighRHand	2	8.9	4.2	0.14	30	128	30	2.77
HopRightLeg	2	1.7	0.8	0.25	75	81	64	0.27
JumpingJack	2	1.5	0.7	0.09	52	50	50	0.19
KickFrontRightFoot	2	3.2	1.5	0.03	38	73	36	0.58
KickSideRightFoot	2	2.2	1.0	0.11	38	46	38	0.34
LieDownFloor	2	15.3	7.2	0.06	20	24	16	4.42
RotateRArmFwd1	2	0.5	0.2	0.48	66	6	5	0.17
SitDownChair	2	16.2	7.6	0.11	20	27	4	3.00
Squat	2	2.2	1.1	0.08	56	55	55	0.33
WalkCrossoverRight3	2	12.9	6.1	0.03	16	137	16	4.33

(b)

Motion Class $\mathcal{C}$	#(kf)	sel (m)	sel (%)	t(kf)	N	$\|H_{0.06}\|$	$\|H_{0.06}^{+}\|$	$t(\Delta_\mathcal{C})$
GrabHighRHand	3	3.2	1.5	0.16	30	59	30	1.08
LieDownFloor	3	6.5	3.1	2.75	20	19	19	2.86
RotateRArmFwd1	3	1.0	0.5	0.33	66	32	32	0.63
SitDownChair	3	3.8	1.8	0.17	20	34	16	1.28
WalkCrossoverRight3	5	4.9	2.3	0.14	16	66	16	2.06

The second to fourth columns indicate the number of keyframes, the size of the preselected dataset in minutes and percent as well as the running time for the preprocessing step. N is the number of relevant motions in $\mathcal{D}_{210}$. $|H_{0.06}|$ and $|H_{0.06}^{+}|$ denote the number of hits and the number of relevant hits, respectively. $t(\Delta_\mathcal{C})$ indicates the running time in seconds required to compute $\Delta_\mathcal{C}$ on the preselected motions.

to improve our ad-hoc method of keyframe selection. To this end, we have conducted first experiments to automatically learn characteristic keyframes from positive and negative motion examples employing a strategy based on genetic algorithms. It would also be interesting to incorporate the keyframe selection methods described by Assa et al. [9].

In a further experiment, we used the 180-min database $\mathcal{D}_{180}^{\mathrm{CMU}}$ containing uncut motion capture material from the CMU database [44]. Similar results and problems can be reported as for $\mathcal{D}_{210}$. Interestingly, our class MT X for "CartwheelLeft" (Fig. 14.4a) yielded no hits at all – as it turned out, all cartwheels in $\mathcal{D}_{180}^{\mathrm{CMU}}$ are right cartwheels. We modified X by simply interchanging the rows corresponding to feature pairs pertaining to the right/left part of the body, see Table 11.1. Using the resulting *mirrored* MT (Fig. 14.4b) four out of the known five cartwheels in $\mathcal{D}_{180}^{\mathrm{CMU}}$ appeared as the only hits. As a second interesting result, no relevant hits were reported as top hits for our "Squat" MT. A manual inspection showed that for all squatting motions of $\mathcal{D}_{180}^{\mathrm{CMU}}$, the arms are kept close to the body, whereas in the training motions

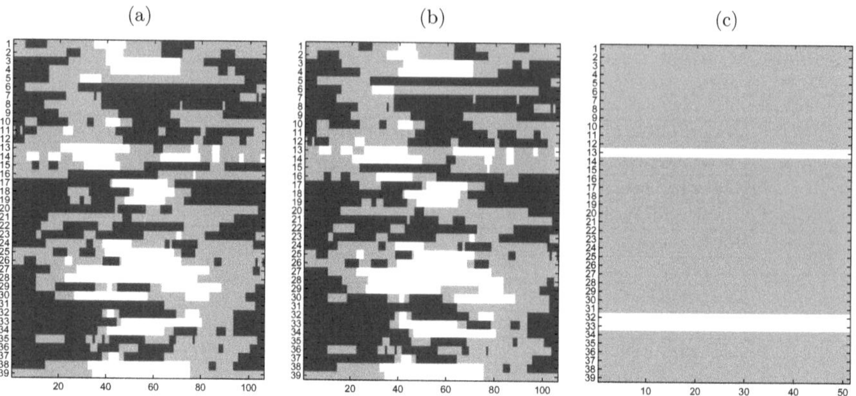

Fig. 14.4. (**a**) Quantized class MT for "CartwheelLeft." (**b**) Mirrored MT of (a). (**c**) Manually designed MT for identifying the motion "sweeping with right hand"

the arms are consistently stretched out in front of the body. Manually modifying the class MT by masking out the features for the arm motions, we were able to retrieve most squatting motions from $\mathcal{D}_{180}^{\mathrm{CMU}}$. Obviously, the class MTs are invariant only under those variations that are reflected by the training motions. However, because of their semantic meaning, class MTs can easily be modified in an intuitive way without any additional training data. Even designing a class MT from scratch (without resorting to any training motions) proved to be feasible. For example, to identify "sweeping with right hand" in $\mathcal{D}_{180}^{\mathrm{CMU}}$, we defined an MT of length 50, setting all matrix entries to 0.5 except for the rows corresponding to F_{13} (right hand fast), F_{32} (spine horizontal), and F_{33} (right hand lowered), which were set to one (Fig. 14.4c). Eight out of ten hits in $\mathcal{D}_{180}^{\mathrm{CMU}}$ were relevant.

14.5 Comparison to Other Retrieval Methods

We compared our MT-based retrieval system to several baseline methods using subsequence DTW on raw motion capture data with suitable local distance measures. It turned out that such baseline methods show little or no generalization capability. The database (3.8 min, or 6 750 frames sampled at 30 Hz) consisted of 100 motion clips: ten different realizations for each of the ten different motion classes. For each of the ten motion classes, we performed motion retrieval in four different ways:

1. Retrieval using a quantized class MT (MT)
2. DTW using the relational feature matrix of a single example motion and Manhattan distance (RF)
3. DTW using unit quaternions and spherical geodesic distance (Q)

Table 14.3. Recall values (r) in the top 5/10/20 ranked hits and separation quotients (s) for different DTW-based retrieval methods: motion templates (MT), relational feature matrices (RF), quaternions (Q), and relative 3D coordinates (3D)

Motion Class	$r^{\mathrm{MT}}_{5/10/20}$	s^{MT}	$r^{\mathrm{RF}}_{5/10/20}$	s^{RF}	$r^{\mathrm{Q}}_{5/10/20}$	s^{Q}	$r^{\mathrm{3D}}_{5/10/20}$	s^{3D}
CartwheelLeft	5/10/10	12.83	5/10/10	1.62	4/6/7	1.63	1/1/2	2.38
Squat	5/10/10	259.5	5/10/10	16.1	5/7/9	2.79	4/6/7	2.52
LieDownFloor	5/9/10	11.65	5/9/10	2.10	4/7/9	1.69	2/3/7	1.29
SitDownFloor	4/6/10	19.33	3/4/8	1.60	2/5/7	2.13	3/5/8	1.56
GrabHighRHand	5/7/9	33.93	5/8/8	9.72	3/5/8	3.39	1/3/4	2.22

4. DTW using 3D joint coordinates (normalized w.r.t. root translation and rotation as well as skeleton size) and Euclidean distance (3D)

For each strategy, we computed a Δ curve as in Fig. 14.2 and derived the top 5, top 10, and top 20 hits. Table 14.3 shows the resulting recall values (note that there are exactly 10 correct hits for each class) for five representative queries. As a further important quality measure of a strategy, we computed the *separation quotient*, denoted by s, which is defined as the median of Δ divided by the median of the cost of the correct hits among the top 10 hits. The larger the value of s, the better the correct hits are separated from the false positives, enabling the usage of simple thresholding strategies on Δ for the retrieval. Only for our MT-based strategy, the separation is good enough. These observations indicate that MT-based retrieval outperforms the other methods. More results can be found at our web site [144].

We also compared MT-based retrieval to adaptive fuzzy querying as introduced in Chap. 12. Recall that the performance of adaptive fuzzy querying heavily depends on the query formulation, which involves manual specification of a query-dependent feature selection. For each query, we carefully selected a suitable subset of features. The resulting precision/recall values on $\mathcal{D}^{\mathrm{MC}}$ are very good and reflect what seems to be achievable by adaptive fuzzy querying, see Table 14.4. For MT-based retrieval, we quote precision/recall values for two quality thresholds, $\tau = 0.02$ and $\tau = 0.06$. Our experiments show that the retrieval quality of our fully automatic MT-based approach is in most cases as good and in some cases even better than that obtained by adaptive fuzzy querying, even after hand-tweaking the feature selection. Hence, our MT-based approach enables us to replace manual, global feature selection by fully automatic, local feature selection without loss of retrieval quality.

14.6 Further Notes

In this chapter, we have applied class motion templates to motion annotation and retrieval. Intuitively, we shifted a specific class MT over the feature matrix

Table 14.4. Comparison of hand-tuned fuzzy queries and MT-based queries on a subset of $\mathcal{D}^{\mathrm{MC}}$

	Elbow-to-knee 	Cartwheel 	Jumping jack 	Hop both legs 	Hop right leg 
Adaptive fuzzy					
Recall	24/27 = 0.89	21/21 = 1.00	51/52 = 0.98	21/36 = 0.58	33/42 = 0.79
Precision	24/26 = 0.92	21/21 = 1.00	51/55 = 0.93	21/34 = 0.62	33/182 = 0.18
Ranking top 5 | 10 | 20	5 | 10 | 20	5 | 10 | 20	5 | 10 | 20	4 | 8 | 16	5 | 10 | 20
MT-based retrieval					
Recall ($\tau = 0.02$)	23/27 = 0.85	12/21 = 0.57	51/52 = 0.98	34/36 = 0.94	40/42 = 0.95
Precision ($\tau = 0.02$)	23/23 = 1.00	12/12 = 1.00	51/51 = 1.00	34/38 = 0.89	40/40 = 1.00
Recall ($\tau = 0.06$)	27/27 = 1.00	19/21 = 0.90	52/52 = 1.00	36/36 = 1.00	42/42 = 1.00
Precision ($\tau = 0.06$)	27/27 = 1.00	19/19 = 1.00	52/52 = 1.00	36/69 = 0.52	42/73 = 0.57
	Hit on head 	Kick right 	Sit down 	Lie down 	Rotate both arms 
Adaptive fuzzy					
Recall	12/13 = 0.92	25/30 = 0.83	16/20 = 0.80	19/20 = 0.95	16/16 = 1.00
Precision	12/14 = 0.86	25/41 = 0.61	16/40 = 0.40	19/21 = 0.90	16/16 = 1.00
Ranking top 5 | 10 | 20	5 | 9 | 12	5 | 8 | 15	5 | 7 | 10	5 | 10 | 18	5 | 10 | 16
MT-based retrieval					
Recall ($\tau = 0.02$)	12/13 = 0.92	14/30 = 0.47	14/20 = 0.70	16/20 = 0.80	16/16 = 1.00
Precision ($\tau = 0.02$)	12/158 = 0.08	14/18 = 0.77	14/14 = 1.00	16/19 = 0.84	16/16 = 1.00
Recall ($\tau = 0.06$)	13/13 = 1.00	26/30 = 0.87	17/20 = 0.85	19/20 = 0.95	16/16 = 1.00
Precision ($\tau = 0.06$)	13/500 = 0.03	26/196 = 0.13	17/21 = 0.81	19/64 = 0.30	16/16 = 1.00
	Walk up staircase 	Walk 	Walk backwards 	Walk sideways 	Walk cross over 
Adaptive fuzzy					
Recall	22/28 = 0.79	15/16 = 0.94	8/15 = 0.53	16/16 = 1.00	10/13 = 0.77
Precision	22/23 = 0.96	15/33 = 0.45	8/22 = 0.36	16/17 = 0.94	10/13 = 0.77
Ranking top 5 | 10 | 20	5 | 10 | 20	2 | 6 | 12	4 | 8 | 8	5 | 10 | 16	5 | 10 | 10
MT-based retrieval					
Recall ($\tau = 0.02$)	22/28 = 0.76	15/16 = 0.94	15/15 = 1.00	16/16 = 1.00	13/13 = 1.00
Precision ($\tau = 0.02$)	22/22 = 1.00	15/45 = 0.33	15/15 = 1.00	16/16 = 1.00	13/13 = 1.00
Recall ($\tau = 0.06$)	27/28 = 0.96	16/16 = 1.00	15/15 = 1.00	16/16 = 1.00	13/13 = 1.00
Precision ($\tau = 0.06$)	27/67 = 0.40	16/88 = 0.18	15/57 = 0.26	16/19 = 0.84	13/48 = 0.27

of the unknown motion data stream while locally comparing the template with data segments. In the comparison, we used a subsequence DTW variant based on a local cost measure that automatically masked out the variable aspects of the motion class. In the resulting distance function, each presence of the specific class within the data stream was indicated by a local minimum close to zero. This strategy allowed us to automatically identify related motions even in the presence of large variations and without any user intervention. Thus, MT-based retrieval is also applicable in a batch mode overcoming some of the limitations discussed in Sect. 12.4.1. Extensive experimental results show that our methods work with high precision and recall for whole-body motions and for longer motions of at least a second. More problematic are very short and unspecific motion fragments. Here, the use of suitably defined keyframes is a promising concept to not only speed up the retrieval process, but also to eliminate false positives.

Regarding a class motion template as a detector or a classifier for motion data, our MT-based approach to motion annotation can also be used for motion classification. However, our classification differs from other techniques in that we work with an explicit motion representation learned exclusively from *positive* training examples describing the respective motion class. In other words, each class MT solely reflects the characteristics derived from a single motion class and is independent of any information implied by other motion classes. This is in contrast to typical supervised learning techniques such as support vector machines (SVM) [60], where classifiers are trained from labeled examples from all classes to be considered in the classification problem. In other words, such a classifier exploits the information that there is a limited, known number of classes. Having this in mind, we review some of the recent work on motion classification, recognition, and annotation.

Arikan and Forsyth [6] propose a technique to synthesize realistic motion sequences from user-specified, textual action commands where automatically selected fragments of a motion database are blended together. In a preprocessing step, each frame of the motion database has been assigned to one or several predefined motion classes. This assignment is performed by a semi-automatic annotation procedure using support vector machine classifiers based on 3D trajectory data. Ramanan and Forsyth [171] also apply this annotation technique for 3D motion data as a preprocessing step for automatic annotation of 2D video recordings of human motion, using hidden Markov models (HMMs) to match the 2D data with the 3D data. Inspired by speech recognition with HMMs, Green and Guan [87] use progressions of kinematic features such as motion vectors of selected joints, so-called *dynemes*, for their video-based motion recognition system. Before a motion class can be recognized, the structure of the HMM for that motion class has to be manually assembled from the set of dynemes and then has to be trained with suitable training data, see also Sect. 9.2 for a discussion of their technique. Wilson and Bobick [216] apply HMMs for video-based gesture recognition. Their main contribution is a parametrization of the HMM's output probabilities that enables certain aspects of gestures to be controlled by continuous parameters. As an example, they quote the gesture of "showing off the size of a fish that one caught." Here, the distance of the hands during the performance of the gesture has an important meaning, which can be captured by their model. Opposed to such HMM-based motion representations, where timing information is encoded in the form of transition probabilities, MTs encode absolute and relative lengths of key events explicitly. The motion recognition system described by Campbell and Bobick [26] is based on a joint angle representation and has been applied to recognize the presence of nine different types of ballet moves from unsegmented motion data streams. Similar to MTs, their technique may also detect the presence of several motion classes at the same time. Bissacco et al. [15] work with state-space (ARMA) models of 3D joint angles derived from 2D video data to roughly distinguish three different types of gait ("walking," "running," "climbing stairs"). Temporal segmentation of motion data, i.e., the process of breaking up continuous mocap data streams into groups of consecutive, semantically related frames, is another important task. Pure segmentation techniques, as opposed to classification/recognition techniques, do not assign class labels or *meaning* to motion subclips. Typical examples are the methods by Fod et al. [67] and Barbic et al. [12].

There are many possible research directions to move forward. For future work, we plan to compare motion templates to HMM-based motion representations and to combine their respective benefits. Furthermore, it would be instructive to compare the classification performance of MTs to machine learning techniques such as SVMs. An interesting research problem is the automated extraction of characteristic keyframes based on our template representation, e.g., by employing genetic algorithms. In collaboration with the HDM school of media sciences (Stuttgart), we shall investigate how motion templates may be used as a tool for specifying animations. Here, similar to approaches such as [6,180], we want to find out if a template-based animation concept is superior to common keyframe-based concepts in view of automated motion synthesis applications. As a further promising application in the field of computer vision, we plan to use motion templates and related motion representations as a-priori knowledge to stabilize and control markerless tracking of human motions in video data [21,181].

References

1. N. Adams, D. Marquez, and G. H. Wakefield, *Iterative deepening for melody alignment and retrieval*, in Proc. ISMIR, London, GB, 2005.
2. R. Agrawal, C. Faloutsos, and A. N. Swami, *Efficient similarity search in sequence databases*, in FODO '93: Proceedings of the 4th International Conference on Foundations of Data Organization and Algorithms, London, UK, 1993, Springer-Verlag, pp. 69–84.
3. E. Allamanche, J. Herre, B. Fröba, and M. Cremer, *AudioID: Towards Content-Based Identification of Audio Material*, in Proc. 110th AES Convention, Amsterdam, NL, 2001.
4. V. Arifi, M. Clausen, F. Kurth, and M. Müller, *Score-to-PCM music synchronization based on extracted score parameters*, in Proc. International Symposium on Computer Music Modeling and Retrieval (CMMR), LNCS 3310, U. Wiil, ed., 2004, pp. 193–210.
5. ——, *Synchronization of music data in score-, MIDI- and PCM-format*, Computing in Musicology, 13 (2004).
6. O. Arikan, D. A. Forsyth, and J. F. O'Brien, *Motion synthesis from annotations*, ACM Trans. Graph., 22 (2003), pp. 402–408.
7. K. S. Arun, T. S. Huang, and S. D. Blostein, *Least-squares fitting of two 3-D point sets*, IEEE Trans. Pattern Anal. Mach. Intell., 9 (1987), pp. 698–700.
8. ASF/AMC, *Motion capture data format*. University of Wisconsin, `http://www.cs.wisc.edu/graphics/Courses/cs-838-1999/Jeff/ASF-AMC.html`, 1999.
9. J. Assa, Y. Caspi, and D. Cohen-Or, *Action synopsis: Pose selection and illustration*, ACM Trans. Graph., 24 (2005), pp. 667–676.
10. N. I. Badler, C. B. Phillips, and B. L. Webber, *Simulating Humans: Computer Graphics, Animation, and Control*, Oxford University Press, 1993.
11. E. M. Bakker, T. S. Huang, M. S. Lew, N. Sebe, and X. S. Zhou, eds., *Proc. 2nd Intl. Conf. Image and Video Retrieval, CIVR 2003, Urbana-Champaign, IL, USA*, vol. 2728 of LNCS, Springer, 2003.
12. J. Barbic, A. Safonova, J.-Y. Pan, C. Faloutsos, J. K. Hodgins, and N. S. Pollard, *Segmenting motion capture data into distinct behaviors*, in GI '04: Proc. Graphics interface, Canadian Human-Computer Communications Society, 2004, pp. 185–194.

13. M. A. Bartsch and G. H. Wakefield, *To catch a chorus: Using chroma-based representations for audio thumbnailing*, in Proc. IEEE WASPAA, New Paltz, NY, USA, 2001, pp. 15–18.
14. ——, *Audio thumbnailing of popular music using chroma-based representations*, IEEE Trans. on Multimedia, 7 (2005), pp. 96–104.
15. A. Bissacco, A. Chiuso, Y. Ma, and S. Soatto, *Recognition of human gaits*, in Proc. IEEE CVPR, IEEE, December 2001, pp. 401–417.
16. E. Blackham, *Die Physik der Musikinstrumente*, Spectrum, Akademischer Verlag, 2. Auflage, 1998.
17. M. Bobrek and D. B. Koch, *Music signal segmentation using tree-structured filter banks*, Journal of Audio Engineering Society, 46 (1998), pp. 412–427.
18. H.-J. Böckenhauer and D. Bongartz, *Algorithmische Grundlagen der Bioinformatik: Modelle, Methoden und Komplexität*, Teubner, 2003.
19. M. Brand and A. Hertzmann, *Style machines*, in Proc. ACM SIGGRAPH '00, Computer Graphics Proc., ACM Press, 2000, pp. 183–192.
20. C. Bregler, *Learning and recognizing human dynamics in video sequences*, in Proc. CVPR 1997, Washington, DC, USA, 1997, IEEE Computer Society, p. 568.
21. T. Brox, B. Rosenhahn, and D. Cremers, *Contours, optic flow, and prior knowledge: Cues for capturing 3D human motion in videos*, in Human Motion—Understanding, Modeling, Capture and Animation, B. Rosenhahn, R. Klette, and D. Metaxas, eds., Springer, Berlin, 2007. Accepted.
22. A. Bruderlin and L. Williams, *Motion signal processing*, in Proc. ACM SIGGRAPH 95, Computer Graphics Proc., ACM Press, 1995, pp. 97–104.
23. S. R. Buss and J. P. Fillmore, *Spherical averages and applications to spherical splines and interpolation*, ACM Trans. Graph., 20 (2001), pp. 95–126.
24. BVH, *Motion capture data format*. University of Wisconsin, `http://www.cs.wisc.edu/graphics/Courses/cs-838-1999/Jeff/BVH.html`, 1999.
25. D. Byrd and M. Schindele, *Prospects for improving OMR with multiple recognizers*, in Proc. ISMIR, Victoria, Canada, 2006, pp. 41–46.
26. L. W. Campbell and A. F. Bobick, *Recognition of human body motion using phase space constraints*, in ICCV, 1995, pp. 624–630.
27. C. Cannam, *Sonic Visualizer*. Centre for Digital Music, Queen Mary, University of London, October 2006. `http://www.sonicvisualiser.org`.
28. P. Cano, E. Battle, T. Kalker, and J. Haitsma, *A Review of Audio Fingerprinting*, in Proc. 5. IEEE MMSP, St. Thomas, Virgin Islands, USA, 2002.
29. M. Cardle, S. B. Vlachos, E. Keogh, and D. Gunopulos, *Fast motion capture matching with replicated motion editing*. ACM SIGGRAPH 2003, Sketches and Applications, 2003.
30. S. Carlsson, *Combinatorial geometry for shape representation and indexing*, in Object Representation in Computer Vision, 1996, pp. 53–78.
31. ——, *Order structure, correspondence, and shape based categories*, in Shape, Contour and Grouping in Computer Vision, Springer, 1999, pp. 58–71.
32. M. Casey and M. Slaney, *The importance of sequences in musical similarity*, in Proc. IEEE ICASSP, Toulouse, France, 2006.
33. ——, *Song intersection by approximate nearest neighbor search*, in Proc. ISMIR, Victoria, Canada, 2006, pp. 144–149.
34. A. T. Cemgil, B. Kappen, and P. Desain, *Rhythm quantization for transcription*, Comput. Music J., 24 (2000), pp. 60–76.

35. J. Chai and J. K. Hodgins, *Performance animation from low-dimensional control signals*, ACM Trans. Graph., 24 (2005), pp. 686–696.
36. W. Chai, *Structural analysis of music signals via pattern matching*, in Proc. IEEE ICASSP, Hong Kong, China, 2003.
37. ———, *Semantic segmentation and summarization of music: methods based on tonality and recurrent structure*, Signal Processing Magazine, IEEE, 23 (2006), pp. 124–132.
38. W. Chai and B. Vercoe, *Music thumbnailing via structural analysis*, in Proc. ACM Multimedia, 2003.
39. K. P. Chan and A. W.-C. Fu, *Efficient time series matching by wavelets*, in ICDE '99: Proceedings of the 15th International Conference on Data Engineering, Washington, DC, USA, 1999, IEEE Computer Society, pp. 126–133.
40. A. W. chee Fu, E. Keogh, L. Y. H. Lau, and C. A. Ratanamahatana, *Scaling and time warping in time series querying*, in VLDB '05: Proceedings of the 31st international conference on Very Large Data Bases, VLDB Endowment, 2005, pp. 649–660.
41. M. Clausen and U. Baum, *Fast Fourier Transforms*, BI Wissenschaftsverlag, 1993.
42. M. Clausen and F. Kurth, *A unified approach to content-based and fault tolerant music recognition*, IEEE Transactions on Multimedia, 6 (2004), pp. 717–731.
43. M. Clausen and M. Müller, *Zeit-Frequenz-Analyse und Wavelettransformationen*, Lecture Notes, Department of Computer Science III, University of Bonn, 2001.
44. CMU, *Carnegie-Mellon Mocap Database*. `http://mocap.cs.cmu.edu`, 2003.
45. J. W. Cooley and J. W. Tukey, *An algorithm for the machine calculation of complex Fourier series*, Mathematics of Computation, 19 (1965), pp. 297–301.
46. M. Cooper and J. Foote, *Automatic music summarization via similarity analysis*, in Proc. ISMIR, Paris, France, 2002.
47. T. H. Cormen, C. E. Leiserson, R. L. Rivest, and C. Stein, *Introduction to Algorithms*, McGraw-Hill Higher Education, 2001.
48. R. Dannenberg, *An on-line algorithm for real-time accompaniment*, in Proc. International Computer Music Conference (ICMC), 1984, pp. 193–198.
49. R. Dannenberg and N. Hu, *Pattern discovery techniques for music audio*, in Proc. ISMIR, Paris, France, 2002.
50. ———, *Polyphonic audio matching for score following and intelligent audio editors*, in Proc. ICMC, San Francisco, USA, 2003, pp. 27–34.
51. R. Dannenberg and H. Mukaino, *New techniques for enhanced quality of computer accompaniment*, in Proc. International Computer Music Conference (ICMC), 1988, pp. 243–249.
52. R. Dannenberg and C. Raphael, *Music score alignment and computer accompaniment*, Special Issue, Commun. ACM, 49 (2006), pp. 39–43.
53. J. W. Davis and H. Gao, *An expressive three-mode principal components model of human action style.*, Image Vision Comput., 21 (2003), pp. 1001–1016.
54. E. de Aguiar, C. Theobalt, and H.-P. Seidel, *Automatic learning of articulated skeletons from 3D marker trajectories*, in Proc. Intl. Symposium on Visual Computing (ISVC 2006), to appear, 2006.
55. J. R. Deller, J. G. Proakis, and J. H. L. Hansen, *Discrete-Time Processing of Speech Signals*, IEEE Computer Society Press, 1999.

56. B. Demuth, T. Röder, M. Müller, and B. Eberhardt, *An information retrieval system for motion capture data*, in Proc. 28th European Conference on IR Research (ECIR), vol. 3936 of LNCS, Springer, 2006, pp. 373–384.
57. S. Dixon and G. Widmer, *Match: A music alignment tool chest*, in Proc. ISMIR, London, GB, 2005.
58. J. S. Downie, *Music information retrieval*, Annual Review of Information Science and Technology (Chapter 7), 37 (2003), pp. 295–340.
59. J. S. Downie, K. West, E. Palmpalk, and P. Lamere. MIREX Audio Cover Song Identification Results, `http://www.music-ir.org/mirex2006/index.php/Audio_Cover_Song`, 2006.
60. R. O. Duda, P. E. Hart, and D. G. Stork, *Pattern Classification*, Wiley Interscience, second ed., 2000.
61. J. W. Dunn, D. Byrd, M. Notess, J. Riley, and R. Scherle, *Variations2: Retrieving and using music in an academic setting*, Special Issue, Commun. ACM, 49 (2006), pp. 53–48.
62. C. Faloutsos, M. Ranganathan, and Y. Manolopoulos, *Fast subsequence matching in time-series databases*, in Proceedings 1994 ACM SIGMOD Conference, Mineapolis, MN.
63. A. C. Fang and N. S. Pollard, *Efficient synthesis of physically valid human motion*, ACM Trans. Graph., 22 (2003), pp. 417–426.
64. O. Faugeras, *Three-Dimensional Computer Vision: A Geometric Viewpoint*, MIT Press, Cambridge, MA, 1993, ch. 9, pp. 341–400.
65. R. Featherstone, *Robot Dynamics Algorithms*, Kluwer, Boston, MA, 1987.
66. N. H. Fletcher and T. D. Rossing, *The Physics of Musical Instruments*, Springer-Verlag, 1991.
67. A. Fod, M. J. Mataric, and O. C. Jenkins, *Automated derivation of primitives for movement classification*, Auton. Robots, 12 (2002), pp. 39–54.
68. G. B. Folland, *Real Analysis*, John Wiley & Sons, 1984.
69. J. Foote, *Visualizing music and audio using self-similarity*, in ACM Multimedia, 1999, pp. 77–80.
70. K. Forbes and E. Fiume, *An efficient search algorithm for motion data using weighted PCA*, in Proc. 2005 ACM SIGGRAPH/Eurographics Symposium on Computer Animation, ACM Press, 2005, pp. 67–76.
71. freeDB.org, *A text-based search engine.* `http://www.freedb.org`, October 2006.
72. C. Fremerey, *SyncPlayer—a framework for content-based music navigation.* Diplomarbeit, Department of Computer Science III, University of Bonn, 2006.
73. C. Fremerey, F. Kurth, M. Müller, and M. Clausen, *A demonstration of the SyncPlayer system*, in Proc. ISMIR, Vienna, Austria, 2007.
74. T. Funkhouser, P. Min, M. Kazhdan, J. Chen, A. Halderman, D. Dobkin, and D. Jacobs, *A search engine for 3D models*, ACM Trans. Graph., 22 (2003), pp. 83–105.
75. M. Giese and T. Poggio, *Morphable models for the analysis and synthesis of complex motion patterns*, IJCV, 38 (2000), pp. 59–73.
76. M. Gleicher, *Operational metric for motion retrieval.* Personal Communication, 2006.
77. R. Göcke, *Building a system for writer identification on handwritten music scores*, in Proc. International Conference on Signal Processing, Pattern Recognition, and Applications, 2003, pp. 250–255.

78. J. Gomes, B. Costa, L. Darsa, and L. Velho, *Warping and Morphing of Graphical Objects*, SIGGRAPH'95 Course Notes #3, ACM SIGGRAPH publication, Los Angeles, 1995.
79. E. Gómez and P. Herrora, *The Song Remains the Same: Identifying Versions of the Same Piece Using Tonal Descriptors*, in Proc. ISMIR, Victoria, Canada, 2006, pp. 180–185.
80. M. Good, *MusicXML: An internet-friendly format for sheet music*, in Proc. XML Conference and Exposition, 2001.
81. M. Goto, *A chorus-section detecting method for musical audio signals*, in Proc. IEEE ICASSP, Hong Kong, China, 2003, pp. 437–440.
82. ———, *SmartMusicKIOSK: Music Listening Station with Chorus-Search Function*, in Proc. ACM UIST, 2003, pp. 31–40.
83. ———, *A real-time music-scene-description system: predominant-F0 estimation for detecting melody and bass lines in real-world audio signals*, Speech Communication, 43 (2004), pp. 311–329.
84. ———, *SmartMusicKIOSK: Music Listening Station with Chorus-Search Function.* Website, June 2004. `http://staff.aist.go.jp/m.goto/SmartMusicKIOSK`.
85. Gracenote, *Gracenote Music Search.* Website, October 2006. `http://www.gracenote.com/music`.
86. F. S. Grassia, *Practical parameterization of rotations using the exponential map*, Journal on Graphics Tools, 3 (1998), pp. 29–48.
87. R. Green and L. Guan, *Quantifying and recognizing human movement patterns from monocular video images: Part I*, IEEE Transactions on Circuits and Systems for Video Technology, 14 (2004), pp. 179–190.
88. A. H. Guest, *Labanotation: the system of analyzing and recording movement*, Routledge, New York, 2004.
89. D. Gusfield, *Efficient methods for multiple sequence alignment with guaranteed error bounds*, Bulletin of Mathematical Biology, 55 (1993), pp. 141–154.
90. A. Guttman, *R-trees: a dynamic index structure for spatial searching*, in SIGMOD '84: Proceedings of the 1984 ACM SIGMOD international conference on Management of Data, New York, NY, USA, 1984, ACM Press, pp. 47–57.
91. K. Hachimura and M. Nakamura, *Method of generating coded description of human body motion from motion-captured data*, in Proc. 10th IEEE Intl. Workshop on Robot and Human Interactive Communication, 2001.
92. J. Haitsma and T. Kalker, *A highly robust audio fingerprinting system*, in Proc. ISMIR, Paris, France, 2002.
93. E. Hsu, K. Pulli, and J. Popović, *Style translation for human motion*, ACM Trans. Graph., 24 (2005), pp. 1082–1089.
94. N. Hu, R. Dannenberg, and G. Tzanetakis, *Polyphonic audio matching and alignment for music retrieval*, in Proc. IEEE WASPAA, New Paltz, NY, October 2003.
95. B. B. Hubbard, *The world according to wavelets*, AK Peters, Wellesley, Massachusetts, 1996.
96. D. M. Huber, *The MIDI manual*, Focal Press, 1999.
97. ISMIR, *Hompage of the international conferences on music information retrieval.* `http://www.ismir.net`.
98. G. Kaiser, *A Friendly Guide to Wavelets*, Birkhäuser, 1994.

99. H. Kameoka, T. Nishimoto, and S. Sagayama, *Harmonic-temporal-structured clustering via deterministic annealing EM algorithm for audio feature extraction*, in Proc. ISMIR, London, GB, 2005, pp. 115–122.
100. E. Keogh, *Exact indexing of dynamic time warping*, in Proc. 28th VLDB Conf., Hong Kong, 2002, pp. 406–417.
101. E. Keogh, K. Chakrabarti, M. Pazzani, and S. Mehrotra, *Locally adaptive dimensionality reduction for indexing large time series databases*, in SIGMOD '01: Proceedings of the 2001 ACM SIGMOD international conference on Management of Data, New York, NY, USA, 2001, ACM Press, pp. 151–162.
102. E. Keogh and M. Pazzani, *Iterative deepening dynamic time warping for time series*, in Proceedings of the Second SIAM Intl. Conf. on Data Mining, 2002.
103. E. J. Keogh, T. Palpanas, V. B. Zordan, D. Gunopulos, and M. Cardle, *Indexing large human-motion databases*, in Proc. 30th VLDB Conf., Toronto, 2004, pp. 780–791.
104. A. Klapuri and M. Davy, eds., *Signal Processing Methods for Music Transcription*, Springer, New York, 2006.
105. H. Ko and N. I. Badler, *Animating human locomotion with inverse dynamics*, IEEE Computer Graphics and Applications, 16 (1996), pp. 50–59.
106. L. Kovar and M. Gleicher, *Flexible automatic motion blending with registration curves*, in Proc. 2003 ACM SIGGRAPH/Eurographics Symposium on Computer Animation, Eurographics Association, 2003, pp. 214–224.
107. ———, *Automated extraction and parameterization of motions in large data sets*, ACM Trans. Graph., 23 (2004), pp. 559–568.
108. L. Kovar, M. Gleicher, and F. Pighin, *Motion graphs*, ACM Trans. Graph., 21 (2002), pp. 473–482.
109. L. Kovar, J. Schreiner, and M. Gleicher, *Footskate cleanup for motion capture editing*, in SCA '02: Proceedings of the 2002 ACM SIGGRAPH/Eurographics Symposium on Computer Animation, New York, NY, USA, 2002, ACM Press, pp. 97–104.
110. B. Krüger, *Dynamikbasiertes Morphing von Bewegungsdaten*. Diplomarbeit, Department of Computer Science II, University of Bonn, 2006.
111. P. G. Kry and D. K. Pai, *Interaction capture and synthesis*, ACM Trans. Graph., 25 (2006), pp. 872–880.
112. F. Kurth, M. Clausen, and A. Ribbrock, *Identification of highly distorted audio material for querying large scale data bases*, 2002.
113. F. Kurth and M. Müller, *Efficient index-based audio matching*, Submitted for publication, (2007).
114. F. Kurth, M. Müller, D. Damm, C. Fremerey, A. Ribbrock, and M. Clausen, *SyncPlayer—an advanced system for content-based audio access*, in Proc. ISMIR, London, GB, 2005.
115. F. Kurth, M. Müller, A. Ribbrock, T. Röder, D. Damm, and C. Fremerey, *A prototypical service for real-time access to local context-based music information*, in Proc. ISMIR, Barcelona, Spain, 2004.
116. M. A. Lafortune, C. Lambert, and M. Lake, *Skin marker displacement at the knee joint*, in Proc. 2nd North American Congress on Biomechanics, Chicago, 1992.
117. M. Last, A. Kandel, and H. Bunke, eds., *Data Mining In Time Series Databases*, World Scientific, 2004.

118. C.-S. Lee and A. Elgammal, *Gait style and gait content: Bilinear models for gait recognition using gait re-sampling*, in Proc. IEEE Intl. Conf. Automatic Face and Gesture Recognition (FGR 2004), IEEE Computer Society, 2004, pp. 147–152.
119. J. Lee and S. Y. Shin, *General construction of time-domain filters for orientation data*, IEEE Transactions on Visualization and Computer Graphics, 8 (2002), pp. 119–128.
120. V. I. Levenshtein, *Binary codes capable of correcting deletions, insertions, and reversals*, Doklady Akademii Nauk SSSR, 163 (1965), pp. 845–848.
121. Y. Li, T. Wang, and H.-Y. Shum, *Motion texture: a two-level statistical model for character motion synthesis*, in Proc. ACM SIGGRAPH 2002, ACM Press, 2002, pp. 465–472.
122. C. K. Liu, A. Hertzmann, and Z. Popović, *Learning physics-based motion style with nonlinear inverse optimization*, ACM Trans. Graph., 24 (2005), pp. 1071–1081.
123. G. Liu, J. Zhang, W. Wang, and L. McMillan, *A system for analyzing and indexing human-motion databases*, in Proc. 2005 ACM SIGMOD Intl. Conf. on Management of Data, ACM Press, 2005, pp. 924–926.
124. Q. Liu and E. Prakash, *The parameterization of joint rotation with the unit quaternion*, in Proc. 7th Intl. Conf. Digital Image Computing: Techniques and Applications (DICTA 2003), 2003, pp. 409–418.
125. B. Logan and S. Chu, in Proc. IEEE ICASSP, Istanbul, Turkey.
126. ———, *Music summarization using key phrases*, in Proc. ICASSP, Istanbul, Turkey, 2000.
127. L. Lu, M. Wang, and H.-J. Zhang, *Repeating pattern discovery and structure analysis from acoustic music data*, in Workshop on Multimedia Information Retrieval, ACM Multimedia, 2004.
128. M. D. Lutovac, D. V. Tosic, and B. L. Evans, *Filter design for signal processing using MATLAB and Mathematica*, Prentice Hall, 2001.
129. N. C. Maddage, C. Xu, M. S. Kankanhalli, and X. Shao, *Content-based music structure analysis with applications to music semantics understanding*, in Proc. ACM Multimedia, New York, NY, USA, 2004, pp. 112–119.
130. A. Majkowska, V. Zordan, and P. Faloutsos, *Automatic splicing for hand and body animations*, in SCA '06: Proceedings of the 2003 ACM SIGGRAPH/Eurographics Symposium on Computer Animation, ACM Press, 2006, pp. 309–316.
131. MATLAB, *High-performance numeric computation and visualization software*. The MathWorks Inc., http://www.mathworks.com, 2006.
132. G. Mazzola, *The topos of music*, Birkhäuser, 2002.
133. J. M. McCarthy, *Introduction to theoretical kinematics*, MIT Press, Cambridge, MA, USA, 1990.
134. MIDI, *Musical Instrument Digital Interface*. Manufacturers Association, http://www.midi.org, 2006.
135. MiniLyrics, *An automatic lyrics display for songs*. Crintsoft, http://www.crintsoft.com, October 2006.
136. H. Miyao and M. Maruyama, *An online handwritten music score recognition system*, in Proc. 17th International Conference on Pattern Recognition (ICPR'04) Volume 1, Washington, DC, USA, 2004, IEEE Computer Society, pp. 461–464.

137. M. Müller and F. Kurth, *Enhancing similarity matrices for music audio analysis*, in Proc. IEEE ICASSP, Toulouse, France, 2006.
138. ——, *Web site accompanying the ICASSP 2006 paper "Enhancing similarity matrices for music audio analysis"*. http://www-mmdb.iai.uni-bonn.de/projects/simmat, 2006.
139. ——, *Towards structural analysis of audio recordings in the presence of musical variations*, EURASIP Journal on Advances in Signal Processing, Article ID 89686 (2007).
140. M. Müller, F. Kurth, and M. Clausen, *Audio matching via chroma-based statistical features*, in Proc. ISMIR, London, GB, 2005.
141. M. Müller, F. Kurth, and T. Röder, *Towards an efficient algorithm for automatic score-to-audio synchronization*, in Proc. ISMIR, Barcelona, Spain, 2004.
142. M. Müller, H. Mattes, and F. Kurth, *An efficient multiscale approach to audio synchronization*, in Proc. ISMIR, Victoria, Canada, 2006, pp. 192–197.
143. M. Müller and T. Röder, *Motion templates for automatic classification and retrieval of motion capture data*, in SCA '06: Proceedings of the 2006 ACM SIGGRAPH/Eurographics Symposium on Computer Animation, ACM Press, 2006, pp. 137–146.
144. ——, *Web site accompanying the SCA 2006 paper "Motion templates for automatic classification and retrieval of motion capture data"*. http://www-mmdb.iai.uni-bonn.de/projects/mocap/SCA2006_results/SCA2006.html, 2006.
145. ——, *A relational approach to content-based analysis of motion capture data*, in Human Motion—Understanding, Modeling, Capture and Animation, B. Rosenhahn, R. Klette, and D. Metaxas, eds., Springer, Berlin, 2007. Accepted.
146. M. Müller, T. Röder, and M. Clausen, *Efficient content-based retrieval of motion capture data*, ACM Trans. Graph., 24 (2005), pp. 677–685.
147. ——, *Efficient indexing and retrieval of motion capture data based on adaptive segmentation*, in Proc. Fourth International Workshop on Content-Based Multimedia Indexing (CBMI), 2005.
148. ——, *Web site accompanying the SIGGRAPH 2005 paper "Efficient content-based retrieval of motion capture data"*. http://www-mmdb.iai.uni-bonn.de/projects/mocap/SIGGRAPH2005_results/SIGGRAPH2005.html, 2005.
149. M. Müller, T. Röder, M. Clausen, B. Eberhardt, B. Krüger, and A. Weber, *Documentation of the macoap database HDM05*. Computer Graphics Technical Report, CG-2007-2, Department of Computer Science II, University of Bonn, 2007.
150. R. M. Murray, S. S. Sastry, and L. Zexiang, *A Mathematical Introduction to Robotic Manipulation*, CRC Press, Inc., Boca Raton, FL, USA, 1994.
151. MusicXML, *A universal translator for common western musical notation*. Recordare, http://www.recordare.com/xml.html, 2006.
152. Mutopia Project, *Music free to download, print out, perform and distribute*. http://www.mutopiaproject.org, 2006.
153. M. Neff and E. Fiume, *Methods for exploring expressive stance*, in Proc. 2004 ACM SIGGRAPH/Eurographics Symposium on Computer Animation (SCA 2004), ACM Press, 2004, pp. 49–58.
154. ——, *AER: aesthetic exploration and refinement for expressive character animation*, in Proc. 2005 ACM SIGGRAPH/Eurographics Symposium on Computer Animation (SCA 2005), ACM Press, 2005, pp. 161–170.

155. J. F. O'Brien, R. Bodenheimer, G. Brostow, and J. K. Hodgins, *Automatic joint parameter estimation from magnetic motion capture data*, in Graphics Interface, 2000, pp. 53–60.
156. A. V. Oppenheim, A. S. Willsky, and H. Nawab, *Signals and Systems*, Prentice Hall, 1996.
157. N. Orio, *Music retrieval: A tutorial and review*, Foundation and Trends in Information Retrieval, 1 (2006), pp. 1–90.
158. E. Pampalk, A. Flexer, and G. Widmer, *Improvements of audio-based music similarity and genre classification*, in Proc. ISMIR, London, GB, 2005, pp. 628–633.
159. B. Pardo, *Music information retrieval*, Special Issue, Commun. ACM, 49 (2006), pp. 28–58.
160. S. Pauws, *CubyHum: a fully operational query by humming system*, in Proc. ISMIR, Paris, 2002.
161. G. Peeters, A. L. Burthe, and X. Rodet, *Toward automatic music audio summary generation from signal analysis*, in Proc. ISMIR, Paris, France, 2002.
162. PhotoScore, *Music scanning software*. NEURATRON, `http://www.neuratron.com/photoscore.htm`, 2007.
163. J. Pickens, J. P. Bello, G. Monti, T. Crawford, M. Dovey, M. Sandler, and D. Byrd, *Polyphonic score retrieval using polyphonic audio*, in Proc. ISMIR, Paris, 2002.
164. K. C. Pohlmann, *Principles of Digital Audio*, McGraw-Hill Professional, 2000.
165. Z. Popović, *Motion Transformation by Physically Based Spactime Optimization*, PhD thesis, Carnegie Mellon University, 1999.
166. I. R. Porteuous, *Clifford Algebras and the Classical Groups*, Cambridge Studies in Advanced Mathematics (50), 1995.
167. J. G. Proakis and D. G. Manolakis, *Digital Signal Processsing*, Prentice Hall, 1996.
168. K. Pullen, *Motion Capture Assisted Animation: Texturing and Synthesis*, PhD thesis, Stanford University, 2002.
169. K. Pullen and C. Bregler, *Motion capture assisted animation: Texturing and synthesis*, ACM Trans. Graph., (2002), pp. 501–508.
170. L. R. Rabiner and B. H. Juang, *Fundamentals of Speech Recognition*, Prentice Hall Signal Processing Series, 1993.
171. D. Ramanan and D. A. Forsyth, *Automatic annotation of everyday movements*, in Advances in Neural Information Processing Systems 16, 2003.
172. C. Raphael, *A probabilistic expert system for automatic musical accompaniment*, Journal of Computational and Graphical Statistics, 10 (2001), pp. 487–512.
173. ——, *A hybrid graphical model for aligning polyphonic audio with musical scores*, in Proc. ISMIR, Barcelona, Spain, 2004.
174. ——, *A graphical model for recognizing sung melodies*, in Proc. ISMIR, London, GB, 2005, pp. 658–663.
175. C. Reinschmidt, *Three-dimensional tibiocalcaneal and tibiofemoral kinematics during human locomotion—measured with external and bone markers*, PhD thesis, Department of Medical Science, University of Calgary, 1996.
176. L. Ren, A. Patrick, A. A. Efros, J. K. Hodgins, and J. M. Rehg, *A data-driven approach to quantifying natural human motion*, ACM Trans. Graph., 24 (2005), pp. 1090–1097.

177. A. Ribbrock, *Effiziente Algorithmen und Datenstrukturen zur inhaltsbasierten Suche in Audio- und 3D-Moleüldaten*. Ph. D. Thesis, Department of Computer Science III, University of Bonn, 2006.
178. C. Roads, *The Computer Music Tutorial*, MIT Press, Cambridge, MA, USA, 1996.
179. O. Rodrigues, *Des lois géométriques qui régissent les déplacements d'un système solide dans l'espace, et de la variation des coordonnés provenant de ces déplacements considérés indépendamment des causes qui peuvent les produire*, Journal de Mathématiques Pures et Appliquées, 5 (1840), pp. 380–440.
180. C. Rose, M. F. Cohen, and B. Bodenheimer, *Verbs and adverbs: Multidimensional motion interpolation*, IEEE Comput. Graph. Appl., 18 (1998), pp. 32–40.
181. B. Rosenhahn, U. G. Kersting, A. W. Smith, J. K. Gurney, T. Brox, and R. Klette, *A system for marker-less human motion estimation*, in DAGM-Symposium 2005, 2005, pp. 230–237.
182. A. Safonova and J. K. Hodgins, *Analyzing the physical correctness of interpolated human motion*, in SCA '05: Proceedings of the 2005 ACM SIGGRAPH/Eurographics Symposium on Computer Animation, New York, NY, USA, 2005, ACM Press, pp. 171–180.
183. Y. Sakamoto, S. Kuriyama, and T. Kaneko, *Motion map: image-based retrieval and segmentation of motion data*, in Proc. 2004 ACM SIGGRAPH/Eurographics Symposium on Computer Animation, ACM Press, 2004, pp. 259–266.
184. S. Salvador and P. Chan, *FastDTW: Toward accurate dynamic time warping in linear time and space*, in Proc. KDD Workshop on Mining Temporal and Sequential Data, 2004.
185. E. D. Scheirer, *Extracting expressive performance information from recorded music*. M.S. thesis, MIT Media Laboratory, 1995.
186. ——, *Music-Listening Systems*, PhD thesis, Program in Media Arts and Sciences, MIT, 2000.
187. N. Sebe, M. S. Lew, X. S. Zhou, T. S. Huang, and E. M. Bakker, *The state of the art in image and video retrieval*, in CIVR, vol. 2728 of LNCS, Springer, 2003, pp. 1–8.
188. E. Selfridge-Field, ed., *Beyond MIDI: the handbook of musical codes*, MIT Press, Cambridge, MA, USA, 1997.
189. W. A. Sethares, *Tuning, Timbre, Spectrum, Scale*, Springer-Verlag, 2005.
190. SharpEye, *Music scanning software*. Recordare, `http://store.recordare.com/sharpeye2.html`, 2007.
191. R. N. Shepard, *Circularity in judgments of relative pitch*, J. Acoust. Soc. Am., 36 (1964), pp. 2346–2353.
192. K. Shoemake, *Animating rotations with quaternion curves*, in Proc. 12th Conf. on Computer Graphics and Interactive Techniques, ACM Press, 1985, pp. 245–254.
193. ——, *Euler angle conversion*, (1994), pp. 222–229.
194. P. Shrestha and T. Kalker, *Audio fingerprinting in peer-to-peer networks*, in Proc. ISMIR, Barcelona, Spain, 2004.
195. SmartScore, *Music scanning software*. MUSITEK, `http://www.musitek.com/smartscre.html`, 2007.
196. F. Soulez, X. Rodet, and D. Schwarz, *Improving polyphonic and poly-instrumental music to score alignment*, in Proc. ISMIR, Baltimore, USA, 2003.

197. G. Strang and T. Nguyen, *Wavelets and Filter Banks*, Wellesley-Cambridge Press, 1996.
198. J. Sullivan and S. Carlsson, *Recognizing and tracking human action*, in ECCV '02: Proc. 7th European Conf. on Computer Vision—Part I, Springer, 2002, pp. 629–644.
199. SyncPlayer, *An advanced audio player for multimodal visualization and content-based retrieval and browsing.* Department of Computer Science III, University of Bonn, http://www-mmdb.iai.uni-bonn.de/projects/syncplayer, 2007.
200. N. F. Troje, *Decomposing biological motion: A framework for analysis and synthesis of human gait patterns*, J. Vis., 2 (2002), pp. 371–387.
201. Tunatic, *Free music identification software.* http://www.wildbits.com/tunatic, October 2006.
202. R. J. Turetsky and D. P. Ellis, *Force-Aligning MIDI Syntheses for Polyphonic Music Transcription Generation*, in Proc. ISMIR, Baltimore, USA, 2003.
203. G. Tzanetakis and P. Cook, *Musical Genre Classification of Audio Signals*, IEEE Trans. on Speech and Audio Processing, 10 (2002), pp. 293–302.
204. G. Tzanetakis, A. Ermolinskyi, and P. Cook, *Pitch histograms in audio and symbolic music information retrieval*, in Proc. ISMIR, Paris, France, 2002.
205. M. Unuma, K. Anjyo, and R. Takeuchi, *Fourier principles for emotion-based human figure animation*, in Proc. ACM SIGGRAPH 1995, ACM Press, 1995, pp. 91–96.
206. B. Vercoe, *The synthetic performer in the context of live performance*, in Proc. International Computer Music Conference (ICMC), 1984, pp. 199–200.
207. W. Verhelst and M. Roelands, *An overlap-add technique based on waveform similarity (WSOLA) for high quality time-scale modification of speech*, in Proc. IEEE ICASSP, Minneapolis, USA, 1993.
208. M. Vlachos, M. Hadjieleftheriou, D. Gunopulos, and E. Keogh, *Indexing multidimensional time-series*, The VLDB Journal, 15 (2006), pp. 1–20.
209. M. Vlachos, G. Kollios, and D. Gunopulos, *Discovering similar multidimensional trajectories*, in Proc. of 18th ICDE, San Jose, CA, 2002, pp. 673–684.
210. R. von Laban, *Kinetografie. Labanotation. Einführung in die Grundbegriffe der Bewegungs- und Tanzschrift*, Noetzel, Wilhelmshaven, 1995.
211. A. Wang, *An Industrial Strength Audio Search Algorithm*, in Proc. ISMIR, Baltimore, USA, 2003.
212. L. Wang and T. Jiang, *On the complexity of multiple sequence alignment*, Journal of Computational Biology, 1 (1994), pp. 337–348.
213. Y. Wang, M.-Y. Kan, T. L. Nwe, A. Shenoy, and J. Yin, *Lyrically: automatic synchronization of acoustic musical signals and textual lyrics*, in MULTIMEDIA '04: Proceedings of the 12th annual ACM international conference on Multimedia, New York, NY, USA, 2004, ACM Press, pp. 212–219.
214. A. Watt and M. Watt, *Advanced animation and rendering techniques*, ACM Press, New York, NY, USA, 1991.
215. Wikipedia. http://en.wikipedia.org/wiki/Motion_capture, 2006.
216. A. D. Wilson and A. F. Bobick, *Parametric hidden markov models for gesture recognition*, IEEE Trans. Pattern Anal. Mach. Intell., 21 (1999), pp. 884–900.
217. A. Witkin and Z. Popović, *Motion warping*, in Proc. ACM SIGGRAPH 95, Computer Graphics Proc., ACM Press/ACM SIGGRAPH, 1995, pp. 105–108.

218. I. H. Witten, A. Moffat, and T. C. Bell, *Managing Gigabytes*, Morgan Kaufmann Publishers, 1999.
219. H. Wolfson and I. Rigoutsos, *Geometric Hashing: An Overview*, IEEE Computational Science and Engineering, 4 (1997), pp. 10–21.
220. M.-Y. Wu, S. Chao, S. Yang, and H. Lin, *Content-based retrieval for human motion data*, in 16th IPPR Conf. on Computer Vision, Graphics and Image Processing, 2003, pp. 605–612.
221. C. Xu, N. Maddage, and X. Shao, *Automatic music classification and summarization*, IEEE Trans. on Speech and Audio Processing, 13 (2005), pp. 441–450.
222. T. Yu, X. Shen, Q. Li, and W. Geng, *Motion retrieval based on movement notation language*, Comp. Anim. Virtual Worlds, 16 (2005), pp. 273–282.
223. V. M. Zatsiorsky, *Kinematics of Human Motion*, Human Kinetics, 1998.
224. V. B. Zordan, A. Majkowska, B. Chiu, and M. Fast, *Dynamic response for motion capture animation*, ACM Trans. Graph., 24 (2005), pp. 697–701.
225. E. Zwicker and H. Fastl, *Psychoacoustics, facts and models*, Springer Verlag, 1990.

Index